Modern Analytical Geochemistry

Modern Analytical Geochemistry
An Introduction to Quantitative Chemical Analysis Techniques for Earth, Environmental and Materials Scientists

Edited by

Robin Gill

LONGMAN

Addison Wesley Longman Limited,
Edinburgh Gate, Harlow,
Essex CM20 2JE, England
and Associated Companies throughout the world

© Addison Wesley Longman Limited 1997

First published 1997

British Library Cataloguing in Publication Data
A catalogue entry for this title is available from the British Library

ISBN 0 582 09944 7

Library of Congress Cataloging-in-Publication Data
A catalog entry for this title is available from the Library of Congress

Set by 16 in 10/12pt Ehrhardt
Transferred to digital print on demand, 2002
Printed and bound by Antony Rowe Ltd, Eastbourne

Contents

Preface

Recent years have seen increasing rigour being applied in many industrialised countries to the regulation of industrial emissions and to the quality of water, air, food and soil. This welcome official concern with the quality of the human habitat will, through the regular monitoring of environmental media that it implies, have important consequences for the training of professional environmental scientists. Monitoring agencies will require more analysts familiar with the special obstacles that compositionally diverse environmental materials can put in the path of reliable analysis. Professionals who interpret such analyses, too, though they may not be trained in the the technicalities of geochemical analysis, will need to recognise the limitations of the analytical methods employed and understand the quality control mechanisms upon which the industry depends, if they are to draw objective and reliable conclusions from their data. Surprisingly, few books currently available provide a comprehensive introduction to this complex and continually evolving area. One of the purposes of writing this book, therefore, has been to fill this niche and provide, both for the student of quantitative environmental analysis and for the analysis user, a clear introduction to the many techniques currently in use. The book is also aimed at a more traditional academic Earth science market, at undergraduate and postgraduate students, at researchers who need a broad overview, and at teachers and supervisers who want an up-to-date entry-level text that will introduce the important principles of quantitative geochemical analysis for the newcomer and non-specialist. At a more advanced level this demand has been admirably met by Phil Potts' *Handbook of silicate rock analysis*, but, having taught the subject to undergraduate and masters students for some years, we are convinced of the need for an introductory text more in keeping with the student budget and a little easier to carry around!

The book thus aims to provide the student, the professional user of geochemical analyses and, we hope, a wider readership with an understanding of the whole spectrum of analytical methods that can be applied to Earth and environmental materials, together with a critical appreciation of their relative merits and limitations. It has been structured to cater for readers with widely varying backgrounds. Firstly there is an extensive glossary (Appendix B), and terms defined there are highlighted in **bold** characters where they first appear in the text. Secondly we have used text/graphic boxes in all chapters in a variety of ways: they provide a useful place to put basic information that might be referred to from more than one chapter, they can be used to provide elementary material that will not be required by all readers, and they are also useful for more specialised information that will be of interest only to a section of readers, or at a second reading. The authorship of each box is indicated by the initials at the end. Finally, for readers not familiar with 'diff pumps' and the like, there is an appendix (A) summarising the elements of vacuum technique.

Special thanks are due to Godfrey Fitton, Anthony Lewis and Philip Rowland, who stepped in at a late stage to write chapters that others had failed to deliver. I am also grateful to the following who have been kind enough to provide expert reviews of individual chapters or who have contributed in other ways: Geoff Abbott, John Bailey, Joel Baker,

Jon Davidson, Tony Fallick, Richard Hinton, Gerry Ingram, Ian Jarvis, Andrew Kinsey, Lotte Melchior Larsen, Susan Parry, Stephen Reed, Nick Rogers, Paul Shand and John Williams. I am indebted to Martin Flower and Don Fraser for suggesting one or two contributors.

Above all I thank my family (especially Mary my wife) for their limitless forbearance!

Robin Gill
Royal Holloway, University of London

Contributors

Louis Brown
Department of Terrestrial Magnetism
Carnegie Institution
Washington DC 20015
USA

Godfrey Fitton
Grant Institute of Geology
University of Edinburgh
West Mains Road
Edinburgh EH9 3JW
United Kingdom

Robin C. O. Gill
Department of Geology
Royal Holloway
University of London
Egham
Surrey TW20 0EX
United Kingdom

Gerry Ingram
Department of Geology
Royal Holloway
University of London
Egham
Surrey TW20 0EX
United Kingdom

Kym E. Jarvis
Natural Environment Research Council ICP–MS Facility
Centre for Analytical Research in the Environment
Imperial College of Science, Technology and Medicine
Silwood Park
Ascot
Berkshire SL5 7TE
United Kingdom

Klaus P. Jochum
Max Planck-Institut für Chemie,
Abteilung Geochemie
Saarstrasse 23
Postfach 3060
D-6500 Mainz
Germany

C. Anthony Lewis

Department of Environmental Sciences
University of Plymouth
Drake Circus
Plymouth PL4 8AA
United Kingdom

David P. Mattey

Department of Geology
Royal Holloway
University of London
Egham
Surrey TW20 0EX
United Kingdom

Susan J. Parry

Centre for Analytical Research in the Environment
Imperial College of Science, Technology and Medicine
Silwood Park
Ascot
Berkshire SL5 7TE
United Kingdom

Michael H. Ramsey

Department of Geology
Royal School of Mines
Imperial College of Science, Technology and Medicine
Prince Consort Road
London SW7 2BP
United Kingdom

A. Philip Rowland

Institute of Terrestrial Ecology
Merlewood Research Station
Grange-over-Sands
Cumbria LA11 6JU
United Kingdom

N. Shimizu

Department of Geology and Geophysics
Woods Hole Oceanographic Institute
Woods Hole
MA 02543
USA

M. F. Thirlwall

Department of Geology
Royal Holloway
University of London
Egham
Surrey TW20 0EX
United Kingdom

J. N. Walsh

Department of Geology
Royal Holloway
University of London
Egham
Surrey TW20 0EX
United Kingdom

Box contributors

Joel A. Baker   Department of Geology
Royal Holloway
University of London
Egham
Surrey TW20 0EX
United Kingdom

John G. Williams   Natural Environment Research Council ICP–MS Facility
Centre for Analytical Research in the Environment
Imperial College of Science, Technology and Medicine
Silwood Park
Ascot
Berkshire SL5 7TE
United Kingdom

Acknowledgements

We are grateful to the following for permission to reproduce copyright material:

Blackwell Scientific for figure 16.2; Chapman & Hall for figures 4.1, 4.2a, 4.2b, 4.5, 4.6, 4.7, 4.10 and 6.8.1; Elsevier Science for figures 4.9, 16.8a and 16.8b; The Journal of Radioanal. Nucl. Chem. and Eilev Steinnes for Table 7.3; Penguin Books for Table 2.1; Plenum Press for figure 14.8; The Royal Society of Chemistry for figure 6.7.2; US Geological Survey for figure 2.2; VCH Verlagsgesellschaft mbH for figure 2.3.

Whilst every effort has been made to trace the owners of copyright material, in a few cases this has proved impossible and we take this opportunity to offer our apologies to any copyright holders whose rights we may have unwittingly infringed.

Chapter 1

What a geochemical analysis means

Robin Gill and Michael H. Ramsey

Why analyse?

The term 'analysis' as used in this book includes any measurement that provides information about the *chemical composition* of a sample. Depending on the application, this information could take various forms. For example:

- the concentrations of certain elements or compounds in the sample;
- the relative amounts (abundance ratio) of two or more isotopes of a particular element;
- what proportion of an element occurs in a specific chemical form (e.g. how much of the sulphur present is in the form of sulphate and how much is sulphide).

Geochemical analyses are those carried out on natural earth or environmental materials such as air, volcanic gas, water, dust, soil or rock, or on processed materials that are relevant to the quality of our environment, such as sewage sludge or industrial effluent.

What purposes do geochemical analyses serve? The following examples illustrate some of the many geological, environmental or industrial applications for accurate chemical analyses:

- *Identifying* or characterising a completely unknown natural material.
- Verifying the quality of a processed product, or testing the contamination of a natural material, against *statutory or recommended limits*.
- Measuring a particular detail of a sample's composition to determine aspects of its *history* (e.g. isotopic dating of rocks, geothermobarometry (Chapter 14), forensic applications).
- Investigating a *geochemical process* (natural or anthropogenic) by following the movement of tracer elements (e.g. Ni) or isotopes (e.g. $^{18}O/^{16}O$), or their distribution in a representative range of samples.
- Determining how composition *varies with time* at a single place (e.g. to determine the residence time of radioactive Cs in Cumbrian soil following the Chernobyl incident).
- Mapping the *spatial distribution* of an element or compound at the present time, perhaps to locate its source (e.g. source of river pollution, or the location of an exposed ore deposit from stream sediment analyses).
- Monitoring the efficiency of an *experimental* or *industrial process* as physical or chemical parameters are varied to determine optimum conditions.

Thus in most cases the motivation for carrying out analyses is not simply to acquire data, but to solve a problem, to locate a source, to test a theory, or to see if a sample satisfies some predetermined quality standard. Most analyses are carried out to determine the concentration of either a particular element of interest (referred to as the *analyte*), or a suite of elements (analytes) sharing a particular concentration range or behaviour. Geochemical analysis involves determining not only concentrations themselves but also the uncertainties ('errors') that are associated with them – as indeed with any scientific measurement – because these restrict the conclusions that can objectively be drawn from the data.

The professional analyst bears the primary responsibility for selecting the least costly technique whose results are sufficiently precise and accurate for the intended application (see Chapter 13). For this reason, he or she needs to be acquainted with a range of alternative techniques that could be utilised according to the circumstances and the demands of the application. But clearly the non-specialist *user* of geochemical analyses, and the *beginner* with a specific problem to solve, also need an up-to-date overview of the relative merits of alternative geochemical methods in order to be able to plan and budget their research effectively, and it is for them, as well as the student of geochemistry or environmental science, that this introductory book has been written.

Types of analysis – terminology

Qualitative versus quantitative analysis

A *qualitative* analysis simply reports a list of elements or compounds that are present at detectable levels in a sample, information which has relatively little use except for identification. This book is concerned primarily with *quantitative* analysis, which measures the *concentrations* of the analyte(s) in the sample. The analytes may be elements, compounds, isotopes or chemical species (e.g. Fe^{2+} as distinct from Fe^{3+} – see Chapter 5).

Major, minor and trace constituents

In dealing with rocks, sediments and minerals, it is useful to distinguish between *major elements* (those present at concentrations exceeding 1% by mass, making up the main minerals of the rock), *minor elements* (concentrations between 0.1 and 1.0%) and *trace elements* (concentrations less than 0.1%). In drawing these distinctions, one should recognise that the same element may be a major element in one type of sample (e.g. sulphur in an ore concentrate) but a trace element in another (sulphur in a fresh basalt).

Bulk analysis versus spatially resolved analysis

Most analyses are designed to measure the overall composition of a homogeneous sample such as a rock powder; they are referred to as *bulk* or (in a geological context) 'whole-rock' analyses. Techniques for carrying out such analyses are described in Chapters 4–12. Certain applications may, on the other hand, require analysis of material that is available only in minute amounts, or which is only one of several materials present in a mixed sample (e.g. analysing individual crystals of a specific mineral *in situ* in a rock section), or the task may be to map the distribution of one constituent in a heterogeneous material. In such cases one must use a technique with *spatially resolved analysis* capability, such as electron probe microanalysis (Chapter 14) or a laser-based method (see Chapters 9, 10, 11).

Chemical form – 'speciation'

The chemical form (e.g. the oxidation state) that an element adopts in a sample, or the relative amounts of the element that exist in alternative forms, may sometimes be of more interest than the element's gross concentration. Sulphur, for example, may exist in a sedimentary rock as sulphide or as sulphate (or even as elemental sulphur), and separate analyses would be required to determine how much sulphur is present in each form. The same is true for other elements that exhibit multiple oxidation states in nature, such as iron, carbon and nitrogen. Another form of speciation is establishing the proportions of carbon present in water in the forms H_2CO_3, HCO_3^- and CO_3^{2-}. These questions are discussed in Chapters 5, 9 and 16.

Isotopic composition of an element

The goal of geological isotope analysis is usually to determine accurately the atomic abundance *ratio* of two or more isotopes of the same element (e.g. $^{143}Nd/^{144}Nd$, $^{18}O/^{16}O$), either in a whole-rock sample or in individual minerals. Such ratios, which can be measured with greater precision than individual isotope concentrations, carry information about the age or derivation of a geological sample. Mass spectrometric techniques are described in Chapters 8–12 and 15.

An environmental geochemist, on the other hand, may wish to measure the concentration or activity of a single radionuclide such as ^{137}Cs.

Spectroscopic analysis

Some analytical techniques are designed to measure slight differences in spectral wavelength that shed light on chemical bonding and the atomic environment in which a particular element resides in a material, rather than on its overall concentration. As information of this kind is of more interest to the structural chemist or mineralogist than the geochemist, this field of *chemical spectroscopy* is not discussed in the present book, except where such measurements can also be used to determine the concentration of a compound.

Units of concentration

In conventional chemical usage, the **concentration**[1] refers to the amount of a particular chemical component (in mass or molar units) present in a unit of *volume* of the substance being analysed, and it may be expressed in such units as grams per litre ($g\,l^{-1}$) or moles per litre ($mol\,l^{-1}$). In geochemistry, on the other hand, the term has a wider meaning, often denoting the amount of analyte per unit *mass* of sample (e.g. % (= g per 100 g), **ppm** (= $\mu g\,g^{-1}$)).

[1] Bold type refers the reader to a definition in the Glossary.

Units such as mol l^{-1} do not comply with the Système International (SI) convention, which does not recognise the litre but regards m^3 as the fundamental unit of volume. The cubic metre is a large volume by laboratory standards, and the inconvenience of units such as kg m^{-3} has led to the widespread adoption of dm^3 (cubic decimetre, equal in volume to the litre) as the laboratory unit of volume. The units and prefixes used in the SI system are summarised in the Glossary.

Ways of reporting analyses

The manner in which an analysis is reported varies according to the type of sample and the chemical assumptions that can be made about it. Table 1.1 shows four examples of quantitative analyses.

Section 1 shows a typical silicate rock analysis. Most rocks consist mainly of silicate minerals, in which the electropositive elements present (metals and metalloids) can be considered as being directly bonded to oxygen. The composition of a rock, or a silicate or oxide mineral, is therefore most conveniently expressed in terms of the weight percentages of the *oxides* of each of the major elements present (i.e. g of each oxide per 100 g of sample), so that oxygen is taken into account in the analysis total, even though it is hardly ever analysed directly. The rock analysis shown includes separate analyses for FeO (iron present in ferrous form) and Fe_2O_3 (iron present in the ferric state). As such speciation analyses are not readily accomplished by rapid instrumental methods, many routine analyses today do not draw this distinction, but present total iron as ΣFeO[2] or ΣFe_2O_3[3].

The concentrations of two volatile constituents, H_2O^+ and CO_2, are included in column 1. The trace of water given off when rock powder is heated at temperatures up to 110°C, by tradition denoted 'H_2O^-', simply represents atmospheric water adsorbed on powder-particle surfaces. Driving off this loosely held water by heating in an oven at temperatures below 110°C is a normal preparatory step before analysing a rock powder. Many rock-forming minerals, however, contain hydroxyl (OH^-) ions bound within the crystal structure, and when the sample is ignited to temperatures up to 1200°C the OH disproportionates and is given off as H_2O vapour:

$$M-OH + M-OH \longrightarrow M_2O + H_2O \qquad [1.1]$$

where M represents the metal to which OH is nominally bound; the mass percentage of water vapour thus evolved is denoted 'H_2O^+'. Any carbonate minerals present will also decompose and evolve CO_2 gas at such temperatures:

$$M-CO_3 \longrightarrow M-O + CO_2 \qquad [1.2]$$

Methods for measuring these volatile constituents are described in Chapters 5, 11 and 16.

Note that the number of decimal places quoted varies from one element to another in column 1. The order of uncertainty in each entry is indicated roughly by the number of significant figures quoted: by tradition, analyses are not shown in published tables with more

---

[2] Total iron content expressed as ferrous oxide.
[3] Total iron content expressed as ferric oxide.

Table 1.1 Illustrative quantitative analyses of four media – assessment of analysis quality

1 Analysis of rock powder ('whole-rock' analysis)*

Dolerite 267904B

Major elements (oxide mass %)	
SiO_2	48.3
TiO_2	2.591
Al_2O_3	13.03
Fe_2O_3	6.84
FeO	7.72
MnO	0.23
MgO	5.46
CaO	10.91
Na_2O	2.34
K_2O	0.51
P_2O_5	0.26
H_2O^+	1.41
CO_2	0.49
Total	100.1

Trace elements (ppm):	
Rb	10
Sr	290
Y	42
Zr	175
Nb	25

2 Published analysis of proprietary mineral water†

Mineral water

	Concentration ($mg\,l^{-1}$)	($mmol\,l^{-1}$)	Charge ($meq\,l^{-1}$)
Ca	32.1	0.803	1.606
Mg	7.3	0.304	0.608
Na	26.5	1.152	1.152
K	0.9	0.023	0.023
Fe	< 0.02	–	–
Al	< 0.01	–	–
Total cation charge:			+3.389
HCO_3^-	152	2.492	2.492
Cl^-	26	0.732	0.732
SO_4^{2-}	11	0.115	0.230
NO_3^-	1.7	0.027	0.027
F^-	0.07	0.004	0.004
Total anion charge:			-3.485
TDS	182		
pH	7.4		

3 Silicate mineral analysis‡

Amphibole

	Mass % oxide	Cations per 24 ox§	Site occupancy‖
Si	57.73	7.786 } Tet¶	Z=8.000
Al	12.04	0.214	
		1.700 } Oct	
Fe(3)	1.16	0.118	
Fe(2)	5.41	0.610	
Mn	0.10	0.011	
Mg	13.02	2.617	Y = 5.056
Ca	1.04	0.150	
Na	6.98	1.825	X = 1.975
K	0.68	0.117	A = 0.117
H	2.27	2.042	OH = 2.042
Total	100.4		

4 Trace element analysis of herbage**

	Moss 201†† ICP-MS ($mg\,kg^{-1}$)†‡‡	INAA ($mg\,kg^{-1}$)
V	2.28	
Cr	0.6	0.79
Mn	243.	
Fe	375.	320.
Co	0.27	0.20
Ni	1.54	
Zn	35.0	34.
As	0.22	0.16
Se		0.38
Br		6.9
Rb	5.9	
Sb	0.086	0.06
Cs	0.54	
Ba	20.1	
La	0.34	
Th	0.080	
U	0.040	

*East Greenland basalt. Data from Gill *et al.* (1988).

†Glendale Spring mineral water, December 1992. Major elements from Nielsen *et al.* cited therein

‡From Gill (1995, Table 8.4)

§Number of atoms of metal per 24 atoms of oxygen (the number oxygen atoms in the amphibole unit cell).

‖The aggregate number of cations occupying each site (per 24 oxygen unit cell). The values calculated from the analysis should approximate to the number of each type of site in the unit cell: 8 (Z site), 5 (Y site) and 2 (X site). The A site in amphiboles is commonly only partially filled. There are two OH sites per unit cell.

¶Calculated partition of Al between tetrahedral and octahedral crystal sites, assuming that eight tetrahedral sites are filled by Si and Al.

**Frontasyeva *et al.* (1994) citing earlier papers.

††*Hylocomium splendens* from Norway.

‡‡7N HNO_3 digest.

significant figures than the precision of the analysis can support (though for the realistic estimation of precision – Box 1.2 – it is necessary to retain at least one 'noisy' digit in working tables). The analysis total in column 1 shows the total percentage of the sample accounted for in the analysis, and provides a rough guide to the completeness and quality of the analysis: a comprehensive, high-quality rock analysis will usually add up to 99.5–100.5%. A significantly lower total would suggest that at least one major constituent had been overlooked (e.g. CO_2 in the case of a carbonate-bearing sample) or that there is analytical bias in some of the measurements (Box 1.3). A total above 100.5% generally indicates an erroneous analysis or assumption (e.g. assuming that an iron-rich sample contains mainly ferric iron when in fact it is ferrous, and thereby associating too much oxygen to each mole of iron). However, a total within the 99.5–100.5% range is never a guarantee of a good analysis, as it may conceal two or more errors of opposite sign that happen to cancel out.

Trace elements, present at levels too low to make a significant contribution to the analysis total, are normally reported in elemental rather than oxide terms, usually as $\mu g \ g^{-1}$ (informally known as 'parts per million' or ppm); for particularly low-abundance elements it may be appropriate to use $ng \ g^{-1}$ ('parts per billion' or ppb).

Section 2 shows an analysis of a typical spring water, taken from the label on a supermarket bottle. The analysis is limited to concentrations of the principal solute species, expressed in $mg \ l^{-1}$, which are assumed to be present as unassociated positive and negative ions. Sulphur is present exclusively in the sulphate form SO_4^{2-} (the solubility of sulphide is far too low to account for the level of sulphur present in solution). Dissolved carbon is represented as bicarbonate ion HCO_3^{-}. The most abundant component of the solution, the solvent H_2O, has not been determined and it is therefore meaningless to quote an analysis total. An estimate of the accuracy of the analysis can, however, be made from its self-consistency: as the solution analysed is electrically neutral, the total positive charge on the cation species (in **milliequivalents**) represented in an accurate analysis should be almost exactly equal to the total negative charge on the anionic species. This can be tested by calculating what is called the *electroneutrality parameter* (EN):

$$EN\% = \frac{\text{sum of cation charge} + \text{sum of anion charge}}{\text{sum of cation charge} - \text{sum of anion charge}} \times 100\%$$

$$= \frac{3.389 + \{-3.485\}}{3.389 - \{-3.485\}} \times 100 = -1.4\%$$

An analysis with an EN value less than 5% (as in this case) is usually considered to be acceptably accurate (see Appelo and Postma, 1994), although, as with the analysis total in column 1, an acceptable value may conceal *two* errors that happen to cancel each other out. An important constituent of aqueous solutions whose concentration is not shown directly in column 1 is the hydrogen ion H^+, whose abundance determines the acidity of the solution. The hydrogen ion activity, measured using a pH meter (Chapter 5), is expressed as pH = 7.4 (a concise way of saying the activity of $H^+ = 10^{-7.4}$).

Section 3 shows an analysis of an amphibole crystal. As amphiboles are silicates, this analysis has many features in common with that in column 1: the concentrations of elements are expressed in terms of mass % oxides, and the quality of a complete analysis such as this may be checked by calculating the oxide total. However, many mineral analyses are carried out today by electron microprobe (Chapter 14), which is incapable of determining OH directly or of distinguishing between Fe(II) and Fe (III). For such analyses the total may have little meaning. Another test of analytical quality, or at least of self-consistency, relies upon the constraint that crystal structure imposes on the number of different types of cation site in the unit cell. The amphibole formula, $AX_2Y_5Z_8O_{22}(OH)_2$, represents a unit cell containing 24 oxygen ions, between which are found eight tetrahedral ('Z') sites, five octahedral ('Y', sometimes known as M_1, M_2 and M_3) sites, two larger ('X' or M_4) sites, and one larger ('A') site. Two of the nominal 24 oxygen ions are actually OH^- ions occupying hexagonal voids in the amphibole structure. Ions enter cation sites according to size and charge: the Z sites accommodate Si and a portion of the Al, Y sites accommodate the remaining Al and Fe, Mn, and Mg, X sites accept Ca and Na, and any K present occupies the A site (many of which may, however, be vacant). Recalculating the analysis, as numbers of each type of cation equivalent to 24 oxygens, allows the total site occupancy to be checked against these geometrical constraints (see Gill, 1995, Chapter 8). That the occupancies of the Z, Y, X and OH sites are close to the expected integer values is a reassurance of a self-consistent analysis, although mutually compensating errors may again remain undetected. The calculation of this 'mineral formula' provides a routine quality check for mineralogists.

Section 4 shows analyses of selected trace elements in a sample of moss from Norway. Moss is widely used as a biomonitor of the atmospheric deposition of trace elements. For such studies it is convenient to use a multi-element technique such as instrumental neutron activation analysis (INAA, Chapter 7). The major elements of moss (C, H, O, N, . . .) have not been determined. In conducting an analysis of this kind, there are no internal constraints equivalent to analysis total or site occupancy to indicate analysis quality, and where possible the prudent analyst seeks to test accuracy by repeating the analysis using another technique (in this case ICP–MS – Chapter 10). The two methods actually measure slightly different things: INAA analyses solid materials directly and therefore gives the total content of a given analyte, whereas the ICP-MS analyses solutions, in this case derived from digesting the moss samples in 7M nitric acid. The good agreement for elements analysed by both techniques suggests that the digestion extracts essentially all of the elements concerned. Note that each method offers some elements that are not available by the other. For any meaningful biomonitoring study, it is essential to analyse alongside the unknown a **reference material** of 'known' composition to provide **quality control**.

Measures of analytical uncertainty

Every measurement is subject to an element of uncertainty, which may be reduced by improving the method or repeating the measurement but can never be entirely eliminated. The magnitude of this uncertainty restricts the uses to which the analysis can be applied, and no quantitative analysis is of much value for scientific purposes without a statement of the uncertainty associated with the values quoted. This uncertainty consists of two contributions: random error and systematic error.

Random error: the *precision* of an analysis

The term precision refers to the closeness of agreement between results obtained by repeating an analytical procedure several times under the same conditions. Saying an analysis is precise is simply a qualitative statement about the **random error** associated with the analysis, not about how close the result is to the true value (see accuracy below). A precise (or high-precision) measurement is one that has a *small* random error. Precision may be expressed in two alternative ways, as **repeatability** or **reproducibility**, according to the circumstances under which the measurements are replicated.

The scatter of random error in chemical analysis usually approximates to a **normal distribution** (Box 1.1). How precise an analysis is can therefore be quantified in terms of the **standard deviation** obtained from replicate measurements: the smaller the standard deviation, the more precise the analysis. For any normal distribution there is a 95% probability (i.e. a 19 in 20 chance) of an individual reading lying within $\pm 2\sigma$ of the mean value.

Box 1.1 The normal distribution I

A variable is said to follow a normal distribution if the values obtained by repeating an analysis many times (say 100 readings) form a symmetric cluster having the characteristic shape as shown in Fig. 1.1.1. This depicts as a series of columns the number (**frequency**) of readings that fall within each interval of Ti concentration. Such a histogram is called a *frequency distribution* for the sample of readings shown in Fig. 1.1.1.

Figure 1.1.1 Sample distribution.

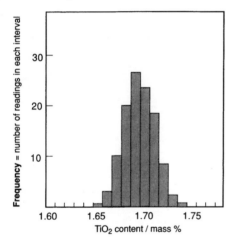

If an extremely large number of readings had been taken, it would be possible to prepare a frequency distribution with a column width of almost zero; the result would effectively be a smooth curve (Fig. 1.1.2), which is known as the *population distribution*. The shape shown in Fig. 1.1.2 is characteristic of a *normal* or *Gaussian distribution*, which can be represented by the equation

$$y = \frac{\exp[-(x-\mu)^2/2\sigma^2]}{\sigma\sqrt{2\pi}}$$ [1.1.1]

Figure 1.1.2 Normal population distribution.

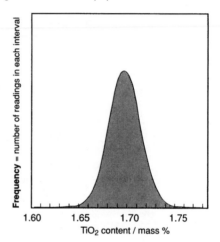

where y represents frequency, x the individual measurements of TiO_2, and μ and σ respectively are parameters describing the location and width of the population distribution in TiO_2 space; they are known as the **mean** and **standard deviation** of the population distribution.

It is a simple matter to calculate the mean and standard deviation for the *sample distribution* shown in Fig. 1.1.1, using formulae given in the Glossary (see **sample mean** and **standard deviation**). *Twice* the standard deviation of a series of replicate analyses, $2s$ (sometimes referred to colloquially, though inappropriately, as '2 sigma'), is the standard way of reporting the precision of an analysis. However, because it represents relatively few readings, the standard deviation s only approximates to the true σ of the underlying population.

Another useful measure of the precision is the **variance** s^2, which is simply the square of the standard deviation s.

Further useful properties of the normal distribution are discussed in Boxes 4.4 and 6.4.

RG

Box 1.2 How to estimate analytical precision

The simplest way to estimate precision is to measure one sample repeatedly and then calculate $\bar{x}$ and $2s$. A reliable estimate of 2σ, however, requires a large number of replicates (Box 1.1), which increases the cost of analysis. Moreover the precision estimated is only representative of one sample at one level of analyte concentration.

In practical analytical terms, the most efficient method for estimating the precision of an analytical method or batch over a range of analyte concentrations and sample compositions is to use duplicated analysis of randomly selected samples. For example, in a batch of 90 soil samples analysed for Pb, nine were selected at random, prepared in duplicate and analysed independently and **anonymous**ly. Each duplicate analysis pair (x_1, x_2) listed in Table 1.2.1 was used to calculate a mean and an absolute difference $|x_1 - x_2|$ (see footnote[1]).

The bottom row expresses the difference of each pair of readings as a percentage of the mean value. The median value of these percentages (final column) provides an estimate of the **coefficient of variation** of the data population (though only if determinations are well above the detection limit of the method ($\times 100$)), providing a useful approximate estimate of the overall precision.

In the more rigorous approach of Thompson and Howarth (1976), the individual means and differences are plotted in a control chart (Fig. 1.2.1), comparing the distribution of these values with that expected if the coefficient of variation were 10%. If that were the case, half of the points should lie below the median (50%) line, 90% below the 90% line and 99% below the 99% line. Six of the nine points do fall below the median line, indicating a **coefficient of variation** better than the 10% model.

Figure 1.2.1 Data from Table 1.2.1 plotted on a control chart for the estimation of precision by comparison with a standard model of 10% precision (after Thompson and Howarth, 1976).

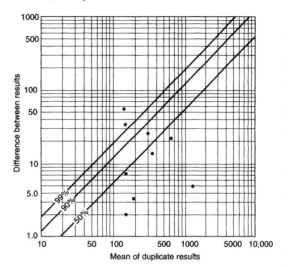

Two points lie above the 99% line, however, representing 22% of the population rather than the 1% predicted by the model. This suggests the data may not be taken from a strictly normal distribution.

MHR

Table 1.2.1   Nine duplicate determinations of Pb ($\mu g\ g^{-1}$) in soils with estimates of precision

Sample	1	2	3	4	5	6	7	8	9	Median
x_1	314.8	634.8	338.4	160.0	198.8	126.4	154.4	1232.0	148.8	
x_2	287.6	612.8	352.4	158.0	195.2	185.2	161.6	1227.0	182.4	
$\bar{x}$	301.2	623.8	345.4	159.0	197.0	155.8	158.0	1230.0	165.6	
$\lvert x_1 - x_2 \rvert$	27.2	22.0	14.0	2.0	3.6	58.8	7.2	5.0	33.6	
$\lvert x_1 - x_2 \rvert \%$	9.0	3.5	4.1	1.3	1.8	37.7	4.6	0.4	20.3	4.1

[1]   This notation signifies the *modulus* of the difference (i.e. its positive value).

The value normally quoted as a measure of precision is twice the sample standard deviation, $\pm 2s$, and one must remember that this is only an approximation to $\pm 2\sigma$ (Box 1.1).

It may be helpful to express the standard deviation as a proportion of the value being estimated; a commonly cited estimate of relative precision is twice the coefficient of variation (i.e. $2 \times 100s/\bar{x}\%$) among repeated measurements, where $\bar{x}$ is the mean value of the repeated measurements. The precision is not constant, but generally improves as the concentration increases.

Practical approaches for estimating the precision of a chemical analysis, together with a numerical example, are discussed in Box 1.2.

Systematic error: bias and accuracy

The **bias** of an analysis refers to the closeness of agreement between the mean of a series of **analyte** concentration measurements and the **true value** of the concentration (in practice, the best available estimate of the true value). Bias is essentially a measure of the **systematic error** of the analytical method at a particular concentration level. 'Bias' was until recently synonymous with the term **accuracy**, which is, however, now reserved for the deviation

---

Box 1.3 How to estimate the bias of an analytical method

The main method for the estimation of bias is by analysing a **certified reference material**. For example, the accuracy of determination of lead in soil can be estimated by analysing a CRM that has both a certified value for Pb and ideally an overall composition closely similar to the unknown samples. Appropriate CRMs may be selected for a given application using compilations such as those of Potts *et al.* (1992) and Govindaraju (1994).

BCR-143 is a CRM prepared from a sludge-amended soil that has a certified Pb concentration (x_c) of 1333 µg g^{-1}. When analysed for Pb five times using an acid digest and ICP–AES (Chapter 4), the Pb values obtained were 1171, 1342, 1374, 1236 and 1312 µg g^{-1}. The mean ($\bar{x}$) and standard deviation (s) of these readings are 1287 µg g^{-1} and 82.6 µg g^{-1} respectively.

Bias $= (\bar{x} - x_c) = 1287 - 1333 = -46$µg g^{-1}
% bias $= 100 \times$ bias$/x_c\% = -3.5\%$

Although bias can be meaured in this way, it may not necessarily be statistically significant. This can be checked by applying a *t*-test[1]:

$$t = \frac{|\bar{x} - x_c|\sqrt{n}}{\sigma} = \frac{|1287 - 1333|\sqrt{5}}{82.6} = 1.245$$

(Refer to Box 1.2 for $|x|$ notation.) The tabulated value of *t* for a probability of 5% is 2.776 (Table 2.1), which the calculated value must exceed if the difference is to be considered significant at the 5% level. As the calculated value (1.246) is less than the tabulated value (2.776), the bias estimated in this example cannot be regarded as having any statistical significance.

It is important to recognise that an analytical method can give analyses that are very precise (*i.e.* reproducible) yet still subject to a large bias. Precision is never a test of bias.

MHR

---

[1] The *t*-test is a statistical test that ascertains the level of significance that can be attached to the difference between two values, by providing an estimate of the probability that the difference is simply the result of random fluctuations in the data. The text involves calculating a value of the parameter *t* (see Box 2.1), and comparing the value obtained with published tables of *t* as a function of probability and number of measurements (Table 2.1). Further explanation may be found in Miller and Miller (1993).

of a single observation from the true value, regardless of whether the error is systematic or random.

The bias (positive or negative) for a particular analytical method can be estimated by analysing a **certified reference material** (CRM) of similar analyte concentration and matrix composition to the samples being analysed (see Potts *et al.*, 1992, Govindaraju, 1994). The ways of doing this, and their limitations, are discussed in Box 1.3.

For a commercial testing laboratory, there is competitive advantage in being formally accredited to carry out particular analytical procedures. This and the related culture of **quality control** are dealt with in Chapter 13.

Chapter 2 Sampling and sample preparation

Michael H. Ramsey

The ultimate objective of most geochemical analysis programmes is to describe the 'true' variation of an analyte's concentration (or the concentrations of several analytes) in some domain of the natural world which is under investigation, such as sediment in streams draining an area thought to contain a mineral deposit, or airborne particulate matter down-wind from an industrial plant. As analysing the entire domain is impractical, the investigator collects a series of **samples** for analysis which are intended to encompass collectively the entire chemical variation present in the domain. Ensuring that the samples collected are sufficiently **representative** of the domain cannot simply be left to chance, but requires the application of statistically rigorous **sampling** procedures.

The laboratory analysis of rocks, minerals or environmental materials thus forms only one stage of a complete geochemical investigation (Fig. 2.1). Though this book is concerned primarily with steps 5 and 6, the tasks of sampling (3) and sample preparation (4) are also inescapable and vital parts of a geochemical investigation that have a substantial bearing on the analytical results. These steps may be closely linked to each other; for example, sample preparation may be deliberately biased so that a specific fraction – such as minerals of

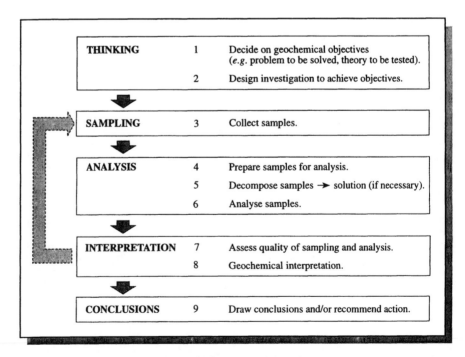

Figure 2.1 Model of a generalised geochemical investigation.

economic importance – is sampled preferentially. The optimum strategy for sampling and sample preparation therefore depends on the geochemical objective.

The process of sampling itself constitutes a source of error, both systematic and random. These errors cannot be avoided entirely, but steps must be taken to reduce them to a level where they do not significantly influence the scientific outcome. The methodologies for estimating sampling errors are not as well established as those for analytical errors, but they are adequate for demonstrating when the magnitude of sampling error has been reduced to an acceptable level for a given application.

Among the decisions to be taken in the design of a sampling scheme are:

(1) the medium to be sampled (e.g. rock, till, water);
(2) what the sample is intended to represent (e.g. the analytes dissolved in water, or those present in suspended solids (particle size $> 0.45\ \mu m$));
(3) the size of sample required (i.e. by mass or volume);
(4) the number of samples required;
(5) the spatial distribution of the sample sites (e.g. sampling density);
(6) whether temporal variation in the sample media is important (e.g. for stream waters);
(7) how the quality of the sampling can be estimated (what methods are applicable for estimating precision/variance and bias of sampling);
(8) what levels of sampling errors will be acceptable;
(9) what scheme of sample identification is to be used (as part of a quality assurance scheme for monitoring sample progress).

Consideration of these factors leads to the design of a detailed sampling protocol, ideally describing all the procedures in an unambiguous way. The protocol is particularly important if sampling is to be carried out by several people or organisations. There is no universal protocol that meets the requirements of every situation; each investigation will require a fresh evaluation of these nine points.

The literature on sampling methodology in geochemistry is sparse. Sampling methods for atmospheres, waters, soils, sediments and plants have been reviewed by Markert (1994), Keith (1988 and 1991) and Kratochvil *et al.* (1984), for industrial products by Smith and James (1981) and for contaminated land DoE (1994).

Theoretical approach to sampling

Sampling errors, like analytical errors, may be random or systematic. One may therefore distinguish between sampling precision (the reproducibility of successive samplings of the same 'population', e.g. at a particular location within a rock body) and sampling bias (a measure of the bias introduced by the sampling procedure). Most sampling protocols are designed to achieve a specified sampling *precision*.

How much of each sample do I require?

Many materials of interest to the geochemist, such as rocks, sediments and airborne particulates, are heterogeneous mixtures. Individual grains are each representative of one component of the mixture, but not of the mixture as a whole. Any sample taken must be large enough for the influence of individual grains on sample composition to be insignificant, and therefore grain size is an important consideration in deciding the minimum mass of sample that will satisfy this requirement: the larger the grain size, the greater the mass of sample needed.

The science of sampling inhomogeneous mixtures was examined in detail by Davis (1954), Gy (1979, summarising earlier work) and Ingamells and Switzer (1973). They developed equations expressing the minimum mass required to achieve the desired sampling precision, in terms of various physical and geometrical properties of the particle population. Such equations have the general form:

$$m = \frac{Cd^3}{\sigma_s^2}$$ [2.1]

where σ_s is the level of sampling precision being sought (expressed as standard deviation in units of concentration), m is the mass of sample required to achieve this reproducibility, and d is the diameter of the largest grains present. C is the product of various physical properties of the mixture and constituent particles. Exact application of such equations to real-life geochemical sampling is generally impractical, however, owing to the detailed information that is required about constituent particles (e.g. size distribution) for the calculation of C, or the assumptions that need to be made in the absence of such information.

There have been several attempts to draw up practical guidelines on good sampling practice. Empirical recommendations were given by Wager and Brown (1960) for the mass required for an acceptably representative sample for homogeneous rocks of various equidimensional grain sizes (Table 2.1).

This rule of thumb, though widely applied, has no statistical foundation. What sampling precision is implied by the term 'representative', for example? Furthermore, some geochemical problems involve the sampling of non-homogeneous materials, such as 'mineralised' rock (in which ore is concentrated in discrete veins), and here Table 2.1 underestimates the mass required for representative sampling. Consider for example a crushed rock (particle size <2.5 mm) containing 5% of sphalerite (ZnS). Table 2.1 suggests that a sample mass of 1 kg would be representative of such material, but Smith and James (1981) showed by rigorous statistical analysis that 70–100 kg would be required to attain an acceptable sampling precision (e.g. ±0.2% ZnS at the 95% confidence limit).

Table 2.1   Mass required to ensure a representative sample of homogeneous rock as a function of grain size. From Wager and Brown (1960) citing an unpublished typescript by E.L.P. Mercy of Imperial College

Grain size of rock (mm)	Minimum sample mass required (kg)
>30	5.0
10–30	2.0
1–10	1.0
0–1	0.5

Another practical illustration is provided by the notorious sampling problem associated with trace gold, which occurs in the crust often in the native (metallic) state. Rather than being randomly dispersed in the host rock, metallic Au typically exists as sporadic flakes 10–100 μm in size. Wildly erratic Au analyses can be obtained if this 'nugget effect' is not allowed for in sampling. Clifton *et al.* (1969) argued that an acceptable sampling precision (say ±50% at 95% confidence) would be obtained from samples containing at least 20 particles of gold, and they proposed a graphical method of determining the sample mass necessary to meet this requirement, based on estimates of Au concentration and the maximum grain size (Fig. 2.2). Thus for an Au concentration of 4 ng g^{-1} (ppb) present as flakes up to 125 μm in diameter (at 95th percentile), samples with a mass of 10 kg would have to be collected and processed to assure acceptable reproducibility.

This analysis was an important step in the quantified approach to sampling protocols, but its limitations need to be recognised. It was necessary to assume that the gold particles are of uniform size, are randomly distributed and make up less than 0.1% of all the particles, and that no other Au-bearing phase is present. If Au were also present in a sulphide mineral, for example, or adsorbed on any organic matter present, then the calculation will not be

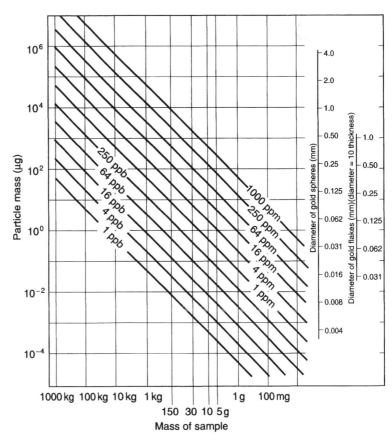

Figure 2.2   A nomogram designed to indicate the mass of sample (rock, sediment) required to contain 20 particles of gold, and hence to give a representative sample (reproduced with permission from Clifton *et al.*, (1960) Sampling size and meaningful gold analysis, *US Geological Survey professional paper* **625-C** 1969). The size of the Au grain is indicated in the vertical axis either by mass or diameter. The expected concentration of gold is selected as one of the lines, and the mass of sample required is read off the horizontal axis.

exact. Nevertheless, Smith and James (1981) suggest that simplified calculations such as this tend to *exaggerate* the sample mass required, so although the calculations may not be entirely accurate they err on the cautious side.

How many separate samples do I require?

The number of samples required for a particular geochemical investigation depends on the scientific objectives, and may also be constrained by cost and time factors. It is therefore impossible to enunciate any universal quantitative principle governing sample numbers, but some general qualitative guidelines apply:

- The greater the number of variables that is to be investigated, or the more classes into which the population is to be divided, the greater the number of samples that will be needed. (However, these quantities may not be known at the start of the investigation.)
- If costly fieldwork is involved, it is prudent to collect too many samples rather than too few, if by doing so a return visit can be avoided. (The same may apply for visits to distant laboratories.)
- Good project design, and sampling and analysis of appropriate quality, are often more important than a large number of samples.
- The collection of several sub-samples from one site, that are combined to form a single composite sample, usually improves sampling precision without increasing the number of samples for analysis.

Quantitative criteria for determining the required number of samples can be devised in specific cases. Consider for example the number (n_1) of soil samples taken in an initial survey of an urban site contaminated with lead. The mean Pb content ($\bar{x}$) in these samples is 1960 µg g^{-1}, with a standard deviation s (including uncertainties in both sampling and analysis) of 200 µg g^{-1}. Suppose the mean is very close to a statutory limit for Pb in soils (x_{stat}) of 2000 µg g^{-1}. How many samples (n_2) would be needed in a subsequent survey to demonstrate (with a 95% confidence limit) that the site did not exceed this statutory limit?

This problem can be treated as an application in reverse of the well-known t-test (Box 2.1). Instead of calculating t for a difference $\{x_{stat} - \bar{x}\}$ and $(n_1 - 1)$ **degrees of freedom** (see Box 2.1), the formula can be rearranged to give the number of samples n_2 required in resampling the site to meet the specified confidence limit:

$$n_2 = \left(\frac{ts}{\{x_{stat} - \bar{x}\}}\right)^2 \qquad [2.2]$$

The tabulated t value used here is based on the prescribed confidence limit, and the mean value and **degrees of freedom** ($n_1 - 1$) from the *initial* survey. From Table 2.2, when $n = 11$, then t is found to be 2.228. Then

$$n_2 = \left(\frac{2.228 \times 200}{2000 - 1960}\right)^2 = 124 \qquad [2.3]$$

The very large number of samples required here reflects both the relatively large measurement errors and the closeness of the initial mean Pb value to the statutory limit.

Box 2.1 Statistical testing – the *t*-test

A clear numerical difference between a measured value and some reference value (e.g. a statutory intervention level for water pollution) may not always be statistically significant, particularly if the measured value has a large random error. Consider repeated measurements of the total As concentration in a sample of drinking water having values of 47, 46, 49, 50 μg ml^{-1}. Assuming there is no analytical bias, is the mean value (= 48 μg ml^{-1}) *significantly* less than the regulatory limit of 50 μg ml^{-1}? With what confidence can we assert that the water *does not exceed* the statutory limit?

The uncertainty on the mean value of *n* measurements is described by a quantity called the **standard error on the mean** (*se*) = $s/\sqrt{n}$. As the number of measurements is increased, the standard deviation (*s*) does not decrease, but the standard error does. It is possible to calculate the number of measurements that must be made, or samples taken, to be able to state with 95% confidence that the water discussed above does not exceed the statutory value 50 mg ml^{-1}. The statistic used for this purpose is called the '*t*' statistic, and it is simply the difference between the mean value (48 μg ml^{-1}) and the regulatory limit (50 μg ml^{-1}) expressed as a multiple of the standard error on the mean:

$$t = \frac{x_{\text{stat}} - \bar{x}}{s/\sqrt{n}} \qquad [2.1.1]$$

The value of *t* that indicates that the difference is statistically significant, t_{tab}, can be looked up in published *t* tables (Table 2.2) for a particular value of *n* (though for reasons that need not be discussed here this is expressed in statistical tables as the 'number of degrees of freedom' (= *n* − 1)) and the required confidence level (e.g. 95%).

For this example the calculated value of *t* is given by

$$t_{\text{calc}} = \frac{50 - 48}{1.826/\sqrt{4}} = 2.19$$

Because we are only interested in the probability of the As level *exceeding* the limit, we look up the tabulated *t* value for the probability of 5% (called a 'one-tail *t*-test'). This tabulated value of *t* is 2.353. The calculated value t_{calc} is less than the tabulated value t_{tab} and therefore the As content of this water is not significantly less than the regulatory limit at the 95% confidence level.

A fuller description of the *t*-test can be found in Miller and Miller (1993).

MHR

Estimation of sampling error

Random error

The standard deviation of replicate analyses of a single sample for an analyte of interest reflects the random errors of laboratory analysis. Where several samples from essentially the same location are analysed for this analyte, the overall standard deviation will be somewhat greater, because in addition to the analytical errors it will include a component arising from errors in sampling. This is easier to understand in terms of variance s^2 rather than standard deviation s, because variances have the useful property of being additive; that is to say, the total variance arising from both analytical *and* sampling errors, the 'measurement' variance s^2_{meas} for the analyte in question, can be written

$$s^2_{\text{meas}} = s^2_{\text{anal}} + s^2_{\text{sampling}} \qquad [2.4]$$

Therefore, once s^2_{anal} has been measured from replicates of a single sample, the *sampling variance* can be evaluated by measuring s^2_{meas} and subtracting s^2_{anal}:

$$s^2_{\text{sampling}} = s^2_{\text{meas}} - s^2_{\text{anal}} \qquad [2.5]$$

Table 2.2   Values of the *t* statistic for $n - 1$ degrees of freedom and selected levels of significance. The significance levels are for one-sided portions of the distribution (for a two-sided test at the 95% confidence interval, for example, a significance level of 2.5% would be used)

Degrees of freedom $(n - 1)$	Significance level (%)					
	10	5	2.5	1.0	0.5	0.1
1	3.078	6.314	12.706	31.821	63.657	318.310
2	1.886	2.920	4.303	6.965	9.925	22.327
3	1.638	2.353	3.182	4.541	5.841	10.215
4	1.533	2.132	2.776	3.747	4.604	7.183
5	1.476	2.015	2.571	3.365	4.032	5.893
6	1.440	1.943	2.447	3.143	3.707	5.208
7	1.415	1.895	2.365	2.998	3.499	4.785
8	1.397	1.860	2.306	2.896	3.355	4.501
9	1.383	1.833	2.262	2.821	3.250	4.297
10	1.372	1.812	2.228	2.764	3.169	4.144
11	1.363	1.796	2.201	2.718	3.106	4.025
12	1.356	1.782	2.179	2.681	3.055	3.930
13	1.350	1.771	2.160	2.650	3.012	3.852
14	1.345	1.761	2.145	2.624	2.977	3.787
15	1.341	1.753	2.131	2.602	2.947	3.733
16	1.337	1.746	2.120	2.583	2.921	3.686
17	1.333	1.740	2.100	2.567	2.898	3.646
18	1.330	1.734	2.101	2.552	2.878	3.610
19	1.328	1.729	2.093	2.539	2.861	3.579
20	1.325	1.725	2.086	2.528	2.845	3.552
21	1.323	1.721	2.080	2.518	2.831	3.527
22	1.321	1.717	2.074	2.508	2.819	3.505
23	1.319	1.714	2.069	2.500	2.807	3.485
24	1.318	1.711	2.064	2.492	2.797	3.467
25	1.316	1.708	2.060	2.485	2.787	3.450
26	1.315	1.706	2.056	2.479	2.779	3.435
27	1.314	1.703	2.052	2.473	2.771	3.421
28	1.313	1.701	2.048	2.467	2.763	3.408
29	1.311	1.699	2.045	2.462	2.756	3.396
30	1.310	1.697	2.042	2.457	2.750	3.385
40	1.303	1.684	2.021	2.423	2.704	3.307
60	1.296	1.671	2.000	2.390	2.660	3.232
120	1.289	1.658	1.980	2.358	2.617	3.160
∞	1.282	1.645	1.960	2.326	2.576	3.090

After Table 21, The Penguin-Honeywell Book of Tables, 1968, Penguin Books

For real sampling programmes, in which multiple samples are collected from various sites in a target area, Ramsey (1994) proposed a method for estimating sampling variance by collecting duplicate field samples, analogous to using analytical duplicates for the estimation of analytical precision (Box 1.2). Each duplicate should be taken at a slightly different position from the original, the separation reflecting the spatial uncertainty in the location of the site; if stream sediment samples are taken every 500 m, for example, then this separation between the duplicates might be ~10 m. The sampling precision can be extracted from the overall vaiance using Equation 2.5, a technique known as analysis of variance (ANOVA – see Miller and Miller, 1988).

*Acceptable levels of
sampling and analytical
error*

The normal purpose of a geochemical survey, for example of a plot of contaminated land, is to document the true geochemical variation $s^2_{geochem}$ in the target area by sampling and analysing from various locations. The data population arising from this exercise will have an overall variance s^2_{total} given by

$$s^2_{total} = s^2_{geochem} + s^2_{sampling} + s^2_{anal} \qquad [2.6]$$

Because the summing of squares is involved, large terms tend to dominate and small terms make a negligible contribution. If meaningful conclusions are to be drawn from the real geochemical variations (represented by $s^2_{geochem}$) and not obscured by measurement artefacts (s^2_{meas}), it follows that the sampling and analytical variances should only account for a small proportion of the total variance. Ramsey (1994) proposed they should together not contribute more than 20% to s^2_{total} for the element of interest if they are not to impair interpretation of the data. Such considerations, which can be presented in the form of a piechart (Fig. 2.3), help in designing the most cost-effective sampling protocol for a given location and geochemical objective. If the measurement variance already accounts for only a small proportion (say 1–5%, for example) of s^2_{total}, little will be gained from attempting to reduce either s^2_{anal} or $s^2_{sampling}$ further. Empirical evidence gathered from a small-scale orientation survey of the proposed sampling site is often the most effective way to optimise the sampling protocol for a more detailed study.

Sampling bias

Systematic sampling error (bias) is more difficult to estimate than random sampling error (precision). Methods for its evaluation depend on the suspected cause of the bias. Where there is a possibility of collected water samples becoming contaminated on site or in transit to the laboratory, for example, high-purity water samples from the laboratory ('field blanks') may be carried to the sampling site with the water sampling equipment. Several field blanks are then processed exactly as normal water samples so that any contamination from the filters, bottles, acid or general field environment can be monitored. The mean concentration of each analyte in the field blank (minus its concentration in the laboratory blank) is an estimate of sampling bias, and is usually subtracted from the concentrations

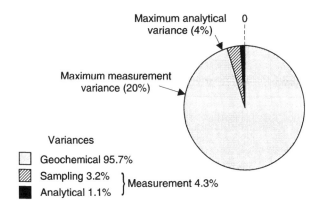

Fig 2.3 A diagrammatic representation of the relative importance of random measurement errors from sampling and analysis (expressed as variance) in a geochemical investigation. When the sum of these two variances is less than 20% (shown by ↓) they will not impair the geochemical interpretation of the data, as shown in this example taken from a survey of Pb in Cornish stream sediments. (Ramsey 1994 from p101 *Environmental Sampling for Trace Analysis* (ed. B. Markert) 1994. Reproduced by kind permission of VCH Publishers, Weinheim, Germany.)

determined in all the samples, if shown to be significantly greater than zero (using the *t*-test – see Box 2.1).

Another approach to the detection of sampling bias is to compare concentration estimates from several totally independent sampling methods. In a sampling test on an area of contaminated land, nine different organisations were recently invited to estimate the mean concentration of lead, using whatever protocol they considered appropriate (Argyraki *et al.*, 1995). The bias between each individual sampler and the overall consensus varied by a factor of 2.5. This demonstrates both the potential magnitude of sampling bias and the benefit of inter-sampler comparison for its detection. It should be possible in principle to establish a 'reference sampling target' for contaminated land or other geochemical media, with a certified concentration value that will allow the direct estimation of sampling bias. This procedure is analogous to the use of reference materials for the estimation of analytical bias (see Box 1.3).

General guidelines for sampling various media

Atmosphere

Sampling of an atmosphere may refer to either the gas phase or the suspended particulate matter. Gas-phase sampling may be required in the open air, or in restricted environments such as within soil. Both types of sampling are susceptible to rapid temporal and spatial fluctuations due to meteorological and social factors, and *in situ* continuous field monitoring of air quality, particularly in urban locations, is replacing laboratory analysis. The optimum positioning and orientation of samplers, together with the timing and duration of the sampling, all need to be investigated in designing the sampling protocol.

Particulate sampling equipment consists of a pump drawing air at a known rate through a filter membrane such as 10 μm glass fibre to collect the suspended particles, although alternative means are available (Markert, 1994, page 125). The duration of the particulate sampling can be calculated to ensure the expected concentration of the analyte is sufficiently high above the analytical detection limit (e.g. × 100) that acceptable precision can be attained (Kratochvil *et al.* 1984).

Water

Fresh water Dissolved elements are traditionally defined as those that pass through a 0.45 μm filter membrane (Chapter 3). Filtration is carried out immediately after the sample is taken, to prevent adsorption of dissolved elements onto the suspended matter. For analysing cations it is normal to acidify the solution to 0.1M (approximately 1% by volume) with high-purity nitric acid, immediately after filtration, to minimise solute deposition onto the container walls; a separate, non-acidified sample should be taken if anions are to be determined, and some analyses (e.g. pH and alkalinity, Chapter 5) may need to be carried out in the field.

Plastic bottles (e.g. polypropylene), soaked overnight in acid (1M analytical grade), are appropriate containers unless organic constituents are to be determined (Chapter 16).

In the sampling of fresh waters, such as springs, lakes, streams and river waters, changes in composition may occur over short time scales, owing to precipitation for example. The effects of sudden peaks of analyte concentration can be reduced by slow-input continuous samplers. Seasonal variability is particularly pronounced in lake water; spatial variation of composition has also been recognised as a serious problem, especially in the surface micro-layer where there may be enrichments of five to several hundred fold in metals such as Pb, Ni and Cu (Barcelona, 1988).

For polluted waters, the sampling protocol must prescribe appropriate safety measures to protect operator health.

Saline waters   Waters with high salinity taken from the sea, an estuary, or some boreholes present similar sampling problems to fresh water, though lower stability may necessitate prompt filtration and acid stabilisation. Rapid refrigeration reduces the rate of alteration, particularly from biological activity which may regulate, for example, concentrations of V and Ni in sea water. Variability of seawater composition on the microscopic scale, such as the surface microlayer, requires a detailed knowledge of marine geochemistry to make reliable interpretations (Chester, 1990).

Solid media

Sediment   A simple polythene scoop is all the equipment that is required for the sampling of surface stream sediment. Wet-strength paper bags, with fold-over closure tabs, are a convenient storage medium in that, unlike polythene bags, they can be transferred directly to a drying oven.

Which part of the sediment to sample cannot be left to subjective judgement. For example, only active steam sediment should be sampled, avoiding any collapsed bank material. It should be sampled only from a specified depth interval (such as the top 2 cm) which should be in the oxidising zone of the sediment. The trace element composition of the sediment changes markedly in this oxidised zone where the precipitation of iron and manganese oxides causes elevation of base metal concentrations due to co-precipitation or adsorption mechanisms. The decision to avoid anthropogenic contamination, or to sample the sediment in a deliberately non-representative manner (e.g. taking only the <170 μm fraction to bias in favour of adsorbed, bio-available trace metals), must be determined by the geochemical objectives of the project.

Soils   A hand-held auger with a thread of 15 by 2.5 cm is ideal for most soil sampling, and wet-strength bags are again a good storage medium. Standardised protocols for sampling soils are in preparation by the International Organization for Standardization (ISO 10381 – Paetz and Crobmann, 1994) . These are appropriate for some regulatory purposes (e.g. environmental site investigations) but not for all scientific investigations. The sampling of soils should ideally be accompanied by a systematic description of the type of soil by a scheme such as that by Munsell (1971). The most appropriate depth or horizon to be sampled can be determined by an orientation survey. The B horizon, for example, is often sampled for mineral exploration in temperate climates. The other sampling parameters, such as the sample weight, the number of sub-samples to be combined into one sample, and whether the surface vegetation should be removed prior to sampling, all need detailed consideration (Webster, 1977; DoE, 1994).

Rocks   For rock sampling the simplest method is to take chip samples of unweathered material from exposures using a hammer. Representative sampling in this situation is problematic as the exposed rock may differ in composition to unexposed rock. Furthermore, rocks will often break along planes of weakness that are liable to have been altered and therefore be chemically unrepresentative. Systematic lithogeochemical sampling may therefore require sampling at a regular interval using a diamond drill core. This method, however, poses its own possibility of bias, in the form of contamination by the drilling materials. Rock chips produced by percussion drilling can be used as an alternative, but information on sample depth is usually imprecise.

Sample bags made of canvas or thick polythene are usually robust enough to prevent cross-contamination between rock samples. It is prudent to split core in half lengthways, sampling one half and storing the other as reserve.

Herbage   The sampling of herbage such as plants, leaves, fruits, bark and roots has the particular problem of the identification of the species required. Furthermore there is the selection of the most appropriate part of the plant, and the optimum time of year for sampling the plant in question (Brooks, 1983). Care must be taken to minimise the effect of soil contamination that can be very pronounced for certain elements that are preferentially excluded by the plant. Careful washing often fails to remove all of the soil particles and the extent of the degree of contamination then needs to be estimated experimentally. This can be achieved using electron microprobe analysis, or by comparison between different acid decompositions (Ramsey *et al.*, 1991).

Sample preparation

The objects of sample preparation are

- to remove unwanted contaminants (e.g. suspended matter from water);
- to preserve the essential features of the sample composition until analysis takes place;
- to convert the form of the sample material into one suitable for chemical analysis;
- in some cases, to separate or concentrate a particular constituent (Box 2.3).

The preparation required varies from the minimal – for waters with no suspended matter – to lengthy multi-step processes such as those employed to separate selected minerals from silicate rocks.

Sample preparation and storage will vary according to the objectives of the study, particularly with regard to the analytes to be determined. For example, agate is preferred for crushing rock samples for trace metal analysis, but is not appropriate for oxygen isotope analysis (Chapter 9). Likewise, samples to be analysed for organic constituents should not be stored in plastic media (Chapter 16).

Waters

Samples filtered and acidified at the time of collection as outlined above need no further preparation, except perhaps for the chemical concentration of certain analytes (see Chapter 3).

Solids

Particulates (filtered from air or water)

For certain instrumental methods of analysis such as neutron activation or X-ray fluorescence, the filter membrane with its particulate load can be introduced directly into the analytical instrument and no preparation other than drying is required.

Other methods require the samples to be presented in the form of a liquid (e.g. ICP–AES (Chapter 4), AAS (Chapter 5)) and the particulate matter together with the filter membrane (if cellulose acetate) need total dissolution, typically in oxidising acids (see Chapter 3). It is often convenient to leave the sample in the filter holder and use this to protect the sample during transport.

Bulk inorganic solids (rocks, sediments and soils)

Most methods of bulk analysis require the sample to be finely divided, firstly on grounds of homogeneity, and secondly to promote dissolution or to facilitate preparation of a machine-compatible solid form (e.g. a powder briquette for X-ray fluorescence analysis). A typical process for preparing solid rock samples for analysis is as follows:

- Removal of weathered or contaminated (e.g. lichen-bearing or mineralised) surfaces using a hammer or a hydraulic rock breaker (Fig. 2.4).
- Crushing of the sample to gravel or small chips (3-5 mm) in a reciprocating steel crusher (Fig. 2.5).
- Grinding of the chips to coarse or fine powder according to the application (Box 2.2). A coarse powder – usually sieved to select a specific particle size range – is required for the mechanical separation of minerals through differences in density, magnetic susceptibility or electrostatic properties (Box 2.3). For bulk analysis, the sample should be ground to a fine powder (consistency of flour).
- Splitting of the powder into representative sub-samples to reduce the volume for storage[1], or to direct into separate analytical streams.
- These physical preparation steps may be followed by one or more stages of chemical processing (Chapter 3).

Sediments require drying (e.g. at $60°C$) before comminution to $<75\ \mu m$ (footnote[2]) using a ball mill or swing mill (Box 2.2). For the purpose of mineral exploration it is usual simply to disaggregate the dried sample, and sieve out the fraction $<170\ \mu m$ (footnote[3]) with a nylon mesh, to concentrate the elements of interest that tend to be partitioned in this size fraction. This powder can then be used directly for analysis if high precision ($<5\%$) is not required.

Soil is defined as material with a natural grain size less than 2 mm. Dried and disaggregated soil should therefore be sieved to <2 mm using a nylon sieve. Air-dried soil, kept at

---

[1]   One usually crushes more material than is needed for the analysis in hand, partly to ensure homogeneity and representivity, and partly to provide for future analytical needs or inter-laboratory exchange.
[2]   Fine enough to pass through sieve mesh number 200.
[3]   Using sieve mesh number 80.

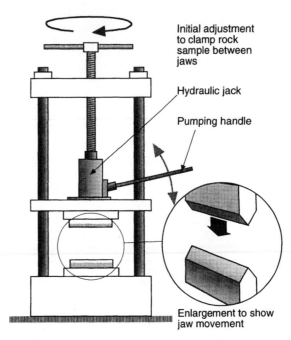

Initial adjustment
to clamp rock
sample between
jaws

Hydraulic jack

Pumping handle

Enlargement to show
jaw movement

Figure 2.4   Hydraulic rock breaker for removing contaminated surface material and breaking rock samples into smaller pieces for crushing. Samples are clamped between the steel jaws, which are compressed by pumping the hydraulic ram. A removable mild-steel housing is placed round the breaker to retain expelled fragments.

this grain size, is most appropriate for the measurement of certain soil properties such as pH and 'extractable' metals (Ministry of Agriculture, Fisheries and Food, 1986).

Spatially resolved analysis   *In situ* analysis of individual mineral grains for elements or isotopes present in an inorganic sample by electron or ion microprobe (Chapters 14 and 15) or laser ablation (Chapters 10 and 11) usually requires a highly polished specimen surface, prepared by slicing the rock sample using a water-cooled diamond saw, grinding it to a flat surface, and polishing on a suitable lap. Microanalysis by laser fluorination (Chapter 9) requires the separated mineral to be in the form of discrete grains (Box 2.3).

Herbage   The inorganic analysis of plant material may be carried out directly on the dried plant tissue, or may require the plant material to be ashed under controlled conditions as described in Chapter 3; this concentrates the analytes by removing the organic bulk of the sample, but results in the loss of volatile elements such as Hg, Se, Cd, As. Freeze drying may be advantageous for some applications. Several types of mill (cross-beater or rotary) are commercially available for comminuting herbage to a specified maximum particle size. Cryogenic milling avoids the Fe or Al contamination associated with standard mills.

The sample preparation required for organic analysis is dealt with in Chapter 16.

Box. 2.2 – Producing fine powder for analysis

Most modern bulk analysis techniques require solid samples to be powdered to the consistency of flour (absence of gritty particles when rubbed between the fingers), either to promote rapid and complete dissolution (Chapter 3) or to ensure that particle size effects do not introduce either poor precision or analytical bias (Chapter 6).

Two types of crushing/grinding equipment are widely used for producing fine powder from hard silicate rocks:

(1) The 'swing mill' consists of a cylindrical pot (with lid) with a free-moving annulus and a puck (Fig. 2.2.1) inside, all fabricated from an appropriate hard material (see below). Coarsely crushed sample is poured in between the puck, annulus and pot walls, the lid is clamped in position, and the assembly is subjected to vigorous horizontal orbital motion on an eccentric drive unit (inside a safety enclosure) for 2–10 minutes. Differential inertial motion of the puck and ring within the pot crush and grind the rock between them to fine powder. The entire assembly is demounted and dismantled for sample recovery and cleaning between samples. A swing mill can typically crush up to 100 g per loading.

(2) The laboratory ball mill (Fig. 2.2.2) consists of a cylindrical mill with removable end pieces. Coarsely crushed sample is placed inside the mill together with 2–3 balls fabricated from the same hard material as the mill. The whole assembly is shaken lengthways by a robust laboratory shaker inside a safety enclosure for up to 20 minutes. Impacts between the balls and end pieces crush sample fragments caught between them. Ball mills cannot handle more than 25 g of sample per loading.

Media used to fabricate such devices need to be hard (to resist abrasion), dense (to crush effectively) and tough (to resist fracture). The commonest materials are:

• hardened steel – cheapest option – acceptable for most non-silicate samples, and for rocks if ferrous metal contamination can be tolerated (Fe, Cr, Ni, Co, Mo, Mn, V);

• tungsten carbide – hard, tough and dense – suitable for most geochemical applications, but contaminates with Ti , Co and Ta;

• agate – chosen for trace element and radiogenic isotope studies because of negligible metal contamination, but agate parts are easily broken, and crushing

times are long because its low density reduces the inertial effect upon which efficient grinding relies;

• alumina ceramic – suitable for crushing non-geological materials, but contaminates with Al and many trace elements (Li, B, Ti, Mn, Fe, Co, Cu, Zn, Ga, Zr, Ba).

In view of health hazards associated with dusts, crushing and powder handling must be carried out only in laboratories in which effective dust-extraction equipment is operating, including extractor ducts and hoods in which fine powder can be safely handled.

Figure 2.2.1 Laboratory 'swing mill'. Elements fabricated from hard grinding medium: A, puck; B, annulus; C, pot lining; D (not shown), lid lining.

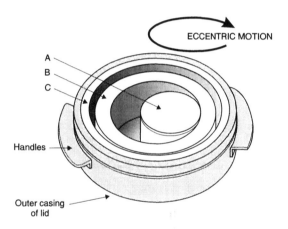

Figure 2.2.2 Laboratory ball mill.

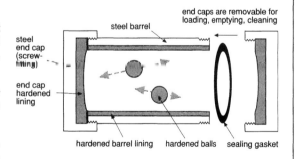

RG

Box 2.3 – Methods of mineral separation

Some geochemical applications (e.g. K–Ar and Ar–Ar dating – Box 9.3) require a pure sample of a selected mineral to be physically separated from the host rock. The first step after removing weathered material is repeated crushing and sieving to reduce the sample to particles of 100–200 μm, the size range in which individual particles are likely to be monomineralic and physical separation methods work most effectively. When crushing is complete the material is repeatedly rinsed in distilled water to remove fine powder (<50 μm).

Most mineral separations employ one or more of the following three methods:

(1) *Hand picking* For isotopic work on mineral separates, where perhaps only 10–20 mg of separated material is required, the quickest route is simply to spread the crushed and sieved sample out on a clean sheet of paper under a binocular microscope, and to hand pick grains of the required mineral using fine tweezers. This work may be carried out in an enclosure flushed by filtered air to minimise contamination.

(2) *'Heavy liquids'* The dried sample is gently agitated (in a separating funnel) in a carefully prepared liquid of **specific gravity** (SG) intermediate between those of the minerals to be separated; thus a liquid of specific gravity 2.87 (such as bromoform – see below) could be used to separate ferromagnesian minerals from quartz and feldspar. The denser fraction settles and can be drained off into a filter to separate the suspension from the heavy liquid (which can be recovered). The light fraction is washed into a separate filter funnel. Allman and Lawrence (1972) and

Hutchison (1974) detail many refinements of this technique.

Two of the heavy liquids traditionally used for mineral separation, the haloalkanes tetrabromoethane (TBE, $C_2H_2Br_4$, SG 2.96 at 20°C) and bromoform (tribromomethane $CHBr_3$, SG 2.89 at 20°C) are now recognised as carcinogenic, and aqueous sodium polytungstate solution (SG 2.96–3.06 at 20°C) is preferred (Gregory and Johnston, 1987). The third, thallium formate–malonate ('Clerici Solution', SG 4.25 at 20°C), is highly toxic and should be prepared and used only by experienced personnel, avoiding all skin/eye contact and dust inhalation.

(3) *Magnetic separation* Minerals that differ in **magnetic susceptibility** are readily separated using an isodynamic magnetic separator (Fig. 2.3.1). The mixed mineral grains fall from a hopper on to a non-magnetic flat-bottomed steel channel passing between the pole pieces of a strong electromagnet. The entire assembly is variably inclined both in the longitudinal direction (so that grains slide down between the pole pieces by gravity, aided by a vibrator – Fig. 2.3.1a) and in the lateral direction (Fig. 2.3.1b). The lateral motion of mineral grains as they progress down the length of the channel is a balance between the force of gravity, acting towards the lower edge of the tilted channel, and the magnetic force pulling them towards the upper edge. High-magnetic-susceptibility minerals therefore tend to descend along the upper edge, whereas lower-susceptibility minerals move along the lower edge. Half way along its length the channel divides in two, each half leading

Errors in sample preparation

Random errors in sample preparation can be detected by processing certain samples in duplicate, in the manner described for determining the precision of analysis (Box 1.2) and sampling (above). Without replicates at the stage of sample preparation, these errors will be included within the overall sampling error, determined from the replicated field sampling.

Possible sources of systematic error (bias) occur at every step in preparation. Unintentional bias may arise by contamination of the sample by the equipment used for preparation (Box 2.2) unless preparation media are carefully selected for the task in hand. Contamination from the preparation equipment can be detected by the processing of **blank** material, such as vitreous silica or clean silica sand, through the grinding procedure. The subsequent analysis of these 'prep–blanks' gives an indication, but not a quantitative measure, of the contamination originating in the equipment for samples of interest.

Residues of preceding samples retained in the equipment may also introduce bias. An extreme example of this can occur when discrete flakes of metallic gold from gold-bearing

Figure 2.3.1 Magnetic separation of minerals using the Franz isodynamic magnetic separator. The open arrow in (b) indicates the differential magnetic force acting on grains with higher magnetic susceptibility.

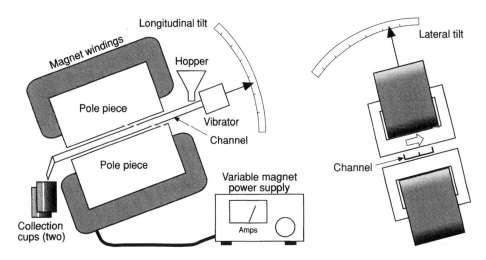

(a) Side view. Mechanism is tilted sideways away from the viewer.

(b) End view. Mechanism is tilted longitudinally toward the viewer.

to a separate chute feeding its own metal beaker in which the separated grains collect.

A mineral's magnetic susceptibility, being composition-dependent, varies quite widely, and therefore the optimum tilt angles and magnet current need to be determined empirically for each application. Magnetic susceptibilities fall into four broad ranges:

magnetite > ferromagnesian silicates > epidote, zircon, sphene, etc. > feldspars

In order to separate olivine, pyroxene or amphibole, it is advisable to remove highly magnetic oxide minerals first using a hand magnet.

RG

rock samples are smeared on the walls of grinding mills, if the grinding time is excessive. Grinding and discarding a small amount of the new sample before the main part is introduced may reduce such errors to an acceptable level.

Volatilisation of certain analytes, such as mercury, can occur in the elevated temperatures caused by the grinding process, or samples may undergo oxidation (Fitton and Gill, 1970). Analytical bias may arise from uneven comminution of a sample due to hardness differences between component minerals.

Most preparation techniques contaminate with some elements, so zero contamination of all elements is impracticable, particularly for multi-element analysis. A more realistic objective is to achieve acceptable levels for error from sample preparation.

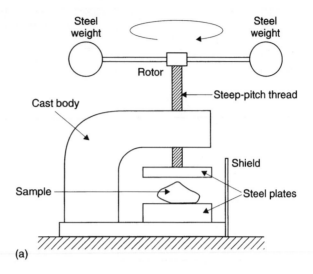

(a)

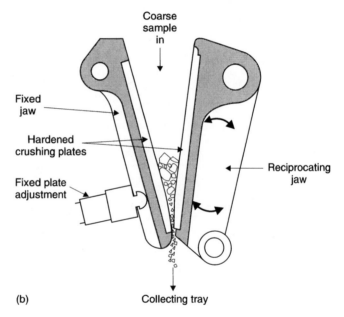

(b)

Figure 2.5   (a) *Fly press.* Pieces of the rock sample are placed between the two horizontal plates (apart) and the rotor is spun vigorously to drive the plates together. The momentum of the weights is sufficient to break most samples into small pieces. This crushing device is easy to clean. (b) *Reciprocating rock crusher.* Pieces of rock are introduced between the reciprocating and fixed jaws, and crushed material falls through to collect in a tray beneath. When the crusher has been switched off and come to rest, the fixed jaw assembly can in some models be swung out to facilitate brushing between samples.
The aperture must be covered immediately after introducing each piece of rock, to prevent fragments of rock flying out and causing injury.
Eye protection must be worn at all times when breaking and crushing rock samples.
Note: The plates between which the rock samples are crushed in both devices are of steel. Samples may therefore become contaminated by occasional small splinters of steel, which may if necessary be removed from crushed material by spreading out the powder on a clean surface and drawing a small permanent magnet across it (covered by a sheet of paper or plastic to keep the magnet clean). This procedure may, however, remove a fraction of any magnetic minerals present as well.

Chapter 3

Dissolution procedures for geological and environmental samples

J.N. Walsh, Robin Gill and Matthew F. Thirlwall

Many of the techniques described in this volume are 'solution' methods of analysis, including atomic absorption spectrometry (AAS, Chapter 5), inductively coupled plasma–atomic emission spectrometry (ICP–AES, Chapter 4) and ICP–mass spectrometry (ICP–MS, Chapter 10). These analytical techniques normally require the sample to be introduced as a stable solution and the analysis obtained can only be as reliable as the dissolution of the sample.

All geological samples can be dissolved quantitatively and there is an extensive literature on the various procedures available for sample dissolution. As silicates are chemically resistant, powerful reagents are needed to attack them and stringent safety measures must be followed (Box 3.1). The texts by Bock (1979) and Sulcek and Povondra (1989) provide information on methods of decomposition for inorganic materials in general, and there are several detailed discussions of sample decomposition and dissolution for geological samples, including Jeffery (1970), Johnson and Maxwell (1981), Potts (1987), Thompson and Walsh (1989) and Jarvis (1992).

It is convenient to group the methods for sample dissolution into four categories according to the type of sample and the results required from them:

(1) Samples already in solution or readily soluble in water.
(2) Samples that are 'leached' with nitric, hydrochloric and/or sulphuric acids.
(3) Samples that are dissolved using hydrofluoric acid (with another mineral acid).
(4) Samples that are dissolved following fusion with an appropriate flux.

Where trace elements are to be determined, reagent purity becomes an important factor (Box 3.2) and a reagent **blank** should always be prepared.

All of the procedures given in this chapter can be carried out safely in a properly equipped laboratory by trained personnel. They should never be attempted without appropriate laboratory training and supervision, nor in inadequately equipped laboratories.

Solution samples and water-soluble solids

Some samples are already available in solution form (groundwater samples, for example) where the only further treatment necessary is to stabilise the solution. The normal analytical protocol will recommend that water samples are first filtered at the point of collection (see below) and then acidified with nitric acid (to a 1% total HNO_3 concentration) in order to stabilise dissolved metals. Nitric acid is now usually preferred to hydrochloric acid for this

Box 3.1 Safety in the geochemistry laboratory

Safe laboratory practice must be scrupulously observed in preparing rock samples for analysis as the reagents used present very specific dangers:

Hydrofluoric acid (HF)

Usually supplied as 40% aqueous solution. Attacks most silicates, including laboratory glassware:

- HF digestions must be carried out in plastic (PTFE) containers or Pt crucibles.
- Excess HF must be removed (by evaporation) or complexed (boric acid) before sample solutions are transferred to glass.
- Operations involving HF must be confined to specialist fume cupboards devoid of glass (polycarbonate sash and baffle) and bare metal (plastic fittings and welded PVC ducts).

Operator safety: Specific training is required for HF operations. HF contact causes no immediate burning sensation or smarting, but it readily penetrates deep tissue and causes intense pain after an hour or two. The user must:

- wear protective clothing and face shield;
- wash hands and fingernails immediately after use;
- carefully clean up even minor spills and trickles;
- label all containers of HF.

Skin contact: Wash copiously with water. Apply glutamate jelly. Seek medical attention.

Perchloric acid ($HClO_4$)

Usually supplied as 60% aqueous solution. Hot concentrated acid is a powerful oxidising agent: reacts vigorously or violently with organic matter. Forms unstable perchlorates with many metals. Acid may accumulate in fume cupboard ducting.

- Must be used only in a dedicated fume cupboard free of organic reagents and residues; for practical reasons this should also be HF compatible (see above).
- Must not be used on samples containing organic matter unless they are first ignited in a furnace.
- Avoid contact with metals.

Operator safety: The user must wear protective clothing and face shield.

Other concentrated mineral acids

All operations involving concentrated acids must be conducted in a fume cupboard. The operator should:

- wear eye protection and protective clothing;
- wash hands immediately after use;
- immediately clean up spills;
- when diluting, always add acid to water, not vice versa.

Box 3.2. Reagent purity and blank values

High-purity reagents and clean apparatus are essential for trace element analysis if the analytical signal is not to be masked by high blank readings.

Water

Distilled water prepared in a traditional metal still is insufficiently pure for trace element analysis and is usually distilled a second time in a silica glass still designed to operate by evaporation below boiling temperature ('sub-boiling distillation'). Alternative approaches employed in many laboratories are (a) to use a high-quality water de-ioniser, which offers greater throughput than a still and allows water quality to be continuously monitored in terms of conductance, or (b) to purify water using proprietary reverse osmosis apparatus.

Acids

Analytical-grade (AnalaR) acids are sufficient for routine trace element work above 10 ppm, but for low-level trace element determinations, for isotope dilution determinations and for radiogenic isotope analyses by TIMS (Sr, Nd, Pb, etc – Chapter 8) it is necessary to purify acids further in the laboratory. Use of a Pyrex glass still leads to high blank values of elements like Sr (typically 30 pg g^{-1}). Use of a silica glass sub-boiling still reduces this to ~5–10 pg g^{-1}. Still higher purity can be obtained (1–2 pg g^{-1}) using a sub-boiling still consisting of two clear 1 litre Teflon® bottles connected (in a 90° configuration) through a threaded Teflon block into which the bottle necks are screw fitted. Analytical-grade acid is placed in one bottle and evaporated from outside by an infrared lamp, while the other bottle is immersed in a cold water bath to condense and collect the distilled product. This design may be used to distil all the comon mineral acids and HF.

Apparatus and laboratory environment

Surfaces and sample containers are usually cleaned by soaking and/or heating in the strongest reagents which will be used in those containers.

Many 'clean' laboratories conducting low-abundance trace element and isotope analysis operate a positive-pressure regime: filtered air is pumped in to elevate the air pressure slightly above atmospheric, such that at all interfaces with the outside world (e.g. doorways) there is a net *outward* flow of air to minimise ingress of airborne particulate contamination. This is usually reinforced by shoe changing or use of over-shoes and a sticky mat at the entrance. More stringent measures are needed for Pb isotope laboratories.

MFT and RG

---

purpose because of possible interference effects from chloride ions affecting ICP–MS and flameless AAS.

Although the analysis of metals in water samples using solution methods of analysis appears straightforward, some consideration should be given to the distribution of metals in natural water samples. Waters may contain metals in true solution, as colloidal material, in suspended particulate material, and adsorbed onto the surfaces of particulate material. Many metals precipitate from solution in alkaline conditions and redissolve when the pH of the solution is reduced. The use of a 0.45 μm nitrocellulose filter (normally used for pre-treatment of water samples) is entirely arbitrary and some 'particulate' matter will undoubtedly pass through this and end up in the sample presented for analysis. The use of 1% HNO_3 will cause some metals to be leached from the very fine particulates. Changing the pH of the solution can cause metals that are in colloidal form to redissolve. Adhering to a recognised water pretreatment protocol thus has the benefit of establishing consistency for the analysis, though its scientific basis may be somewhat doubtful.

Most solid geological samples cannot be dissolved in water and strong acids are required as a preliminary to using solution methods of analysis. However, a limited number of sample types can be dissolved directly in water (evaporites are the most obvious example). It is also worth noting that two widespread accessory minerals in sedimentary rocks, halite and gypsum/anhydrite, are dissolved in cold water.

Acid leaching

Samples can be selectively attacked by treatment with the standard mineral acids (hydrochloric, nitric or sulphuric acids, for example). This method of sample digestion makes no attempt to dissolve the silicate matrix present in most geological samples. The elements to be analysed are 'leached' from the sample. This method is well suited to carbonate-rich samples (where acetic acid may be used to dissolve carbonates without attacking the silicate matrix), to geological samples being analysed for exploration purposes (where the elements of interest are in readily dissolved sulphide minerals, for example), and to many environmental monitoring programmes where interest is focused on adsorbed contaminants rather than the insoluble substrate. In this last case it is reasonable to assume that, if the elements do not dissolve in these strong mineral acids, they cannot be regarded as being biologically available and are unlikely to pose any environmental hazard.

Aqua regia is commonly used as the leaching acid, although nitric acid alone, and occasionally other acid mixtures, may be recommended for specialised applications. Aqua regia (three parts HCl + one part HNO_3) will leach many metals, notably base metals (Cu, Cd, Mn, Pb, etc.) from rocks with considerable efficiency; for example, Cu recoveries are almost always close to 100%. The sample is normally treated with the acid at 80–95°C for 1–3 hours. The effectiveness of aqua regia for sample attack and dissolution is thought to be due to the complexing power of the chloride ion and to the catalytic effect of Cl_2 and NOCl and of the chloride ion present as the reaction takes place (Bock, 1979). Occasionally 'reversed' aqua regia (three parts HNO_3 + one part HCl) may be preferred. One advantage with using aqua regia, compared with evaporations using HF or fusion techniques, is that volatile components (As, Hg, S, Sb, Se etc.) can be retained. Another advantage is that the total concentration of dissolved solids in the resulting filtered solution is minimised. This can be important for low-level trace element work with some analytical methods. The use of hydrochloric acid has some disadvantages in ICP–MS and flameless AA where it can cause analytical **interferences**, and for these techniques nitric acid alone is preferred. However, this is considerably less effective at leaching metals from many minerals, and the analysis obtained for an aqua regia leach of a geological sample will be very different to that obtained following a nitric acid leach.

A further, very important problem encountered with analyses based on partial acid leaching of the material is that there are few recognised **reference materials** available that can be used for method validation. Where standards are available, considerable care is needed in following the sample pretreatment **protocol** to ensure **reproducibility** in the analytical determinations.

Acid leaching may also be useful for selectively removing unwanted components of rock samples to eliminate bias, for example to remove hydrothermally introduced Sr from a basalt being analysed for original Sr isotope composition. In such cases the leachate is discarded, and the solid residue is analysed.

Analysis of environmental samples - digestion procedures

Acid digests are widely used in the analysis of environmental samples, including sample matrices such as soils, sewage sludges, air particulates collected on filters, domestic, urban and industrial dusts, refuse, and plant and animal tissue. These samples are normally prepared by dissolving out the elements of interest with mineral acids under prescribed conditions. Analytical protocols are available (e.g. US EPA 9010 for total cyanide) and should be selected according to what the analysis is attempting to show. Dissolution in cold dilute nitric acid may be a preferred preparation, for example, where the analysis is intended to determine the amount of analyte that is biologically available. Increasing acid strength and temperature will release increasing amounts of elements, even partially leaching many elements from silicate minerals, notably from clay minerals. A 'complete' analysis will, however, require the complete dissolution of the sample and for many environmental samples this will involve using hydrofluoric acid to break down any silicate minerals present.

Another example of selective dissolution is extracting a particular *species* of an element. For example, treating a soil with petroleum ether can solubilise any Si that may be present as the organosilicon pollutant siloxane. This procedure, however, depends on the sample preparation not altering the speciation.

Wet or dry 'ashing'

When samples containing organic matter are to be analysed for inorganic constituents by solution analysis, it is normal to remove the organic component quantitatively by 'ashing'. One of two alternative approaches can be employed. The organic material may be 'burnt off' by heating the sample in air – 'dry ashing' – or the untreated sample may be digested in an acid (or an acid mixture) that is strongly oxidising – so-called 'wet ashing'. Both of these procedures are widely used and each offers advantages and disadvantages.

Dry ashing is relatively simple: heating the sample in air at temperatures in excess of $500°C$ will remove organic material and the ignited residue can then be treated as any inorganic sample, digesting it in nitric acid, aqua regia, etc. The limitation is that several volatile elements are likely to be lost during ignition. Hg, Se and As are elements of environmental concern that are likely to be almost entirely removed. Other elements may be partially lost, for example Pb and Cd. Loss of volatile elements during ashing may be minimised by adding 'ashing aids' before ignition (magnesium nitrate has been used with some success) but this approach has not been widely used as it introduces into the sample a large excess of unwanted material, with evident potential for contamination.

Two mixtures of acids have been extensively used for 'wet ashing' digestion of samples with a significant organic content:

(1) mixtures of sulphuric and nitric acids;
(2) mixtures of perchloric and nitric acids.

Samples are heated with the acid mixture for a number of hours in open vessels, and the organic material is oxidised gradually as the digestion proceeds. There is some safety hazard associated with both of these techniques (Box 3.1) and care must be taken not to overheat the samples. Wet ashing does not suffer from the limitation of loss of volatile elements. However, the process can be time consuming, may not fully dissolve the sample and needs constant supervision. The methods were discussed in some detail by Jolly (1963)

Hydrofluoric acid digestion

Dissolution using hydrofluoric acid (HF), in combination with another strong mineral acid having a higher boiling point, is a widely used procedure for 'opening up' silicate-based materials. Most of the common silicate minerals are attacked by hydrofluoric acid, especially when warmed, although a few refractory minerals will fail to dissolve quantitatively.

The most widely used method for dissolving rock samples is an open evaporation of the sample with hydrofluoric acid together with nitric (HNO_3) or perchloric ($HClO_4$) acids in a **Teflon**® beaker or platinum crucible (Fig. 3.1a) on a hotplate and/or under an infrared lamp. The evaporation must be done in the presence of an acid with higher boiling point than HF to ensure that insoluble fluorides are converted to more soluble salts (nitrates or perchlorates) on completion of the evaporation; for this reason hydrochloric acid (which boils at a lower temperature than HF) cannot be used. HF attacks the silicate matrix of the sample and excess acid is lost as the evaporation proceeds. Silicon is quantitatively removed as the volatile SiF_4. The remaining solids are dissolved in nitric or hydrochloric acid. Perchloric acid is preferable to nitric for the evaporation: it has a higher boiling point and is more effective at dissociating fluorides, some of which are sparingly soluble. However, perchloric acid requires special precautions because of the danger of explosive reaction with organic material (Box 3.1). It is also more expensive than nitric acid and the residual chloride ions may cause interferences in ICP–MS.

Some refractory minerals fail to dissolve quantitatively in the HF–$HClO_4$/HNO_3 attack, including cassiterite, chromite, tourmaline, rutile and zircon. In practice, the failure to dissolve zircon fully is the most serious limitation for most rock compositions, and quantitative Zr data cannot be obtained with this type of sample preparation.

The loss of Si will normally be an advantage as the total content of background-enhancing dissolved solids is reduced and trace element detection limits will thus be improved, but the disadvantage for major element analysis is obvious. In addition to Si, other volatile components will be lost in the evaporation (see comments in preceding section). A further disadvantage of open evaporation is the possibility of airborne contamination, either from the atmosphere or from the boiling of an adjacent beaker; this may be significant at

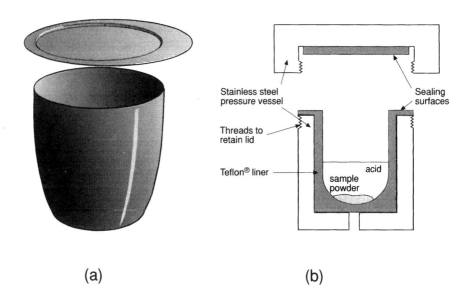

(a)   (b)

Figure 3.1   (a) Platinum crucible. A snug-fitting lid is provided to retain heat during fusions but is not used for evaporations. A hot Pt crucible should be handled only with Pt-tipped tongs to prevent degradation
through alloying with iron. Crucibles are also available in other precious metals and alloys. (b) A simple
form of Teflon bomb with steel pressure vessel. In more elaborate versions, the top of the container is
machined to a sharp edge which forms a seal with a V-shaped groove in the lid.

the 1–2 ppm level for Pb, with implications for Pb isotope studies (Chapter 8). To avoid
contamination, the evaporation may be carried out in an acid-resistant (usually Teflon)
evaporation cell flushed by filtered air to carry away fumes, or under a clean air hood.
Another alternative to the open evaporation method is to use a sealed digestion vessel.
This may be a fully sealed Teflon container in a stainless steel pressure vessel (a 'bomb',
Fig. 3.1b), a sealed Teflon container that can be placed in a microwave, or simply a sealed
Teflon vessel that can be heated on a hotplate. These procedures overcome some. but
not all the limitations in the method. They certainly can be used to reduce atmospheric
contamination and retain volatile components. It is possible to determine Si using a
closed-vessel hydrofluoric acid dissolution at temperatures up to 180°C; the capsule
must be cooled to room temperature before unsealing, and the excess HF is usually ren-
dered unreactive as the HBF_4 complex by adding boric acid (as saturated solution), to
minimise damage to glassware in subsequent handling.

However, closed-vessel digestions are not entirely successful at dissolution of refractory
phases. High temperatures in the region of 200°C are required to dissolve refractory minerals such as zircon. This can be achieved in bombs where the Teflon liner is enclosed in a
stainless steel pressure vessel, a technique successfully used for dissolving zircon for Pb isotope analysis (Krogh, 1973). However, such bombs are inconvenient to use if a large number
of samples are to be analysed. The microwave-based systems that are available commercially
contain Teflon liners in plastic retaining vessels. These will not withstand the pressures
that the steel liners can achieve and they cannot be used with perchloric acid because of the
risk of explosive reaction with plastic. When used with nitric acid and hydrofluoric acid,
the microwave vessels cannot reach the necessary temperatures as the boiling point of the
HNO_3–HF mixture is too low. Consequently analytical experience has demonstrated that
microwave systems, as currently available, are no more successful at dissolving refractory

phases than open evaporations. The same limitation will apply to the closed unlined Teflon vessels that are heated on a hotplate as temperatures here will be no higher.

It is possible to use hydrofluoric acid in combination with sulphuric acid. This has a much higher boiling point and used in a microwave digestion system it is capable of greater success in dissolving zircon. However, the sulphuric acid presents other analytical difficulties and the sulphates produced are much less easy to dissolve.

Digestion by fusion

Fusing a silicate sample with an appropriate **flux** is the most complete and rigorous method for sample dissolution. There are virtually no known silicate minerals that cannot be brought into complete solution when fused with the appropriate flux. The range of fluxes in common usage is considerable. Most are salts of the alkali metals: sodium carbonate and sodium hydroxide were widely used but have now been largely replaced by lithium metaborate ($LiBO_3$) and lithium tetraborate ($Li_2B_4O_7$).

Lithium metaborate is probably the most widely used flux in current use for the dissolution of geological samples. It has several advantages: it will attack almost all the major and accessory minerals (Ingamels, 1970; see Bock, 1979; Cremer and Schloker, 1976); it introduces only Li and B into the dissolved solution; it can be used at favourable flux:sample ratios (three parts flux to one part sample by mass); and fusions can be carried out in platinum, zirconium or graphite crucibles. The normal procedure (Thompson and Walsh, 1989) is to mix 0.5 g of sample carefully with 1.5 g of flux in a platinum crucible, and heat to $900°C$ in an electric muffle furnace or over a Meker burner for approximately 30 minutes. The sample is cooled and the crucible and fusion mixture are placed in 300 ml of water containing 10 ml HNO_3. A stirring bead is placed inside the crucible, the solution placed on a magnetic stirrer, and the solution stirred until the fusion mixture has fully dissolved. It is important to ensure that stirring is continuous and that as the sample dissolves it does so into a dilute solution, otherwise there is a risk that the silica will polymerise and fail to dissolve.

An alternative is to carry out the fusion in a graphite crucible and whilst the mixture is still molten pour it into a dilute nitric acid solution. This should then be stirred vigorously without delay. This is a more rapid procedure (samples should fully dissolve in 10 minutes), and avoids the risk of mineralised or high-organic-content samples attacking and damaging expensive platinum crucibles. However, it is a somewhat slow and labour-intensive procedure.

Alternative fluxes are available and should be considered. Lithium tetraborate is an alternative to the metaborate, but offers little advantage and the fused mixture is slower to dissolve. Boric acid has also been used as a flux and has the advantages of introducing only B into the analyte solution. However, the fused mixture forms a viscous melt that is difficult to remove from the crucible.

Sodium carbonate (Na_2CO_3), peroxide (Na_2O_2) and hydroxide (NaOH) are also potential fluxes. Obviously they can only be considered if Na is not required in the analysis. Sodium carbonate requires a high flux:sample ratio (six parts flux to one part sample) and does not fully attack some refractory minerals. Sodium peroxide can be used at comparatively low flux:sample ratios (4:1) and only low temperatures (less than 600°C) are needed to carry out the fusion or sintering. Sodium hydroxide fusions require the highest flux:sample ratio (10:1) but they can be carried out in inexpensive nickel crucibles, are completed rapidly (approximately 2–5 minutes) and will attack quantitatively most silicates including many refractory phases.

Dissolution of a geological sample using a fusion procedure should ensure complete quantitative attack and subsequent dissolution of all the elements present in the sample. Where Si is to be determined in the solution, fusion methods are normally preferred. Fusion methods have the significant disadvantage, however, of introducing substantial amounts of additional solid into the solution, and greater dilution of the sample may then be required (see below) and careful matching of standards is likely to be necessary. This can be a serious limitation in trace element analysis using some analytical methods.

Adjustment of analyte concentration

In all forms of sample dissolution for quantitative analysis, a known mass m of solid sample is weighed into the digestion vessel, and on completion of the digestion the resulting solution is made up (diluted) with distillied water to a known volume, V, in a volumetric flask of appropriate volume. (In laboratories where high sample throughput is required, however, solutions may instead be made up to a predetermined weight in the digestion vessel before transfer to a storage tube; this obviates time-consuming quantitative transfer techniques and expensive graduated glassware.)

What is determined in subsequent analysis is the concentration of analyte i in the solution, C_i^{soln}. The mass concentration in the solid C_i^{solid} (see Chapter 1) can be calculated using the formula

$$C_i^{solid} \text{ ppm} = C_i^{soln} \frac{V}{m} \qquad [3.1]$$

where C_i^{solid} is in ppm ($\mu g \ g^{-1}$), C_i^{soln} is in $\mu g \ ml^{-1}$, V is in ml and m is in g. Alternatively

$$C_i^{solid} \text{ mass \%} = C_i^{soln} \frac{V}{m} \ 10^{-4} \qquad [3.2]$$

Dilution

For multi-element techniques such as ICP–AES where major and trace elements are to be determined in the same solution (Chapter 4), the digested sample should be made up to the minimum practicable volume that is consistent with the constraints of the technique (e.g. the limit on total dissolved solids in ICP–MS – Chapter 10).

Where analytes differ widely in abundance or in analytical **sensitivity**, as for example in atomic absorption spectrometry, it may be necessary to prepare more dilute solutions for specific analytes in order to ensure their concentrations lie within, or close to, the linear portion of the calibration (Chapter 5). A predetermined volume of the sample solution is pipetted into a volumetric flask and made up to the mark using distilled water or dilute acid. A reagent blank of the same dilution should also be prepared.

Preconcentration procedures

Low-abundance analytes may have concentrations in the initial sample solution that lie near to or below the detection limit of available analytical methods. For example, the heavier **rare earth elements** (HREEs) occur in many igneous rocks close to or below the ppm level. When 0.2 g of rock is digested and made up to 20 ml (a typical dilution factor), these elements will be present in solution at concentrations around 0.01 μg ml^{-1}, unacceptably close to the ICP–AES detection limits which vary from 0.002 to 0.05 μg ml^{-1} (Potts, 1987). To gain precise analyses of such elements, it is necessary to increase quantitatively the analyte's concentration in the solution prior to analysis.

Simply evaporating the solution fails to solve the problem because it concentrates major elements as well as trace elements, increasing the **background** signal and the potential for **interference**. There are three commonly used methods for *selective* preconcentration of a specific analyte:

(1) *Ion exchange separation* The sample solution is allowed to percolate slowly down a vertical glass column (Fig. 3.2a) packed with **ion exchange resin** (in the form of minute beads). Different solute ion species are attracted to varying degrees by active sites on the resin, and accordingly migrate down the column at different rates as carrier solution (the 'eluant') is passed through the column. The analyte(s) can be recovered at the bottom of the column by collecting the predetermined fraction in which it is concentrated. Other solutes emerge in separate fractions which can be discarded. The fraction containing the analyte(s) – the 'eluate' – can be evaporated to dryness and the solids redissolved in a few millilitres of dilute acid.

Ion exchange separation is widely used for separating all elements of interest in **radiogenic** isotope studies, for example Sr, Sm, Nd, U, Th, Pb (see Chapter 8), and for the separation of **rare earth elements** for analysis by ICP–AES (Chapter 4) or isotope dilution–mass spectrometry (Chapter 8). Two main types of procedure are used: chromatographic and ion-selective procedures.

- *Ion exchange chromatography* is based on the distribution of cations between reagent and resin. The nature and volume of both resin and reagents determine how long an element of interest is retained on the resin: since each cation species has a specific distribution coefficient, a determinable volume of reagent is required to elute the cation of interest. This volume may be simply determined by passing a solution containing the element of interest through the resin, collecting successive eluted fractions and identifying the fraction containing the element by a colour-change indicator, or using ICP–AES (Chapter 4). The reagent volume required changes with reagent molarity and resin volume, and thus care must be taken to keep these parameters constant. In separating an element for mass spectrometric analysis the critical factor is selecting a reagent molarity that eliminates overlap into the collected volume of **isobaric** interferents or of elements that may cause ionisation suppression. The chromatographic method has commonly been used for Sr separation, with Sr being eluted after passing some 45 ml of 2.5M HCl through about 30 ml of cation exchange resin. The only isobaric interferent, Rb, is eluted much earlier, but if higher acid molarities are used, the smaller reagent volumes required tend to

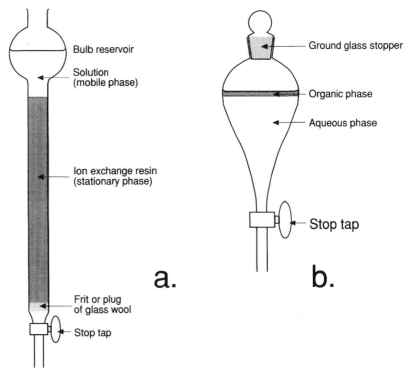

Figure 3.2   (a) Ion exchange column; (b) separating funnel

lead to overlap between Rb and Sr. Lower HCl molarities require excessive reagent volumes and consequently larger Sr contributions from the HCl.

- *Ion-selective* procedures rely on the element of interest forming an ion complex, usually anionic, that adheres to the resin whilst a particular reagent is eluted, but breaks down when the reagent is changed, allowing collection of the required element. For this to be effective, interfering elements should not form a similar anion complex. Complexes used in ion-selective procedures in isotope geochemistry include the lead bromide anion formed in 1M HBr solution, the uranium complex generated by dissolution in 8M HNO_3, and the REE–acetate complex anion utilised in the REE separation procedure of Hooker *et al.* (1975). These procedures can produce very pure separates of Pb, U and REEs with relatively small volumes of reagent being used, although the REE separate may also contain substantial Ti and Zr in Ti-rich rocks.

(2)   *Solvent extraction* A number of elements form stable complexes that partition strongly into the organic phase when an aqueous sample solution is shaken with a smaller volume of an immiscible organic solvent such as methyl isobutyl ketone (MIBK, 4-methylpentan-2-one). After shaking in a separating funnel (Fig. 3.2b) and being allowed to separate, the denser aqueous fraction (containing the bulk of the solute) may be drained away leaving behind the organic fraction enriched in the analyte complex, which may for example be analysed directly by flameless AAS (Chapter 5). Extracting Au from a bromide solution into MIBK yields a detection limit of $\sim$1 ppb by electrothermal AAS (Meier, 1980).

These preconcentration methods lower the **detection limit** in two ways: by concentrating the analyte in a smaller volume of solution, and by removing the bulk of the dissolved solids which elevate the **background** signal and may cause **interference**.

(3)   *Fire assay* Certain elements, notably the platinum group metals, have a strong affinity for the metallic state: when a silicate sample is fused in intimate contact with a molten metal (usually Pb or Sn), such elements partition strongly into the metallic melt. In Pb fire assay, the sample powder is mixed with a flux containing PbO and a reducing agent (e.g. flour) and heated to 1000°C. Reduction of PbO during heating yields molten Pb, which scavenges platinum group metals and sinks to the base of the silicate melt where it is readily separated on cooling. A second stage of concentration is achieved by heating the Pb bead to 900°–1300°C and removing the Pb as oxide, leaving behind a metallic residue of noble metals which can be dissolved in aqua regia for analysis by, for example, flameless AAS (Chapter 5) or neutron activation (Chapter 7). Details of this time-hallowed technique are given by Potts (1987).

Other chemical separation techniques include co-precipitation (e.g. Potts, 1987, page 492) and electrodeposition (Arden and Gale, 1974).

Chapter 4

Inductively coupled plasma–atomic emission spectrometry (ICP–AES)

J. N. Walsh

Introduction

Inductively coupled plasma–atomic emission spectrometry (ICP–AES) is now established as one of the most widely used techniques of elemental analysis. ICP–AES was developed as an experimental technique during the 1960s, commercially available instruments became available in the 1970s, and during the 1980s it was established as the method of choice in a great diversity of application fields.

The inductively coupled plasma was originally devised as a medium for growing crystals at high temperatures (Reed, 1961), but its potential as a source for emission spectrometry was soon recognised. The outstanding potential of the ICP as a source for elemental analysis was apparent from the earliest days. Specifically, the very high temperature of the plasma was noted, and it was appreciated that many spectral lines for the elements could be produced, with atom and ion lines available for even the most refractory elements. Chemical bonds could not survive in the ICP; thus traditional chemical interference (as found in atomic absorption spectrometry) should be eliminated. The early work also demonstrated that the background signal from the plasma itself was comparatively low, ensuring good detection limits. The second distinguishing feature of the ICP as a spectroscopic source (which was not appreciated in the early days) was its ability to give linear calibration lines over several orders of magnitude. This feature was only recognised after the pioneering work of Greenfield and co-workers who developed an appropriate sample introduction method. Linearity of calibration lines is an essential feature for routine analysis that permits simultaneous measurement of major and trace components.

As the name suggests, ICP–AES measures the atomic spectra of the elements being determined. The origin, element–specific nature and utilisation of atomic spectra are dealt with in Box 4.1.

Instrumentation

There are three essential components in an ICP–AES instrument (Fig. 4.1): the source unit (the ICP torch); the spectrometer; and the computer. The source unit provides the energy to generate the emission spectral lines. The spectrometer separates and resolves these lines and measures their signal strength. The computer enables the analyst to convert the signal into a numerical measurement of the concentration of the analyte elements. There is a

Box 4.1 Atomic spectra

Different types of atom emit and absorb **electromagnetic waves** (such as visible **light**) at their own characteristic wavelengths. This fact allows the analyst to detect and analyse individual elements without having to separate them chemically. *Spectral* techniques that exploit this element-specific nature of atomic spectra form the foundation of today's rapid multi-element methods of chemical analysis.

How does an atom absorb or emit electromagnetic waves, and why does the wavelength vary between one element and another? The answers to both questions lie in the **quantised** energy levels occupied by electrons in atoms. These are determined by wave mechanical considerations; readers unfamiliar with this subject will find a simple introduction in the books by Gill (1995, Chapters 5 and 6) and Shriver *et al.* (1994, Chapter 1).

Ground and excited states

An atom in which every electron occupies the lowest energy 'slot' available to it is said to be in its *ground state*. Such a situation is illustrated in a simplified way for the potassium atom in Fig. 4.1.1. The valence electron of potassium occupies the 4s orbital. Higher-energy orbitals are vacant. The ground state is the most stable configuration the atom can have, the one normally encountered at low temperatures.

Injecting energy into the atom, by means of heat, incoming photons or energetic particles (e.g. electrons), may cause an electron to 'jump' temporarily to a vacant orbital at a higher energy, or be ejected out of the atom altogether. Figure 4.1.1(a) shows a valence electron being promoted, but when the energy input is large (e.g. an X-ray photon) the promoted electron may come instead from a deeper energy level (not shown). In either case, a temporary vacancy is created at a lower energy than the promoted electron. This *excited* state of the atom is unstable, and almost immediately an electron will fall spontaneously from a higher level to fill the vacancy. In doing so it dissipates an amount of energy (the energy difference between these two states) which is *radiated* as an electomagnetic photon. Figure 4.1.1(b)

shows a number of possible transitions and the wavelengths of the associated atomic emission lines.

At low (room) temperatures most atoms in a substance are in their ground states. As the temperature is increased, however, the proportion of excited atoms in the substance rises (the reason why flames emit light).

Wavelength and atomic number

Planck showed that a simple relationship exists between the frequency v (in hertz) of any **monochromatic** beam of electromagnetic radiation and the kinetic energy E_q (in joules) of the constituent **photons**:

$$E_q = hv \qquad [4.1.1]$$

where the coefficient h is Planck's constant (6.626×10^{-34} J s). Since $\lambda v = c$ (the speed of light), it follows that

$$E_q = hc/\lambda \qquad [4.1.2]$$

where λ represents the wavelength of the electromagnetic wave. If all the photons under consideration are derived from the same type of electron transition in atoms of the same type, the photon energy is the same for every photon, and equals the energy difference between the initial and final electron energy levels between which the transition takes place (Fig. 4.1.1b):

$$E_q = E_{\text{initial}} - E_{\text{final}} = \Delta E$$

Thus

$$\Delta E = hc/\lambda \qquad [4.1.3]$$

This equation indicates that the wavelength emitted by electron transitions between two energy levels in an atom is inversely proportional to the energy difference between them. As electron energy levels, and the energy differences between them, are related to the charge on the nucleus (= the **atomic number**, Z – see Gill, 1995, Chapter 6), it follows that the wavelength of a particular transition will depend on the *identity* of the atom emitting it. This fact is fundamental to all spectral methods of chemical analysis. The λ–Z relationship is particularly straightforward for X-ray spectra (Box 6.1).

wide diversity of instrument designs and configurations, but all use the same basic principles. Much emphasis has been placed in the past on 'optimisation' procedures for ICP instruments and such investigative work is undoubtedly valuable. Nevertheless, one must not overlook the fact that one of the main reasons for the success and widespread acceptance of ICP–AES is its ease of use. It is straightforward to install and set up, and routine elemental analysis for a whole suite of elements is easily accomplished.

Figure 4.1.1 Ground state and excited states illustrated for the potassium atom. The S, P and D columns refer to the total orbital angular momentum of the atom; for an explanation, refer to Atkins (1994).

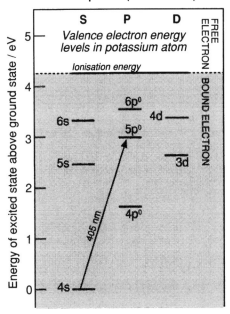

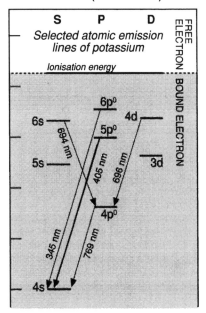

Emission and absorption spectra

Most analytical applications of atomic spectra measure the *emission spectrum* of the sample, the wavelengths present in the light it gives out. The energy required to excite the sample atoms may be introduced by heating (e.g. ICP–AES, Chapter 4), by photon irradiation (e.g. XRFS, Chapter 6) or by particle bombardment (e,g. electron microprobe, Chapter 14). The excited atoms emit radiation when their electrons fall from higher energy to fill the vacancies created at lower levels (relaxation). An emission spectrometer measures the wavelengths given out and their intensities.

The wavelengths of light that an atom can *absorb* (its absorption spectrum) are also **quantised** (Fig. 4.1.1a) and are as characteristic of that atom as its emission spectrum. One or two analytical methods utilise atomic or molecular **absorption** spectra, measuring the intensity of absorption to determine the analyte concentration (Chapter 5). Absorption atomic spectra incidentally provide the only means of analysing the compositions of the Sun and stars.

RG

Source unit

The excitation system used in virtually all commercial ICP–AES systems is the argon plasma. The essential principles of the operation of the argon ICP are illustrated in Box 4.2.

Generator   The power for the plasma is supplied by the radiofrequency generator and several designs of generator are currently used. The two most important variables in RF generator design are

Box 4.2 The inductively coupled plasma

A plasma can be defined as a luminous volume of gas that has some of its atoms or molecules ionised (Ebdon, 1982). In ICP–AES the plasma is normally generated within a 'quartz glass' (vitreous silica) 'torch' (Fig 4.2.1) by inductive coupling to a radio frequency power source – a 'load coil' surrounding the torch.

The two- (or three-) turn hollow copper coil (internally cooled by water flowing through) is placed around the 'quartz glass' torch and connected to a radio-frequency generator. The ICP torch assembly that is most commonly used in commercial systems consists of three concentric tubes. The outermost one carries the plasma gas, and the central tube carries the analyte sample as an aerosol with argon gas. The intermediate tube carries the 'auxiliary' gas supply. This is not always used but may be required to lift the plasma higher and slightly away from the tip of the injector gas tube. The argon gas supporting the plasma enters the torch through a 'T' junction and flows up through the torch in a spiral pattern (see Fig. 4.2.1). The conduction in the argon gas is initiated with a high-voltage electrical (Tesla) spark, allowing energy to be transferred from the coil to the torch. The radiofrequency magnetic field produced by the coil causes energetic oscillatory motion of free electrons within the argon gas. Further ionisation is caused by collisions between free electrons and argon gas atoms. The motions of free electrons in response to the radio-frequency magnetic field constitute eddy currents that bring about intense **ohmic** (resistive) heating of the argon gas to a maximum temperature of about 10,000°C.

This method of heating is unusual in spectroscopic sources: there are no electrical contacts and no combustion of inflammable gases. The temperature of the argon gas closest to the load coil is much cooler than the main part of the plasma; this is essential to prevent the quartz glass reaching its melting temperature of approximately 3,000°C. The temperature distribution within the spectroscopic source is also different to other sources (Fig. 4.2.2). There is a volume of relatively gentle temperature gradient, decreasing progressively above the load coil. The sample that is to be analysed is introduced into the plasma through a central injector tube and forms a tunnel through the plasma. This tunnel is cooler than the surrounding plasma in the lower parts of the ICP and the sample is thus heated from the outside 'inwards'. The region of the plasma that is normally observed for spectroscopic analysis is higher in the tail flame of the plasma where the temperature is of the order of 6000°C (see Fig. 4.2.2).

The ICP is used as a spectroscopic source in emission spectrometry, generating intense emission spectral lines for many elements. The high operating temperature eliminates most chemical interferences, and the general absence of self-absorption and self-reversal in the plasma gives calibration lines that are linear over several orders of magnitude. The ICP is also used as an ion source in ICP–MS (Chapter 10).

JNW

the operating frequency and the power output. Many ICP source units are operated at either 27 MHz or 40 MHz. Increasing the operating frequency improves the signal to background ratio of the emission line, but using too high a frequency can lead to increased analytical interferences. There has been some trend in recent years to move from the 27 MHz band to the 40 MHz band for commercial systems. Most commercial ICP source units are now operated at 'low' power – a typical working level would be 1.0–1.2 kW power delivered to the plasma. The early instruments sometimes used much higher power – several kilowatts was not uncommon. Practical experience has demonstrated that best results can be obtained in ICP–AES at lower power levels, and good compromise operating conditions for a wide range of elements may readily be achieved. Some manufacturers have moved to even lower power levels (600–800 watts) operated with a 'mini-torch' plasma. A potential advantage of this approach is that less expensive solid-state generators can be used.

ICP torch Box 4.2 also shows the geometry of the ICP torch. This is constructed from 'quartz glass' (vitreous silica) and normally consists of three concentric tubes. The outermost tube carries

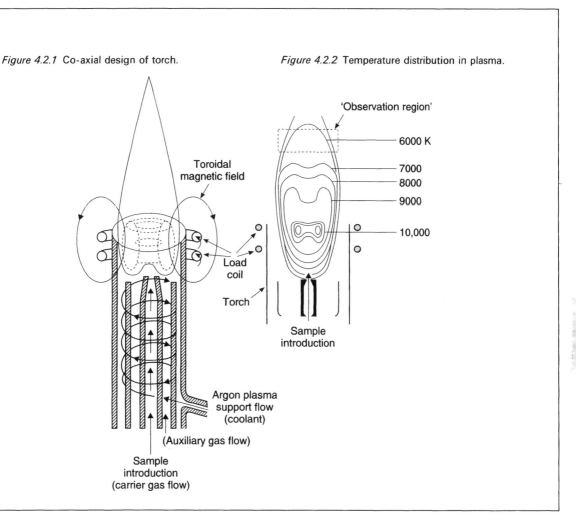

Figure 4.2.1 Co-axial design of torch.   *Figure 4.2.2* Temperature distribution in plasma.

the *coolant* (or 'plasma') gas and, as shown in Box 4.2, this enters the tube at right angles and swirls upwards in a spiral flow pattern. The plasma is produced as the gas flows through the two–turn (occasionally three) coil. The coil (known as the load coil) is connected to the generator and the plasma is generated by **ohmic** heating of the gas as it passes through the coil. Cooling is achieved by water flowing through the coil, which is hollow. The plasma is generated in the centre of the tube and the outer part of the gas flow cools the glass torch; hence the term 'coolant gas flow' is sometimes used.

Solutions to be analysed are introduced into the ICP through the central *injector* gas tube. The other gas flow is the *auxiliary* gas flow and the function of this gas inlet is simply to lift the plasma. This is sometimes desirable to avoid damage to the tip of the injector gas tube when corrosive materials are introduced into the plasma, for example high levels of alkali metal salts that might well fuse into the injector gas tip at the high operating temperatures of the ICP

The method of sample introduction and the torch geometry shown in Box 4.2 are critical for the successful operation of the ICP as an analytical source. The sample is aspirated

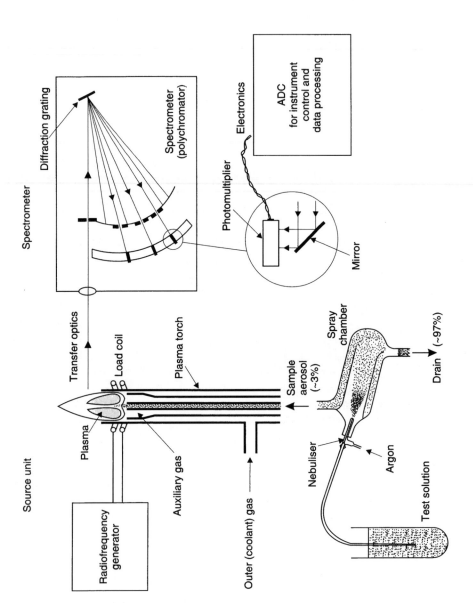

Figure 4.1   Schematic diagram showing the major components of a typical multi-element ICP–AES system. The analyte solution is introduced into the plasma as an aerosol suspended in argon gas. Emitted light from the elements present in the plasma is resolved into its component wavelengths by a multi-channel optical spectrometer. **Photomultiplier** tubes or **charge-coupled devices** (CCDs) measure the light intensity at each designated wavelength, and comparison with standard samples allows element concentrations to be measured. Reprinted with kind permission of Chapman & Hall from *Handbook of Inductively Coupled Plasma Spectroscopy*, M. Thompson and J. N. Walsh, 1989, Blackie and Son Ltd.

through the central injector tube and it 'punches' a hole into the plasma. The effect of this can be seen in Box 4.2, Fig 4.2.2 – the temperature profile through the plasma. The plasma forms a doughnut or toroidal shape around the solution as it is introduced, and heating of the sample occurs from the outside inwards. The ICP has two main attributes as an analytical source. It operates at a high enough temperature to eliminate most chemical interferences and produces a wide range of spectral emission lines for most elements. The emission signal increases linearly as concentration is increased over four or five orders of magnitude producing linear calibration lines. This is a direct consequence of the unique method of sample introduction. Heating the sample progressively from the outside reduces the levels of self-absorption and self-reversal that are a common feature of most 'flame' methods of analysis. In the ICP, the sample is heated to temperatures in excess of $7000°C$ and desolvation, vaporisation, dissociation, ionisation and excitation are almost instantaneous.

The normal observation region in the ICP for emission spectrometry is higher in the tail flame – some 14–18 mm above the induction coil. The temperature in this region of the plasma is lower, but still in excess of $6000°C$. The background signal from the argon plasma is very low and excellent signal to background ratios are achieved. The flow rate used for the coolant gas is normally $10–15 \, l \, min^{-1}$ with $0.6–1.4 \, l \, min^{-1}$ for the injector gas. Argon is the preferred gas for almost all commercial ICP–AES systems. It has the advantages of optical transparency, reasonably low cost, chemical inertness and low thermal conductivity.

The basic mechanism by which atomic emission spectral lines are generated can be represented by the process shown in Box 4.1. When energy (supplied here by the ICP) is transferred to a ground-state atom, electrons may be promoted to vacant higher-energy levels. As the atom moves to a lower-energy state (higher in the ICP tail flame) this electron may drop back to the lower-energy level and energy will be radiated in the form of a photon (Box 4.1). For each specific transition, light of one specific wavelength is generated (i.e. an emission spectral line). However, if the energy supplied is sufficient, electrons may move to more than one vacant energy level and may return to the ground state via several intermediate levels. Consequently, a series of spectral lines may be produced covering a range of different wavelengths – the characteristic emission 'lines' for an element. Where a high-energy source (such as the ICP) is used to generate the spectral lines, both atomic and ionic species may generate distinct series of emission lines.

Wavelengths are normally quoted in nanometres (10^{-9} metres), but angstrom (Å) units (10^{-10} metres) are still used in some of the older literature. Every element will have its own set of emission lines, and samples that are mixtures of several elements will emit a range of spectral lines contributed by each element present. Useful spectral emission lines are in the wavelength range 165–800 nm, with the majority lying between 175 and 450 nm. There are several compilations of emission lines that are useful for ICP–AES (Boumans, 1980, Parsons *et al.*, 1980, Winge *et al.* 1984).

Sample introduction   The most widely used method for introducing the sample is the pneumatic nebuliser, and one type of system is shown in Fig. 4.2. This is a simple 'cross-flow' nebuliser. The sample to be analysed is pumped through a capillary tube and converted to an aerosol by a flow of argon gas introduced into the nebuliser at right angles to the solution capillary. The aerosol then travels up the central (injector) tube of the ICP torch. Details of two alternative types of ICP–AES nebuliser are shown in Fig. 4.3: (a) the concentric (Meinhard) nebuliser and (b) the Babington nebuliser. Cross-flow, concentric and Babington-type nebulisers are all used in commercial systems and each has advantages. Concentric nebulisers are claimed to offer excellent reproducibility, but may be less tolerant of high dissolved solids and are diffi-

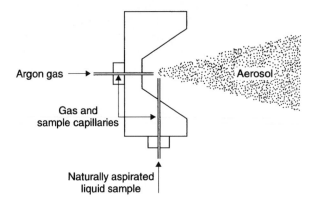

Figure 4.2   A simple cross-flow nebuliser can be used to produce an aerosol of the solution to be analysed with two capillary tubes accurately aligned and set at right angles.

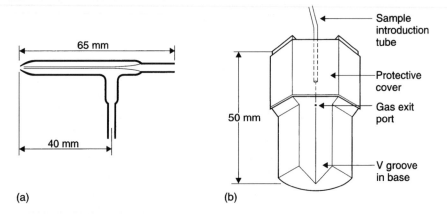

Figure 4.3   Two alternative nebuliser designs for ICP–AES: (a) the Meinhard nebuliser is constructed entirely from glass and the capillary that carries the solution must be very accurately aligned with the surrounding tube that carries the argon gas; (b) the Babington–type nebuliser, in which the aerosol is produced as the analyte solution runs down a narrow groove and over a small hole through which argon gas flows at high pressure. Reprinted with kind permission of Chapman & Hall from *Handbook of Inductively Coupled Plasma Spectroscopy*, M. Thompson and J. N. Walsh, 1989, Blackie and Son Ltd.

cult to clear when blockages occur. Babington-type nebulisers are the least likely to block, but may not offer the same high reproducibility as other types of nebuliser. Cross-flow nebulisers are a compromise with many satisfied users, although the nebuliser must be constructed with great care; accurate alignment of the capillary tubes is critical.

The efficiency of ICP–AES nebulisers is low: only 1–2% of the solution is converted to useful aerosol, most of the solution being lost via the drain (Fig. 4.1). The spray chamber is used to separate the larger droplets from the fine aerosol mist that continues up to the plasma. The droplet size distribution produced by a typical nebuliser is shown in Fig. 4.4. It is only the finer-sized droplets that travel up to the plasma and are used for analysis. The efficiency and the reproducibility of nebulisation of a sample is a critical part of quantitative ICP analysis. The smallest change in conditions will substantially alter the proportion of sample reaching the plasma; with only such a small proportion of the sample reaching the ICP, any changes will have a dramatic effect on the emission signal.

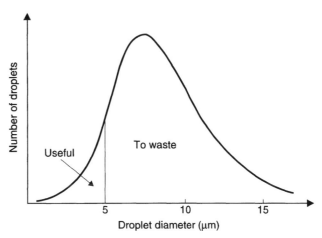

Figure 4.4   The typical distribution of droplet sizes produced by a conventional ICP nebuliser. Only the droplets of less than 5 μm are analytically useful (small enough to reach the ICP and be dissociated in the plasma). Increasing the proportion of small droplets is possible with some nebulisers (ultrasonic nebulisers, for example) but the increased quantity of water causes problems in the plasma.

Another part of the sample introduction system that may be used in some systems is a peristaltic pump, which is used to pump the analyte solution into the nebuliser. It is not essential for all systems (notably for concentric nebulisers). The peristaltic pump controls the flow rate of the solution uptake (typically 1–3 ml min^{-1}) although experience shows that deliberately changing the flow rate has a limited effect on the analytical signal. A more important factor is the role the peristaltic pump plays in stabilising the nebulisation rate when solutions of different viscosity are aspirated.

An important design parameter of the ICP–AES sample introduction system is the 'clean-out' rate between the analysis of samples. It is important that the spray chamber does not have any 'dead' space and is not made too large, otherwise there is a risk of memory effects from one sample to the next. A typical analysis will require a flush time of 30–45 seconds between the analysis of samples. If the samples are of very different compositions, this time may have to be increased. Some elements (boron is one example) do appear to take longer to clear from the sample introduction system than others when introduced at high ($>1000 \mu$g ml^{-1}) levels.

Spectrometer

The role of the sample introduction part of an ICP–AES system is to convert the dissolved sample into an aerosol. The ICP torch converts the aerosol first into particulate material (as the water is evaporated), then into atomic (and in some cases ionic) form and finally energises the atom or ion to emit photons of energy. The role of the spectrometer is to separate the emitted radiation into discrete spectral lines, each of which can be attributed to one of the elements present. By measuring the intensities of selected spectral lines, the spectrometer provides a quantitative measure of the concentration of each analyte.

The spectrometer resolves the light emitted in the tail flame of the ICP. Normally a 4 mm window of emitted light between 14 and 18 mm above the load coil is observed. Fig. 4.5 shows diagrammatically the linking of the ICP to the spectrometer. The transfer optics

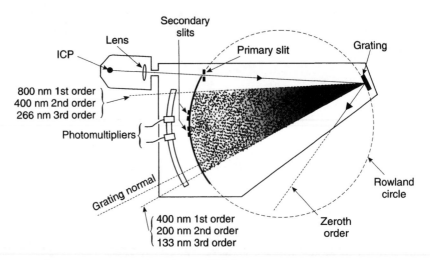

Figure 4.5   Paschen–Runge spectrometer design – a widely used spectrometer design for multi-element simultaneous analysis, employing a concave grating. Reprinted with kind permission of Chapman & Hall
from *Handbook of Inductively Coupled Plasma Spectroscopy*, M. Thompson and J. N. Walsh, 1989,
Blackie and Son Ltd.

directs the light emitted in the ICP onto the diffraction grating of the spectrometer and
ensures each diffracted beam is brought to a focus at the appropriate exit slit. Accurate alignment and maximum mechanical stability are essential for achieving good analytical detection limits and reproducibility.

Two categories of multi-element spectrometer are available to meet the needs of different
applications: these are simultaneous (polychromator) and sequential (monochromator)
spectrometers. A polychromator offers 20–30 fixed detector channels that can measure different wavelength components simultaneously. A monochromator detects only one wavelength at a time, but is more versatile for scanning regions of the spectrum. Both classes of
spectrometer utilise a diffraction grating (Box 4.3) to disperse the incoming radiation into
separate wavelengths.

Simultaneous   One of the outstanding features of ICP–AES analysis is the potential for rapid, simultaneous
spectrometers   determination of a broad suite of elements. Therefore, simultaneous spectrometers (illustrated in Fig. 4.5) are widely used by instrument manufacturers when configuring ICP–
AES systems. Although several other designs of spectrometer have achieved commercial
popularity, it is appropriate to begin by considering the important features of the
Paschen–Runge polychromator design shown in Fig. 4.5. Light from the ICP tail flame is
directed onto a concave diffraction grating through a primary entrance slit assembly. This
primary slit would typically be 20 µm wide and in most systems it can be moved (usually
under computer control) through a limited traverse. The diffraction grating resolves the
incident light into its component wavelengths which are diffracted onto accurately located
photomultipliers. The principles of how a diffraction grating operates are shown in Box
4.3. The photomultipliers are mounted behind narrow secondary (exit) slits, and aligned
to measure specific wavelengths (the emission spectral lines for each analyte element). The
intensity of the signal recorded by the photomultipliers can be used as a measure of the concentration of the element.

Box 4.3 How a diffraction grating works

A diffraction grating is an optical device using diffraction and **interference** to disperse a polychromatic (multi-coloured) beam of light into its constituent wavelengths.

Diffraction refers to the scattering of light by objects similar in size to the wavelength of the light itself. A diffraction grating consists of a reflecting aluminised surface upon which are engraved thousands of regularly spaced fine grooves, the spacing (*a* in Fig. 4.3.1a) typically being in the range 0.5–3.0 μm. Each groove scatters incoming light over a wide angle. Because the spacing of the grooves is similar to the wavelength of the light, however, rays scattered in the same direction from adjacent grooves generally differ in **phase** and will *interfere* with one another: they produce a combined wave whose amplitude may have any value between zero and the sum of the two contributing amplitudes, depending on the phase difference between them. Figure 4.3.1(b) shows two waves scattered at the same angle θ_1 from equivalent points on adjacent grooves. In this case the geometry is such that the waves are **in phase** (compare bars) and will reinforce each other. Similar rays diffracted at an angle θ_2, on the other hand, are **out of phase** (Fig. 4.3.1c) and will cancel each other out.

The combination of such interference effects from thousands of grooves, *acting on incident light comprising a single wavelength*, is to produce a series of sharp bright lines, occurring at specific angles (θ in Fig. 4.3.1a),

that depend on the wavelength and the groove spacing:

$$a(\sin \theta + \sin \phi) = n\lambda \qquad [4.3.1]$$

The pattern of lines is called a *diffraction pattern*. The integer *n* (0, 1, 2, ...) signifies that reinforcement may arise with phase differences of 0, 1λ (as in Fig. 4.3.1b), 2λ or any multiple of λ, and is called the *order* of the diffraction peak concerned.

When the incident light contains several wavelengths, as when a spectrum is analysed in a grating spectrometer, the line for each wavelength present in the spectrum occupies a different angular position around the grating, allowing each to be measured separately by means of an appropriately positioned detector. The phenomenon of diffraction from a grating will be familiar to many people in the form of the coloured patterns produced by the microscopic tracks on a compact disc when viewed in natural light. Several orders of colour can be seen, though the situation is complicated by overlap between the colours of one order and those of the next, and by the circular geometry of the tracks.

Visible and **ultraviolet** spectra require gratings with 300–2000 grooves per millimetre (spacings of 500–3000 nm). **X-rays** on the other hand ($\lambda = 0.003 - 3$ nm) may be diffracted by the regularly repeated atomic structure of a crystal (typical repeat distance 0.1–2 nm), which acts as a three-dimensional diffraction grating (see Box 6.2).

RG and JNW

Figure 4.3.1 (a) Construction of a diffraction grating (greatly enlarged); (b) waves scattered from adjacent grooves such that they are in phase; (c) waves scattered **out of phase**.

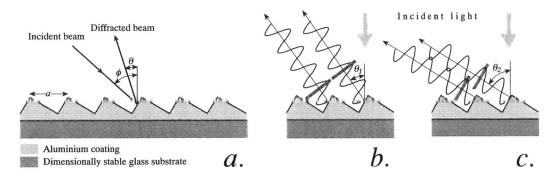

The spectrometers used with ICP sources come in a wide range of shapes and sizes; there is no ideal configuration and, in the final analysis, all systems are a compromise, particularly between spectral resolution and transmittance.

Good optical resolution is essential for all ICP–AES systems. The more effective the spectrometer is at separating the spectral line to be measured from adjacent lines, the lower the interferences will be. This is especially important in analysing a complex matrix, where there are many elements that may potentially cause spectral interferences. There are several procedures that can be used to increase the resolution of an ICP spectrometer, including;

- using narrower slits;
- increasing the size of the spectrometer;
- using a higher-dispersion diffraction grating;
- measuring spectral lines in higher orders (Box 4.3).

However, these procedures may also have an adverse effect on other desirable features of the spectrometer (see discussion below).

The intensity of the emission lines from some elements in the ICP is considerable (the alkaline-earth elements, for example). However good the design of the spectrometer, some stray light from these intense lines is bound to be scattered from internal surfaces inside the spectrometer and fall onto the photomultipliers for other elements, causing higher background readings and degrading detection limits. Improvements in the details of spectrometer design have done much to reduce stray light problems in modern systems. Light transmission through the spectrometer will ultimately influence detection limits. Reducing slit widths, using longer-focal-length spectrometers and higher-dispersion gratings will all reduce light throughput (see discussion of resolution above).

The spectral range that an ICP spectrometer might need to measure would cover spectral lines from 852.12 nm (Cs) through to 167.08 nm (Al). It is unlikely that a commercial ICP–AES can achieve this using first order spectral lines alone. Invariably some compromise must be adopted; for example, if Cs is not to be determined (its performance in the ICP–AES is poor), the highest wavelength measured can be reduced to 780.02 nm (Rb) or to 766.49 nm (K). Low-wavelength lines can be measured in the second order (Box 4.3). For example, the Mo 202.03 spectral line will give a second-order reflection at 404.06, and although this is a less intense line it can be used for analysis. It is also possible to use even higher-order spectral lines (see Fig. 4.5). The advantage of using higher-order spectral lines is that the wavelength range the spectrometer is required to measure can be reduced, with a consequent reduction in capital cost. In Fig. 4.5 the wavelength range of the spectrometer is from 400 to 800 nm. Emission spectral lines in this range are measured in the first order, lines between 200 and 400 nm are measured in the second order and lines between 133 and 266 nm could be measured in the third order. Careful design of the diffraction grating can limit the loss of transmission that occurs when higher-order spectral lines are used.

Spectral lines below 200 nm are absorbed by air and below 190 nm the amount of absorption is substantial. Consequently, if lines below 190 nm are to be measured, it is necessary to use a vacuum spectrometer or to replace the air with nitrogen or argon. Some spectrometers do both, purging the system with argon or nitrogen and reducing the pressure.

An important alternative design of spectrometer is the 'echelle' configuration (Fig. 4.6). These were originally used for compact designs where space was a major constraint. The spectrometer grating has widely spaced grooves (30–300 mm^{-1}) and is illuminated at a high angle by the incident light from the ICP source. The echelle spectrometer grating gives high-order spectra over a narrow angle. An additional prism is introduced into the spectrometer and placed in front of the diffraction grating. The prism acts as an 'order-sorting' prism and is mounted at right angles to the grating dispersion plane. In consequence the spectrum is produced as a two-dimensional array of 'dots' rather than lines. An exit slit cassette rather than a series of narrow line slits is therefore used and the detectors are mounted behind this cassette. The echelle spectrometer has been available for many years and it offers advantages of flexibility, compactness and possibly economy. Some designs have suffered from the disadvantage of lack of stability, critical for the echelle design because of the smaller size of the final spectra. There is reason to believe that the stability problems

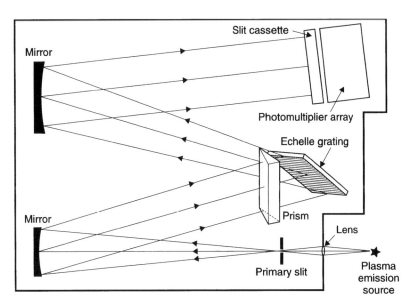

Figure 4.6   The echelle spectrometer configuration. Reprinted with kind permission of Chapman & Hall from *Handbook of Silicate Rock Analysis*, P. J. Potts, 1987, Blackie and Son Ltd.

may have been overcome in the most recent systems and ICP–AES systems using echelle spectrometers are gaining in popularity.

Wavelength stability is essential for quantitative ICP–AES analysis. In addition to mechanical stability there is also the need for temperature stability and many modern ICP spectrometers are heated (typically to 30°C or 35°C) and maintained at this temperature ±0.1°C.

Ideally, the analyte signal would be measured after subtracting the background underneath the emission line for each sample, but in practice it is not possible to measure backgrounds directly. The most widely used method for measuring **background** in routine analysis is to run an instrument **blank** solution and use the emission signal given by the blank as the background. Clearly this is less than ideal as it is unlikely that the blank will contain the same matrix elements as the analyte sample. An alternative procedure to measure background is to offset the entrance slit by a small amount and measure the background on either side of the signal. Interpolation between these two background signals should give the background beneath the signal. This is what most instrument manufacturers refer to as background correction. Although this procedure can work well, it is less successful when the matrix of the samples being analysed changes as there may be spectral interferences from matrix elements at the wavelengths where background signals are being measured.

Sequential spectrometers   Sequential spectrometer systems have achieved increasing popularity in recent years, and now represent the most widely used type of ICP–AES analytical system. Unlike the polychromator, they cannot offer simultaneous analysis but they do offer a number of other advantages. There are many different designs of sequential spectrometers used by instrument manufacturers, and Fig. 4.7 shows one of these, the Czerny–Turner spectrometer. The light emitted in the ICP source is focused onto the entrance slit of a monochromator containing a single photomultiplier, the diffraction grating being mounted so that it can be rotated under computer control. By rotating the grating, any spectral line within the wave-

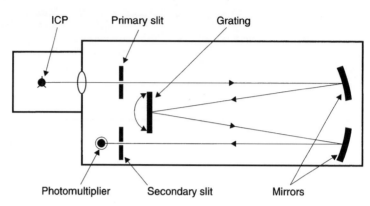

Figure 4.7   Sequential ICP–AES – Czerny Turner spectrometer. There are several designs for sequential spectrometers and this is one simple arrangement. Reprinted with kind permission of Chapman & Hall from
Handbook of Inductively Coupled Plasma Spectroscopy, M. Thompson and J. N. Walsh, 1989, Blackie
and Son Ltd.

length range of the spectrometer can be accessed. Elements must be measured one after
another, but there is no limit on the number of lines that could be used for analysis.

A significant limitation for this type of sequential spectrometer is the accuracy with which
it is able to locate the exact spectral line required for analysis. Several procedures have been
used to improve the accuracy of line location. These include the use of a reference line (a
line from an Hg discharge lamp, for example), stepping motors to calibrate the drive mechanism, and peak location by performing a series of scans across the anticipated peak position.

Several ingenious designs for sequential spectrometers have been used by instrument
manufacturers to overcome some of the limitations of the early systems. These include a
design using a Paschen–Runge-type spectrometer which has a series of precalibrated positions on the exit slit assembly block. A photomultiplier is mounted behind this and can be
driven to the nearest position to the analyte spectral line with a high degree of accuracy.
The exact peak position can then be found by making quite small movements of the
entrance slit. This arrangement is rapid and more accurate than the Czerny–Turner design
shown in Fig. 4.7. Many other types of spectrometer design can be used, each having its
own advantages and disadvantages. Descriptions of these can be found in the manufacturers' literature and in Thompson and Walsh (1989).

Sequential spectrometer systems benefit from the flexibility they can give to the analytical programme, and the reduced cost of the hardware. Simultaneous spectrometers have
fewer moving parts and should give greater stability. Higher-precision analysis is possible
with internal standards being used where the standardising element is measured simultaneously to the analyte element and this is only possible with a simultaneous spectrometer.
Simultaneous spectrometers have a much greater throughput of samples, particularly
where a wide range of elements is sought in each sample. This increase in speed of analysis
will usually more than offset the higher initial cost of the instrumentation.

Computer

The third part of any ICP–AES system is the computer and associated electronic interface
equipment. Most modern systems will be controlled from a PC-type computer, and recent

improvements in performance and reductions in prices have brought real benefits to ICP analysis. The computer used with a commercial system will now represent only a small part of the total purchase price of the instrument and service costs should be minimal. However, the computer is probably the first part of the system that will become obsolete. The interface between the spectrometer and the computer has a relatively simple function – to convert the voltage produced by the photomultipliers into a digital signal that can be processed by the computer. It will contain analogue to digital converters (**ADCs**) and power supply boards. In detail the electronics is complicated, and many of the boards are manufactured solely for the instrument supplier. Repairing or replacing them can be the most expensive part of servicing an ICP system. It is also important that amplifiers and ADCs have sufficient **dynamic range** to cover the range of concentrations measured by modern ICP systems – at least five orders of signal magnitude.

Analytical performance

ICP–AES is now an established and mature analytical technique in geochemistry. It is also the solution method of choice for a wide range of routine analytical applications. Modern ICP–AES instruments are very reliable and simultaneous systems can generate a 20–30 element analysis for a prepared sample in approximately 1 minute, with more than 200 samples being analysed in a typical working day. It is a very cost-effective technique and ideally suited to geochemical projects requiring the analysis of large numbers of samples. There is no shortage of texts that present application studies and detailed 'recipes' of analytical procedures for the analysis of geochemical samples (Thompson and Walsh, 1989). Most analytical applications for ICP–AES are dominated by the sample preparation (i.e. dissolution) procedure to be adopted. The sample preparation techniques used for geochemical solids can be broadly categorised into: (a) extracting the analyte elements without attempting a complete dissolution of the sample; (b) dissolving the sample using hydrofluoric and another mineral acid; (c) fusing the sample with an appropriate flux and dissolving the resulting fused mixture. Details of the main sample dissolution procedures used in geochemical and environmental analysis have been discussed in Chapter 3. Each sample preparation technique has its merits, and choosing the most appropriate technique will depend upon the information sought from the analysis.

The range of elements that can be determined in geological samples by ICP–AES is considerable (Fig. 4.8). All of the 10 major elements (Si, Al, Mg, Fe, Ca, Na, K, Ti, P and Mn) normally determined can be measured, in addition to most of the common trace elements. The rare earth elements are measured routinely using a cation separation and preconcentration procedure (see discussion below). There are, however, some limitations to the elements that ICP–AES can measure. The halogen elements, the inert gas elements, O, N and C cannot realistically be measured. It is difficult to measure some trace elements that occur at very low concentrations (less than $1 \mu g\, g^{-1}$) in geological samples, although it is possible to measure some of these if separation and preconcentration techniques are employed (platinum group metals are a possibility). ICP–AES is not successful at measuring the heavy alkali metals (Rb and Cs) at trace levels; they are too readily ionised in the

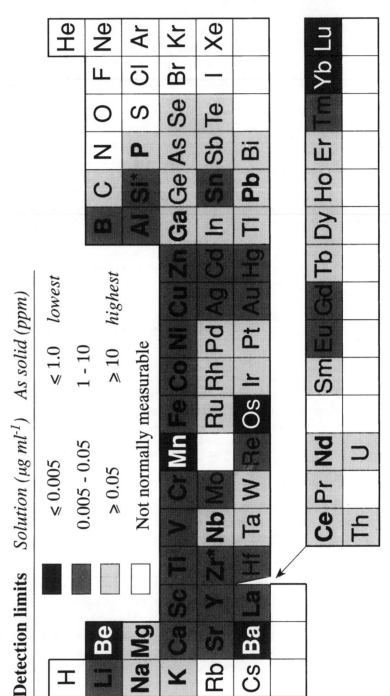

Figure 4.8 Periodic table showing elements that can be determined by ICP–AES and their working detection limits (6s data from Potts, 1987 – see Box 4.4). Bold symbols indicate elements that can be determined in typical silicate rocks without preconcentration.
* Requires fusion digestion.

high temperature of the plasma and only the atom lines can be used for analysis. U, Th, W and Ta are normally below the detection limits for ICP–AES.

Instrument calibration can be carried out with either natural or synthetic standards. Synthetic standards are relatively simple to prepare if the range of elements to be determined is small, but when a large suite of major and trace elements is to be determined, natural standards will be preferred. The main limitation with natural standards is the availability of standards that cover the range of elements to be determined and the reliability of the certified values for the standard samples. **Matrix effects** are a relatively minor problem in ICP–AES, but it is desirable to make an approximate matching of standards and samples. This is easier to achieve with natural standards than with synthetic standards.

Spectral interference (spectral line overlap) is more serious. The ICP is an efficient mechanism for generating a large number of spectral lines from most of the elements of the periodic table. In a solution containing a dissolved rock sample, the complex mixture of elements is likely to produce very complex emission spectra, with emission lines from one element overlapping the main emission line from another element. Careful selection of the analysis line and increasing the resolution of the ICP spectrometer can help considerably to eliminate or reduce spectral line interference. In most cases acceptable emission lines can be found. However, there are cases where this is not possible and an attempt has to be made to calculate the extent of an interference mathematically. Providing the interfering element is also measured during an analytical programme, it is then possible to apply a calculated interference correction to the analytical result, and experience shows that this process works well if the interference is not too great. Nevertheless, spectral interferences remain a significant limitation in ICP–AES and degrade **detection limits** for trace element analysis in natural materials.

Sequential ICP–AES instruments offer greater flexibility for selection of the emission line that is used in an analytical programme. This may be advantageous in eliminating spectral line overlap interferences, especially where the bulk composition changes significantly from one sample to another. However, with multi-element simultaneous ICP–AES instruments it is easier to measure any potentially interfering element in all samples and apply appropriate corrections.

Applications

Water analysis with ICP–AES

One of the most abundant, and widely analysed, environmental samples is water. The routine elemental analysis of natural waters is now a major application area for ICP–AES. Most water authorities and most organisations concerned with the assessment of water quality will operate at least one ICP–AES instrument as part of their water evaluation process.

As ICP–AES is essentially a solution method of analysis, and is capable of measuring a wide range of major and trace elements simultaneously, it is in many respects ideally suited to the routine determination of many elements in waters. Normally, the only sample pretreatment required will be filtration (to remove particulate material) and acidification (to stabilise metals present in solution). An analytical protocol for water analysis based on

Box 4.4 The normal distribution II: What is meant by detection limit?

The smallest analyte concentration that can be reliably detected by an instrument under given operating conditions is called the **detection limit** for that analyte. It represents the smallest analyte (peak) signal that can be distinguished from the background signal with a specified statistical confidence. In order to define detection limit more rigorously, it is necessary to examine the statistics of background measurement, and in particular the properties of the normal distribution.

For a series of measurements conforming to a normal distribution, it is a simple task to predict the probability of an individual reading lying within a specified range of the mean. The dark band in Fig. 4.4.1 highlights that part of the distribution lying within $\pm 1\sigma$ of the mean. One can show (by integrating Equation 1.1.1 (Box 1.1) or by reference to tables) that this band accounts for 68.72% of the total area under the Gaussian curve. Accordingly the probability of an individual reading falling within this range of the mean is 68.72%; conversely the probability of the reading lying *outside* it must be $100.00 - 68.72 = 31.28\%$. The probability of a reading lying within the range $\pm 2\sigma$ from the mean is 95.44%, and the probablity of it lying *beyond* $\pm 2\sigma$ is 4.56%. For $\pm 3\sigma$ of the mean the corresponding figures are 99.74% and 0.26%.

Figure 4.4.2(a) depicts 160 replicate readings of the spectral background in the vicinity of an analyte peak. In ICP–AES analysis, these background readings would be made by measuring a reagent blank solution at the peak wavelength itself; in other methods such as XRF (Chap-

Figure 4.4.1. Normal population distribution divided into $\pm 1\sigma, \pm 2\sigma$ and $\pm 3\sigma$ ranges

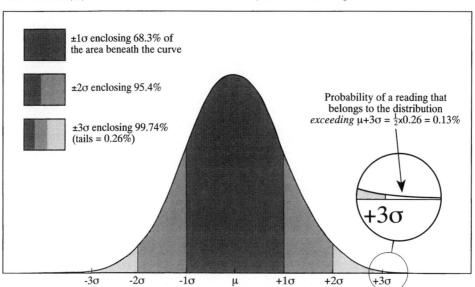

$\pm 1\sigma$ enclosing 68.3% of the area beneath the curve

$\pm 2\sigma$ enclosing 95.4%

$\pm 3\sigma$ enclosing 99.74% (tails = 0.26%)

Probability of a reading that belongs to the distribution *exceeding* $\mu + 3\sigma = \frac{1}{2}\times 0.26 = 0.13\%$

$+3\sigma$

-3σ -2σ -1σ μ $+1\sigma$ $+2\sigma$ $+3\sigma$

filtration of the sample through a 0.45 μm filter and acidification to 1% with nitric acid will therefore be normal practice. Most ICP–AES laboratories will be set up to carry out this work.

ICP–AES suffers from some limitations when the natural levels of many trace elements are sought. This can be seen in Tables 4.1(a) and 4.1(b). Table 4.1(a) gives some typical concentration levels for the major elements in river waters, and compares these with $3s$ **detection limits** (see Box 4.4) for ICP–AES. Although concentration levels in natural waters will vary considerably, and the concentration levels suggested in Table 4.1(a) are only guidelines, it is clear that the concentration levels for many trace elements in natural

ter 6) and neutron activation (Chapter 7), the background may instead be measured at a fixed offset from the peak. Figure 4.4.2(b) shows the sample distribution defined by the readings (mean $\bar{x}_b$ and standard deviation s_b), which approximates to a normal population distribution (Box 1.1). From Fig 4.4.1 the probability of an individual reading lying more than $3s_b$ above or below the mean $\bar{x}_b$ is approximately 0.26%. This represents the area beneath the two tails of the distribution beyond $\pm 3s_b$. Since the ideal curve is symmetrical and its tails have equal areas, the probability of a background reading specifically *exceeding* $\bar{x}_b + 3s_b$ is $0.5 \times 0.26\% = 0.13\%$. In other words, any apparent peak that exceeds the mean background reading by more than this margin has only a 0.13% chance of representing an outlying background signal, and therefore we can conclude with $100.00 - 0.13 = 99.87\%$ confidence that it represents a 'real' analyte signal (signifying the presence of analyte). For an

apparent peak *smaller* than this figure, however, there is a greater probability of the signal representing no more than an outlying background value generated in the absence of analyte. The value of $3s_b$ divided by the *sensitivity* (transforming it into units of concentration) is therefore referred to as the $3s_b$ *(or '3 sigma') detection limit* for this analysis; some analysts call this the *lower limit of detection* or 'LLD' (Potts, 1987).

The $3s_b$ detection limit provides a basis for *detecting* the presence of a peak, but at this level the 'peak' is too small to be quantitatively *measured*. The threshold for confidence in a quantitative analysis of a trace element is usually set at $6s_b$/sensitivity (sometimes known as the *limit of determination* or 'LoD'), or sometimes even at $10s_b$/sensitivity (the *limit of quantitation* or 'LoQ').

RG

Figure 4.4.2 (a) Repeated background measurements and (b) sample distribution of readings showing mean and $\pm 1s$ and $\pm 3s$ ranges.

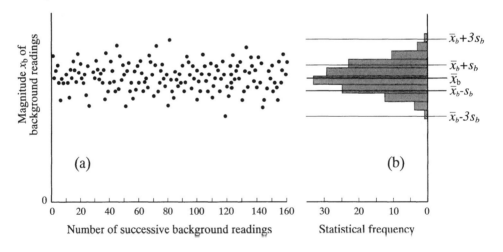

waters are very low and are beyond the working range of ICP–AES. Preconcentration by simple evaporation (up to some 20 ×) is a routine operation for water analysis. Even this is insufficient to bring the concentration levels for many elements into the working range of the ICP–AES technique (see data in Table 4.1b). Some attempts have been made to improve detection limits even further (using ultrasonic nebulisation, hydride generation for appropriate elements, or ion exchange preconcentration), but the limitations of ICP–AES should be recognised and other analytical techniques such as ICP-MS (Chapter 10) considered.

Table 4.1   Typical element concentrations in river water and ICP–AES detection limits for: (a) elements determined directly; and (b) after 20 × preconcentration

(a)	$\mu g\ ml^{-1}$*	$3s$ detection limit in solution ($\mu g\ ml^{-1}$)
Na	6300	30
K	2300	90
Mg	4100	105
Ca	15,000	60
Sr	50	1.5
Ba	20	3
Fe	100	15
B	10	1.5
S	3700	120
Si	1300	45

(b)	$\mu g\ ml^{-1}$*	$3s$ detection limit in solution ($\mu g\ ml^{-1}$)
Li	3	3
Be	0.4	0.15
V	0.9	3
Cr	1	4.5
Mo	1.5	6
Mn	15	15
Ni	1.5	12
Cu	3	3
Zn	20	12
Al	20	60
P	20	15

*Anticipated level in a 'typical' river water

Table 4.2   A sample preparation protocol for the analysis by ICP–AES of water samples with high total-dissolved-solid contents

100 ml of mixed sample + 5 ml conc. HNO_3
Heat in 250 ml tall-form beaker
Evaporate to approx. 50 ml
Cool and add 5 ml conc. HNO_3
Cover with watch glass
Heat at 80°C for 1 hour
Cool – dilute to 100 ml
Filter or centrifuge
Dilute aliquot 1:10

Analysis of sea water and saline formation waters using ICP–AES   To avoid degradation of nebuliser performance, a dissolved solid content of approximately 1% is normally regarded as an upper limit for ICP–AES sample introduction systems, although many nebulisers will cope with higher levels. Saline formation waters (which are similar in composition to sea water) have 3–4% dissolved solid content – largely as NaCl.

Table 4.3 Effective ICP–AES detection limits (3*s*) for saline waters taking account of spectral interference and matrix effects

Elements	Expected level in sea water (Rosler and Lange 1972) (μg ml^{-1})	Working detection limit (μg ml^{-1})
Na	10,561	0.1
Ca	400	0.01
Mg	1272	0.02
K	380	0.2
	(ng ml^{-1})	(ng ml^{-1})
Ag	< 0.002	0.01
Al	0.16–1.9	0.01
As	< 0.03	0.1
B	4.6	0.01
Ba	0.05	0.01
Co	< 0.002	0.02
Cu	0.001–0.09	0.01
Fe	0.002–0.02	0.01
Hg	< 0.002	0.1
Li	0.1	0.01
Mn	0.001–0.01	0.01
Mo	< 0.002	0.01
Ni	< 0.002	0.02
P	> 0.001–0.1	0.1
Pb	0.004–0.005	0.05
S	884	0.1
Si	0.02–4	0.1
Sn	0.003	0.1
Sr	13	0.01
Th	< 0.002	0.1
V	< 0.002	0.02
Y	< 0.002	0.005
Zn	0.005–0.014	0.01

This may cause instability (and poor analytical precision) in the ICP and will certainly influence the analyte signal; this effect is variable from element to element. The signal for most elements shows some degree of suppression, but for the alkali metals the signal will increase. It is probable that these changes are due to adding the high salt loading to the plasma. This will have the effect of lowering the power delivered to the argon ions that sustain the plasma. It is possible to analyse saline and sea waters directly by ICP–AES but careful matrix matching of analyte solutions with calibration standards is necessary. A further problem is that sample pretreatment is necessary; it is not realistic to analyse the sample as collected. The analytical protocol described in Table 4.2 has been used successfully on a routine basis for such samples. It removes any particulate material and ensures that elements to be determined have been put into true solution (colloidal and very fine particulate material may be present in the sample when collected). The procedure should also ensure that elements are stabilised in an acid solution (many analyte elements will precipitate out from alkaline solution).

The four most abundant cations in saline and formation waters will normally be Na ($\sim$10,000 μg ml^{-1}), Ca ($\sim$500 μg ml^{-1}), Mg ($\sim$500 μg ml^{-1}) and K ($\sim$200 μg ml^{-1}). At these levels, severe curvature of the calibration lines is likely (especially for the alkali metals), so it is advisable to dilute the samples by 1:10 prior to ICP–AES analysis. The dilution also reduces the severe matrix effects, and indeed the diluted solutions may also be used for the determination of Li (a reasonably abundant trace element in such samples). All other elements are measured in undiluted samples using calibration standards that have a reasonable matrix match (mostly NaCl concentrations) to the analyte solutions. Table 4.3 compares the detection limits that can be realistically achieved with some published abundance levels for sea water. These are given as working detection limits, making appropriate allowance for the spectral interference and matrix effects for these difficult samples. Table 4.3 indicates that several elements can be determined successfully: Na; Ca; Mg; K; Al; B; Ba; Li; S; Sr. Other elements are marginal, or will only be satisfactorily determined when present at abnormally high levels. Several elements are better determined with flame atomic absorption spectrometry (Chapter 5), notably Pb, Cd and possibly Cu and Zn. Although ICP–AES has good detection limits for these elements, the presence of spectral interferences in ICP–AES causes a deterioration in attainable detection limits and offers some advantage for AA.

Analysis of rocks, soils and minerals

ICP–AES is now recognised as one of the preferred methods for the elemental analysis of major and trace elements in geological materials. The advantages that it offers include the wide range of elements that can readily be determined, the speed and comparatively modest cost of analysis, the diversity of sample types that can be analysed, absence of serious interferences, good trace element detection limits and acceptable precision for major element analysis.

The analysis of solid samples using an analytical technique that is essentially a solution method assumes that appropriate methods of sample dissolution are available. The methods for dissolving geological samples have been considered in Chapter 3 and these must form the basis for the analysis of geological samples by ICP–AES.

Analysis of the sample following dissolution in hydrofluoric and perchloric (or nitric) acids is arguably the most cost-effective and widely used method for the analysis of geological samples using ICP–AES. Details of the procedures used are given in Chapter 3 together with some of the advantages and limitations of the method. All major elements, excluding only Si, can be determined, a broad range of trace elements can be measured, and both natural and synthetic standards can be analysed, allowing precision and accuracy to be evaluated. Instrument calibration is normally done with natural reference materials, as the preparation of multi-element synthetic standards can be time consuming. A large selection of standard reference materials is available. By using a simultaneous ICP–AES and preparing dissolved samples in bulk the method has reduced the cost of a quantitative multi-element analysis to substantially lower levels than other analytical techniques.

The HF/HClO$_4$ dissolution method is also used routinely for atomic absorption analysis of geological samples (Chapter 5). It is possible, therefore, to add to the ICP–AES analysis selected elements that are better determined by AAS (Rb, Pb and possibly Cd are examples).

The main limitation of the HF/HClO$_4$ dissolution method is that Si (usually the most abundant electropositive element in a geological sample) and Zr cannot be determined. For a complete silicate analysis by ICP–AES, it is necessary to resort to a fusion method of

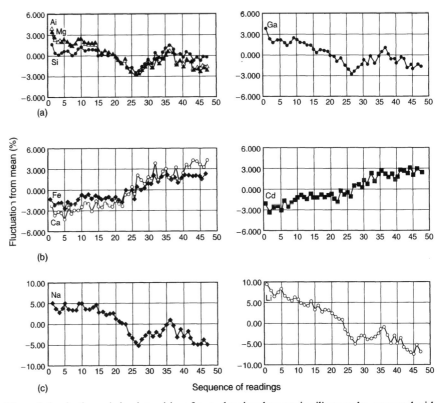

Figure 4.9   The variation in the emission intensities of several major elements in silicate rocks, compared with potential internal standardisation elements.

Co-variation plots for several elements from 47 sequential measurements made over a 3 hour period. (a) Ga shows good time correlation with the major elements Si, Al and Mg, and was selected as the internal standard for these elements. (b) Cd shows good correlation with Fe and reasonable correlation with Ca. There is clearly a very different drift pattern shown by these elements when they are compared with Si, Al, Mg and Ga. (c) Li can be used to monitor Na signal with reasonable confidence. Reprinted from *Chemical Geology* **95**, 113–121 (1992) with kind permission of Elsevier Science–NL, Sara Burgerhartstraat 25, 1055 KV Amsterdam.

sample dissolution (Chapter 3), which as well as retaining Si also ensures the complete take-up of Zr.

High precision analysis of major constituents: Commercial ICP–AES instruments have been available for 15–20 years. Many improvements and refinements in the instrumentation have been introduced during that time, but one aspect of the analytical performance that has shown only marginal improvement is the precision attainable, which remains no better than 1–3%. It is comparable with **AAS** and better than **ICP–MS**, but it is not as good as **XRFS**. Many parts of the ICP–AES system can contribute to the fluctuation of the analyte signal, but it is generally recognised that the sample introduction system, especially the nebuliser, is the main source of noise. Of the numerous attempts to improve the ICP sample introduction system, very few of the improvements have enhanced signal stability. Indeed some improvements (the use of non-blocking Babington-type nebulisers, for example) may lead to lower-precision performance. Consequently, ICP–AES as routinely used fails to satisfy the increasing demand for high-precision major element rock analysis.

An alternative approach to attaining high signal stability for major element analysis is to accept the fluctuation in the analytical signal, but to monitor and compensate for it using multiple internal standards to allow for the disparate analytical response of the elements in the ICP (Walsh, 1992). Several potential internal standardisation elements were evaluated for a high-precision major element analysis technique based on fusion dissolution (Chapter 3). Careful matching of each major element analyte to the most appropriate internal standard element demonstrated that by using three separate internal standardisation elements, the analytical precision could be reduced to less than 0.5%. It can be seen from Fig. 4.9 that Ga is an excellent internal standard element for Si, Al and Mg, but would be quite inappropriate for Fe or Ca (for the spectral lines used in this evaluation). The time variation in the Ca and Fe analytical signals is closely imitated by Cd, and this is used as the internal standard element for these two elements. The alkali metals invariably behave differently to other elements, and to monitor Na (and K if necessary) another alkali metal (Li) is used.

Current developments

ICP–AES can now be regarded as a mature analytical technique, with many years of experience of analysis of real samples, and many laboratories routinely process many thousands of samples per year. Advances in instrumentation have produced small systematic improvements rather than spectacular leaps forward in instrument capabilities. There have, however, been a number of important developments in recent commercially available instruments

- The use of axial (rather than radial) plasma viewing geometry has been developed, as used in ICP–MS (Fig. 10.1). This is claimed to produce a lower background signal from the plasma, more intense spectral emission, minimisation of some interferences and greater stability, with consequent improvements in sensitivity and detection limit. However, not all manufacturers have adopted this approach, and it remains to be seen how successful it will become.
- There has been a trend towards the replacement of multiple photomultiplier tubes with continuous **charge-coupled device** (CCD) detectors and other solid-state detector systems.
- A trend in recent instruments has been greater acceptance of the echelle spectrometer configuration, which is more compact. The stability problems of early echelle spectrometers appear to have been overcome.
- The use of echelle spectrometers in conjunction with CCDs has led to several new instruments capable of measuring very large numbers of emission spectral lines simultaneously. This can offer real advantages in the analysis of samples where **spectral interference** is a problem. The best (most interference-free) line can be selected for each element without the limitations of the conventional Paschen–Runge spectrometer with limited photomultiplier tubes.

The volume of data that such ICP–AES instruments generate has, however, necessitated an order of magnitude increase in the data storage and data manipulation capabilities of the associated computing equipment.

Rare earth element analysis in geological samples

The diagnostic value of rare earth element concentrations in geological samples is widely accepted, and there is a continuing requirement for the reliable and rapid determination of this suite of elements. ICP–AES provides one of the most widely used methods (Walsh *et al.*, 1981) for REE analysis. Samples are dissolved with an HF/HClO$_4$ attack and a fusion of any resistant material not dissolved (which may retain a significant proportion of the rare earths, e.g. in minerals such as zircon). The diluted sample solution is loaded on an ion exchange column containing Dowex 50 resin. Major and unwanted trace elements are eluted using 1.7M HCl and discarded. The REE fraction is then eluted with 4M HCl. After separation, the rare earths are concentrated into a small volume of solution and then determined simultaneously on the ICP–AES system. The separation, summarised in Fig. 4.10, will not remove Sc, some of the Sr and Ba, and a small proportion of the Ca in the sample. These will accompany the REEs in the solution analysed; they should not cause serious analytical difficulties, although the levels of spectral interferences should be checked and appropriate corrections made. Standardisation of the ICP–AES system will normally be carried out with multi-element rare earth solutions. These can be prepared directly by dissolution of the separate rare-earth oxides or from commercially available REE stock solutions. The precision attained in REE analysis using ICP–AES is better than 5% unless working close to detection limits (Walsh *et al.*, 1981). Table 4.4 compares 3*s* detection limits with chondritic rare earth abundance levels; for most elements these detection limits are comparable with or better than chondritic abundances, and in most geological samples REE abundances will be considerably higher.

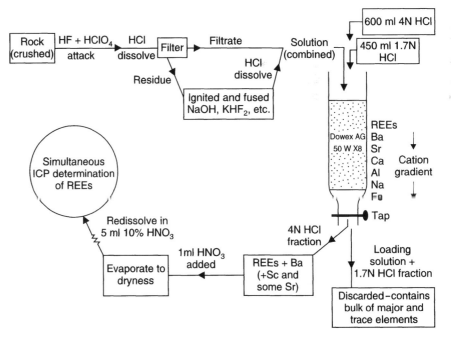

Figure 4.10   Schematic diagram showing the stages in the rare earth element preparation and preconcentration procedure for ICP–AES determination. Reprinted with kind permission of Chapman & Hall from *Handbook of Inductively Coupled Plasma Spectroscopy*, M. Thompson and J. N. Walsh, 1989, Blackie and Son Ltd.

Table 4.4   Detection limits for REEs determined in rock samples by ICP–AES after ion exchange preconcentration

Element	3s detection limit* (ppm)	Chondritic abundance[†] (ppm)
La	1.125	0.330
Ce	1.29	0.865
Pr	0.24	0.122
Nd	1.005	0.630
Sm	0.165	0.203
Eu	0.03	0.077
Gd	0.21	0.275
Dy	0.075	0.342
Er	0.12	0.225
Yb	0.06	0.220
Lu	0.015	0.034

*From Walsh *et al.* (1981).
[†]From Nakamura (1974).

Chapter 5

Atomic absorption spectrometry and other solution methods

A. P. Rowland

This chapter covers an assortment of widely used 'wet-chemical' analytical methods that target a single analyte or a specific group of analytes. Though lacking the multi-element capability and speed of ICP–AES, they entail relatively low capital outlay and can in some cases offer detection limits that are significantly lower, or can tackle analytes (e.g. anions) not attainable by ICP–AES. The chapter concludes with one or two miscellaneous analytical techniques not covered elsewhere.

Atomic absorption spectrometry

In atomic absorption spectrometry, a light beam consisting of the emission spectrum of the element of interest is passed through an absorption cell containing the sample in an atomised state. Analyte atoms in the cell, by absorbing light at their characteristic wavelengths and re-emitting in random directions, cause a net attenuation in the beam intensity in proportion to their abundance. The optimum analyte line in the beam spectrum is isolated by a monochromator (Chapter 4) and detector and the absorbance (the degree of attenuation at that wavelength) is measured. The use of a light source specific to the element of interest reduces **spectral interference** compared with emission methods and allows the use of a relatively simple spectrometer.

Atomic absorption spectrometry (AAS) is a well-established technique for metal determination, being the instrumental method of choice for aqueous samples in geochemical laboratories for 20 years between the mid 1960s and mid 1980s. For the majority of laboratories, atomic absorption offers the facility for single element analysis, although the more recent instruments may allow sequential analysis with batches of samples being processed for each element in turn. Recently, there have been commercial instruments developed with the capability of simultaneous multi-element analysis.

Atomic absorption analysis is applicable to the analysis of a wide range of samples from geochemistry studies following dissolution or preparation of slurries. Fusion with sodium carbonate or lithium borate or acid digestion using HF and/or aqua regia may be used for dissolution of rocks, soil or sediments, for example for geochemical mapping applications. Similarly, acid digests of plant material following digestion with nitric and perchloric acids have recently been reported as an application used to indicate mineralisation and pollution (Farago and Mehra, 1994). Geochemical cycling studies use atomic absorption to determine dissolved components in water. Standard works are available for water analysis, many giving general information for AAS and specific details of flame conditions, detection limits, sensitivity, etc. (e.g. Greenburg *et al.*, 1992).

AAS still has a place in the modern analytical geochemistry laboratory providing a complementary technique to ICP–AES and ICP–MS. AAS instruments entail much lower capital cost and, being commonly under-utilised, provide valuable flexibility available at short notice. Methods are well documented, not requiring the operator to be an expert to produce quality results. Flame methods are simple and with a few notable exceptions relatively free from interferences. Atomic absorption is less dependent on variations in flame temperature conditions than atomic emission. Instruments usually have a facility to operate in the emission mode to cater for elements which are more sensitive by this technique.

Application Reviews on Geological and Inorganic Materials appear biennially in *Analytical Chemistry* (Jackson *et al.*, 1995); *Atomic Spectroscopy* provides a comprehensive listing annually (Lust, 1994); the *Journal of Analytical Atomic Spectrometry* contains comprehensive annual update reviews of AAS (Hill *et al.*, 1994) and annual reviews of environmental materials which include the application of AAS to geochemical materials (Cresser *et al.*, 1995). New applications are relatively rare, being confined mainly to increasing throughput, for example simultaneous multi-element AAS, or to isolating specific analytes. Many new papers are appearing in the Far East, reproducing work from an earlier period in the Western world. A detailed theoretical treatise has recently been published by the Ontario Geological Survey (Vander Voet and Riddle, 1993). Students who require a more detailed tutorial study of flame spectrometry should refer to the very readable book by Cresser (1994).

What is atomic absorption?

An atomic absorption spectrometer consists of a light source, an atomisation cell and a monochromator–detector system (Fig. 5.1). In the most commonly used configuration, the atomisation cell is an air–acetylene flame, into which a sample solution is introduced as an aerosol. The high temperature of the flame dissociates the sample into a cloud of ground-state atoms, which can absorb energy (at specific wavelengths characteristic of the elements present) from the light beam passing through the flame. Though the excited atoms re-emit at the same wavelengths (Box 4.1), they do so in random directions, resulting in a net attentuation in beam intensity *along the beam path* (Box 5.1). The attentuation of the beam at a specific analyte wavelength, to which the monochromator is tuned, reflects the number of ground-state analyte atoms in the optical beam path, and thus provides a measure of the concentration of the analyte in the sample.

The light beamed through the atomisation cell comprises not white light but *the emission line spectrum of the analyte element*. It consists of the very wavelengths at which analyte atoms in the flame absorb, and the degree of specific beam attenuation is therefore much greater than would occur with white light. This key feature lies behind the analytical advantages that AAS offers: high sensitivity, freedom from spectral interference, and low capital cost. A separate analyte-specific lamp is employed for each element to be determined.

Radiation source   All forms of AAS require a light beam consisting specifically of the emission spectrum of the analyte element (in the visible-UV wavelength range). The most widely used light source is the hollow-cathode lamp, whose construction is illustrated in Fig. 5.2. The key component is a cylindrical cathode composed largely of the element of interest, commonly as a coating on the inside of the cylinder. Applying power to the lamp generates an electric discharge which ionises the low-pressure inert gas with which the lamp is filled. Positive gas ions are accelerated towards the cathode and, on impacting, sputter ions of the element of interest from the cathode surface. These ions, excited by the discharge, emit the element's

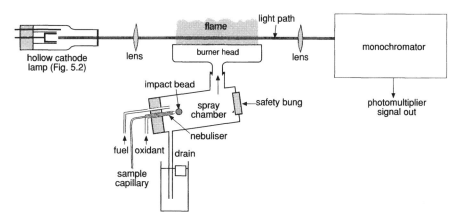

Figure 5.1   Schematic layout of a flame atomic absorption spectrometer.

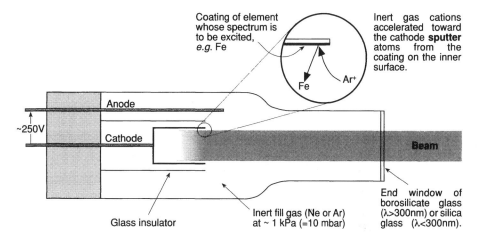

Figure 5.2   The construction of a hollow–cathode lamp as used in atomic absorption spectrometry. Applying a DC potential of several hundred volts between anode and cathode causes an electrical discharge and ionisation of the fill gas. When positive ions, accelerated towards the cathode, impact on its coated inner surface, they *sputter* atoms of the element of interest into the gas, where they are excited by the discharge and radiate their characteristic emission spectrum.

characteristic emission spectrum. A separate hollow cathode lamp is required for each ana lyte to be determined; for convenience, several may be mounted together in a lamp turret. The alignment of an individual lamp can be adjusted to optimise the beam intensity on the monochromator.

Because it is not practicable to operate an acetylene flame in a light-tight enclosure, an atomic absorption spectrometer must be designed to tolerate ambient light conditions. In order to filter out ambient light electronically, lamp output is pulsed, usually at a frequency of 50 or 60 Hz. The detector output is processed synchronously to register the *difference* signal between on and off parts of the cycle and reject the unmodulated signal component. By this means it is possible to eliminate ambient light (and emission from the flame itself) from the analytical signal.

The optimum operating current varies from one lamp to another (refer to the instrument manual or the method screen on the computer). Higher lamp current will produce brighter

Box 5.1 Atomic fluorescence analysis

When a population of ground-state analyte atoms is exposed to radiant energy of precisely the right wavelength (whether from an atomic line source like a hollow-cathode lamp, or from an intense continuum source), electrons in the atoms are promoted by the absorbed energy to a higher energy level. Before long each atom relaxes back to the ground state by emitting a photon of the same or slightly longer wavelength, a process known as fluorescence. The wavelength of fluorescent radiation is characteristic of the element emitting it.

Fluorescent photons are emitted in random directions. Absorption and fluorescence attentuate the light intensity in the *original beam direction* (the basis of atomic absorption analysis), but increase the intensity of analyte-characteristic radiation in *other directions*. In atomic fluorescence analysis, light from a high-intensity hollow-cathode lamp is beamed at an 'atom cell' (typically a hydrogen flame) containing the analyte in atomic form. A photomultiplier is positioned at right angles to the original beam direction (Fig. 5.1.1), behind an optical filter which isolates the waveband of interest. The intensity of the fluorescent radiation is proportional to the abundance of the emitting analyte over several orders of magnitude (typically 0.01–100 µg l^{-1} in aqueous solution).

Atomic fluorescence was initially unpopular as an analytical technique on account of its susceptibility to **matrix effects**, but for volatile elements like Hg, or elements forming volatile hydrides like As, Se, Sb and Te (where hydride generation eliminates matrix differences), it is proving very effective. It is claimed to have detection limits one to two orders of magnitude lower than cold-vapour AAS (CV–AAS) (Stockwell and Corns, 1994). Other elements can be determined by forming volatile alkyl derivatives (e.g. diethyl cadmium). In all such cases the volatile compound dissociates in the heat of the flame, providing neutral analyte atoms in the vapour state.

Fluorescence can also be used in gas analysis, for example for determining trace SO$_2$ levels in monitoring air quality.

RG

Figure 5.1.1 Schematic layout of an atomic flurescence spectrometer (after Stockwell and Corns, 1994).

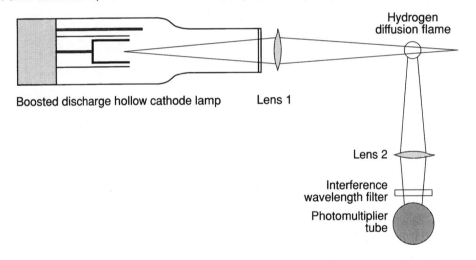

emission and lower baseline noise, but will result in spectral broadening (with a consequent reduction in sensitivity) and a shorter lamp life. Lamps should be allowed to warm up before analysis is begun.

Multi-element hollow-cathode lamps are available for up to six elements, though the range of specific elements that can be combined successfully is limited. These lamps allow capital savings for rarely analysed elements, but are not recommended if optimal performance is required (Cresser, 1994) as they offer lower sensitivity than single-element lamps.

An alternative form of light source, the electrodeless discharge lamp, provides a more intense spectral line and offers increased sensitivity for some of the more volatile elements such as As, Sb, Bi, Se and Te. A salt of the element of interest, sealed in a vial with an inert gas at low pressure, is heated by a radiofrequency induction coil, causing atomisation of the salt and emission of the element's characteristic atomic spectrum. Such lamps give less stable emission than hollow-cathode lamps and have a shorter operating life.

Drift in lamp emission can be compensated by using a 'double-beam' instrument configuration where the source beam is split, alternately passing through or by-passing the flame/atomisation cell (Fig. 5.3). The two beams are recombined before entering the monochromator and subsequently the two signals are resolved in the amplifier to compensate for source drift. Where this is not available, a longer lamp warm-up time is required, and more frequent zero and calibration checks may be necessary.

Monochromator   The monochromator isolates from the lamp spectrum the atomic line of interest (normally in the wavelength range 180–900 nm) for absorption measurement. In AAS the spectral resolution of the method is determined by the line-width of the lamp spectrum, not by the resolving power of the monochromator receiving slit (generally much wider than the line bandwidth – see Fig. 5.3). Because the spectrum is simple (consisting of the emission lines of a single element, or of a few non-interfering elements in the case of a multi-element lamp), line overlap of the kind encountered in atomic emission spectrometry (Chapter 4) poses less of a problem, and a relatively simple low-resolution grating spectrometer is sufficient (see Box 4.3 for principles). The minimum spectral performance required is determined by the spectrum of iron, whose emission lines are very close together.

The detector used is normally a **photomultiplier** (Box 6.3).

Flame atomic absorption

The sample solution, drawn up through a capillary tube, is dispersed as a fine aerosol in oxidant gas using a pneumatic nebuliser as discussed in Chapter 4 (Fig. 4.3a). The aerosol passes through a spray chamber where larger droplets (those larger than 5 µm) fall to the bottom or condense on the walls of the spray chamber and run to waste through the drain, leaving a uniform mist of finer droplets that gives a more stable atomic absorption signal. Some nebulisers incorporate an impact bead to increase the proportion of fine droplets transported into the flame and thereby improve sensitivity. The oxidant-sample aerosol is mixed with fuel in the chamber directly beneath the burner head (Fig. 5.1). For most geochemical applications it is important to ensure that nebuliser and spray chamber are fabricated from acid-resistant materials.

Acetylene is the standard fuel used in modern AAS analysis. For most applications compressed air is used as the oxidant. The mixture should normally be adjusted to give an *oxidis-*

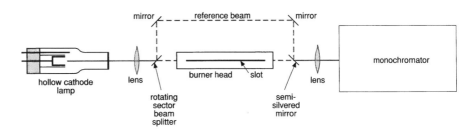

Figure 5.3   Schematic plan view of a double-beam atomic absorption spectrometer.

ing flame, that is non-luminous with a hazy blue inner cone. For some metals, however, particularly Cr, Mo and Sn, a white luminous *reducing* flame gives greater sensitivity. Ideal flame conditions may be found for each element by adjusting the gas flows and burner position until maximum absorbance is obtained while aspirating a standard solution that is known to give about 0.4 absorbance. The burner most commonly used for the air–acetylene flame has a single slot about 100 mm long and 0.3 mm wide. The beam is directed along the length of the flame giving optimum absorption. Solutions containing a high dissolved solids content, for example from a borate or carbonate fusion, may clog the standard burner, and a three-slot 'Boling burner' is preferred for such solutions, greatly reducing burner blockage.

Using nitrous oxide (N_2O) as the oxidant in place of air provides a hotter flame that is preferred for refractory analytes such as Si and Al. The nitrous oxide–acetylene flame provides improved sensitivity and greater freedom from **interference** for a number of elements, notably Ba, Ca, Cr, Mo, Sr and Sn. Flame composition and burner position are more critical. Generally a slightly fuel-rich flame is required with a red zone approximately 2mm high. Because the nitrous oxide–acetylene flame has a faster flame speed than air-acetylene, a modified burner with a shorter (50 mm) slot must be used to prevent flash-back. Most instruments are fitted with an automatic gas control unit to sense that the correct burner is fitted, and the spray chamber incorporates a blow-out plug to prevent explosive pressure build-up (Fig. 5.1). Devices may also be provided to detect power failure, gas flow failure or drain problems, facilitating safe shut-down of the system. The manufacturer's instruction manual should be consulted on the safe use of compressed gases and cylinders, acetylene and flammable solvents.

Interferences   The formation of ground-state atoms can be inhibited by incomplete dissociation of compounds or by ionisation. The extent of the interferences depends on the flame conditions and the burner height. Refractory compounds may not be completely dissociated at the temperature of the flame, for example calcium phosphate in an air–acetylene flame. The addition of an excess of a 'releasing agent' (e.g. La or Sr for the determination Ca), overcomes the interference from phosphate by displacing the analyte in the refractory compound. Alternatively one may use the higher-temperature nitrous oxide–acetylene flame to ensure dissociation of the refractory compounds. Higher-temperature flames may, however, cause appreciable ionisation of the analyte element, reducing the population of ground-state atoms and therefore absorption sensitivity. To control ionisation it is necessary to add a suitable cation having a lower ionisation potential than that of the analyte, for example Na, K, Cs (typically 0.2 to 0.5% w/v). Matrix interference may be appreciable if the physical characteristics of the sample are very different from the calibration standard. This may be controlled through sample dilution, matrix matching or through the use of **standard addition**.

Calibration/optimisation   The degree of beam attentuation (absorption) recorded by the detector is expressed by the absorbance

$$\text{Absorbance} = \log_{10}(I_0/I) \qquad [5.1]$$

where I_0 is the intensity of the incident light beam and I is the intensity of light emerging from the flame. The relationship between absorbance and concentration is usually linear up to an absorbance value of about 0.4; above this value the calibration response becomes increasingly non-linear. For some elements, typically the more volatile elements such as

As, Cd, Se, Te and Zn, non-linearity (resulting from self-absorption) is more pronounced. Instrument manuals provide information on the linear range for optimum working conditions for each element/wavelength/flame. For example, for Ca analysis using a reducing flame at 422.7 nm, the optimum working range is between 1 and 4 $\mu g\ ml^{-1}$. With a less sensitive line at 239.9 the working range is 200 to 800 $\mu g\ ml^{-1}$.

It is not necessary to confine analysis entirely to the linear region of the calibration. Curvature-correction software is provided with most current AAS instruments, usually dividing the curved region into segments and using parabolic curve fitting segment by segment. The elements which may be determined routinely by flame AAS are shown in Fig. 5.5.

Electrothermal atomisation

Trace metal concentrations in geological materials and natural waters are too low for quantification by flame AAS (F–AAS) or ICP–AES. Substituting an electrothermal atomiser in place of the flame ('ET–AAS') significantly increases the sensitivity of AAS, as the sample is vaporised directly into the instrument beam. This technique is used for the determination of trace elements, usually after their chemical separation from the bulk matrix (Cresser *et al.*, 1992). ET–AAS is the most sensitive established atomic spectrometric method for a large number of elements, except for those forming thermally stable oxides or carbides (B, Hf, Ti, lanthanides) and for non-metals (Farago and Mehra, 1994). Detection limits in absolute terms are in the region of $0.1–10^{-3}$ ng for flame AAS and $10^{-3}–10^{-5}$ ng for ET–AAS. An illustration of comparative values between flame and electrothermal AAS is contained in Table 5.1.

In ET–AAS, a transient atomic absorption signal is measured in contrast to the continuous signal observed in flame AAS. In electrothermal AAS analysis, the sample solution is pipetted into the electrothermal atomiser (usually a graphite tube) and subjected to a four-stage controlled heating process:

(1) drying – to evaporate the solvent (up to a minute at 100–150°C);
(2) ashing – to destroy the sample matrix while retaining the analyte (up to a minute at 600–1800°C);
(3) atomisation – to atomise and vaporise the analyte (typically 5 seconds at 900–2700°C);
(4) burn-off – to clean off the sample residue (a few seconds at a few hundred degrees above the atomisation temperature). The furnace must be allowed to cool for 20 s before injecting the next sample.

Table 5.1 Illustration of 3s detection limits ($\mu g\ l^{-1}$) for F–AAS, ET–AAS and CV–AAS

	Wavelength (nm)	F–AAS ($\mu g\ l^{-1}$)	ET–AAS ($\mu g\ l^{-1}$)	CV–AAS ($\mu g\ l^{-1}$)
Al	309.3	100	3	–
As	193.7	1000	2	1
Ba	553.6	30	2	–
Cu	240.7	10	1	–
Hg	253.7	500	–	1
Mn	279.5	10	0.2	–
Mo	313.3	100	1	–
Pb	283.3	50	1	–
Se	196.0	1000	2	1
Sn	224.6	100	5	1

The instrument beam passes along the axis of the tube and undergoes transient attenuation by the analyte vapour evolved during the atomisation stage. The graphite tube must be flushed with inert gas (nitrogen or argon) during the heating cycle to prevent oxidation of the graphite. Modern instruments provide for automated control of the heating cycle.

Types of ET–AAS   Most commercially available electrothermal atomiser systems are based on a resistively heated graphite tube or rod. **Pyrolytically coated graphite** is recommended by most manufacturers as the impervious pyrolytic layer retains the sample solution on the surface, it offers less opportunity for elements like Mo, Si, Ti, V to form stable carbides that resist volatilisation and cause memory effects, and it is more resistant to oxidising mineral acids. The tube configuration has the advantage of improved sensitivity (confined by the tube walls, analyte atoms spend longer in the beam) and longer lifetime (more effectively flushed by inert gas). Other tube materials are available, such as W, Ta, Mo, but are generally regarded as inferior to graphite, being brittle and prone to attack by acid matrices. Yao and Huang (1986) compared pyrolytic graphite, electrolytic graphite and tantalum-foil-lined graphite tubes for the direct determination of trace Be in rocks and found all three tubes gave satisfactory results. Sen Gupta (1993) used pyrolytically coated graphite tubes for the determination of the elements Sc, Y, Nd, Sm, Eu, Dy, Ho, Er, Tm and Yb in rocks. Later he found that a tantalum-foil lining increased the sensitivity almost 10-fold, whilst at the same time allowing the determination of more refractory elements such as La, Ce, Pr, Gd, Tb and Lu.

An analyte may be vaporised and atomised from the surface of the graphite tube before the gas in the tube has reached a uniform temperature. It then experiences cooling which may lead to significant chemical interferences. Placing a pyrolytic graphite platform inside the graphite tube (known as a L'vov platform after its inventor), onto which the sample is injected, delays the atomisation process until the gas has been heated throughout, thus reducing analytical problems.

Operating conditions   Nitrogen is most commonly used as the flushing gas, except for the analysis of nitride-forming elements such as Ba, Mo or V for which argon is preferred. Optimum gas flows are usually specified by the manufacturer. The flow of gas may be automatically stopped or reduced during the atomisation stage, reducing inter–element effects and increasing sensitivity.

Typically sample volumes between 2 and 50 μl are injected into the graphite tube at room temperature. It is possible to use multiple drying stages to increase the sample loading. Injection of excess liquid may result in spreading of liquid to the ends of the tube introducing memory effects. It is strongly recommended that an autosampler is used for accurate injection as operators may find difficulty in injecting accurately into the graphite tube. One manufacturer nebulises the sample directly into a heated graphite tube.

Control of the heating cycles during the drying, ashing, atomisation and burn–off cycle is by setting the temperature and duration of each phase on the controller. Rates of change of temperature between phases may be controlled by a rapid temperature rise (stepped change) or via a more gradual and controlled rise (ramped). The user will need to establish the temperature and ramp rates for each analyte and method, as instrument manufacturers only provide general guidelines for analysis conditions. It is essential to ensure complete evaporation of the solvent during the drying phase. Drying too rapidly, however, results in spitting leading to poor reproducibility. Maximum controlled volatilisation of the matrix is required for the dry ashing stage without significant loss of the analyte. If the matrix is not completely removed, a build-up of matrix will produce a gradual loss in sensitivity. The rate of heating during the atomisation phase can be quite critical. However, it is difficult to provide particular advice for all situations. When using a L'vov platform, it is important to use the maximum ramp rate in order to minimise interference effects.

Background correction   Background absorption interference, due to broad-band molecular absorption or physical scattering of the beam by smoke or salt particles, is far more significant in ET–AAS, particularly in the ultraviolet wavelength range. It is therefore almost certain that background correction will be required. In addition it may be necessary to minimise the interference by optimising the conditions and by adding a matrix modifier. Instruments are equipped with correction systems of deuterium arc, Zeeman effect or Smith–Hieftje type.

A deuterium arc lamp emits a continuum spectrum (mainly in the ultraviolet). For background correction, a deuterium lamp beam is accurately aligned along the same optical path as the hollow-cathode lamp; a rotating sector-silvered mirror alternates the two beams through the graphite tube at a frequency of about 10 Hz, and the photomultiplier output samples each beam synchronously. As non-analyte sources of absorption (e.g. smoke emitted during ashing) vary through the heating cycle, the lamp beam and deuterium beam will be affected in equal proportion. The time-varying correction required to zero the absorbance of the deuterium beam at each stage of the heating cycle is automatically applied as well to the hollow-cathode signal, effectively eliminating the background interference. The bandwidth of the hollow-cathode beam is very narrow (that of the emission line alone), and it therefore responds strongly to analyte-specific absorption, whereas the deuterium continuum fills the much larger bandwidth of the monochromator receiving slit and, being relatively insensitive to the narrow-band analyte absorbance, responds mainly to background absorbance (Fig. 5.4).

In Zeeman effect background correction, a magnetic field is applied to the graphite tube, splitting the spectral line into two polarised beams of slightly different wavelength. The π line is absorbed by both analyte and background, whilst the σ line is only absorbed by the background (see Potts (1987) for a full explanation). Zeeman offers more accurate correction for high background and for 'structured' background (consisting of discrete lines). This method is restricted to electrothermal atomisation and may result in reduced sensitivity for some elements and reduced linear range.

One manufacturer offers Smith–Hieftje correction (Smith and Hieftje, 1983) in which the hollow-cathode lamp is pulsed with a very high current interspersed with the normal current value. High current results in line broadening and self-reversal of the emission line (i.e. splitting of the atomic line). The normal line undergoes analyte *and* background absorption, whereas the 'wings' of the split line lie beyond the absorption bandwidth of the analyte atoms, and therefore experience only background absorption, providing the means for an automated background correction. This method offers accurate correction at higher absorbance levels, and structured background error is virtually eliminated. However, broadening of the emission line profile causes a reduction in sensitivity.

Calibration and   Calibration in ET–AAS is by the method of **standard addition** to overcome the likelihood
minimisation of inter-   of severe interference effects. Aliquots of sample are spiked with known concentrations of
element effects   the analyte; the resulting spiked samples and the original sample are then measured. Sample concentration is determined by extrapolation. At least two spiked samples (i.e. three measurements) are required and the calibration must be linear. This may not overcome all types of interference, and may create problems in correctly assessing the blank.

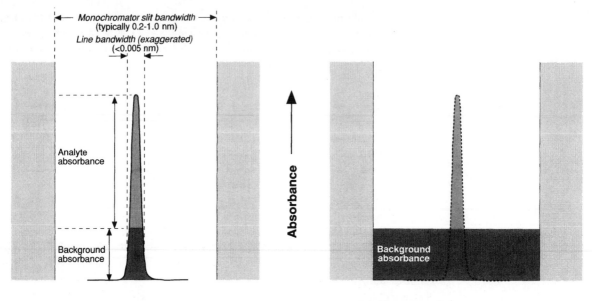

(a) Absorbance in the hollow cathode lamp beam

(b) Absorbance in the deuterium arc lamp beam

Figure 5.4   The principle of deuterium-arc background correction.

Luecke (1992) recommends the use of HNO_3 for the determination of K and Na in silicate samples. Nitric acid is the preferred matrix for ET–AAS; HCl or H_2SO_4 should be avoided in the final solution for analysis. The degree of interference observed may also be dependent on the design of the graphite tube atomiser and may also be age dependent. It should be stressed that interference tests should be carried out at various stages of the graphite tube lifetime.

Matrix modification   The addition of certain reagents may significantly reduce the interference effects. Ammonium nitrate added as a modifier aids the evolution of ammonium chloride from HCl solutions during the dry-ashing step and considerably reduces both the background absorption and the chemical interference. The modifier serves to increase the volatilisation temperature of the analyte. Other commonly used matrix modifiers are diammonium hydrogen phosphate, magnesium nitrate, nickel nitrate, Pd or nitric acid. Ammonium fluoride removes severe matrix effects in the determination of Co and Ni in a variety of standard rocks, and orthophosphoric acid reduces loss of Zn in the analysis of gallium arsenide.

Cold vapour techniques

An alternative atomisation technique is the cold vapour technique for mercury. Hg compounds are reduced to elemental mercury with tin (II) chloride, and the Hg vapour is then flushed by an inert gas (nitrogen or argon) into a silica–glass tube aligned in the beam path of the AAS instrument. Cold vapour generation (CV–AAS) is also routinely used for Hg in a variety of sample types in spite of interferences in complex matrices.

Another low-temperature AAS method is applicable to the determination of elements like As, Sb, Sn and Bi (Fig. 5.5) that form volatile hydrides on being reduced with sodium boron hydride ($NaBH_4$) or lithium aluminium hydride ($LiAlH_4$). The hydrides are swept by nitrogen into a silica glass tube mounted above the conventional burner. Sensitivity by hydride generation is 100–300 times greater than by flame atomisation. For the determination of Se, CV–AAS remains the method of choice for all but the lowest concentrations (Haygarth *et al.*, 1993).

Cold vapour and hydride generation techniques have been successfully used with atomic fluorescence spectrometry (Box 5.1).

Current and future developments of AAS

In reviewing analytical methods used for environmental media, Cresser *et al.* (1995) note that nearly 40% of the references use AAS for the analysis of geological materials, supporting the view that despite the introduction of ICP–AES (Chapter 4) and ICP–MS (Chapter 10), AAS still has a role to play in the analysis of rocks, minerals and sediments. New publications report the application of AAS for the analysis of Au, Ag, Ca, Ni and Sn. Direct analysis of solid material continues to be an active area of development, with the emphasis on the introduction of slurry samples to both flame and electrothermal AAS (cf. Chapter 10).

Flow injection may be combined with AAS as a means of sample introduction to improve sensitivity, for on-line sample dilution and reagent mixing or for calibration. The sample is injected, via a liquid injection valve, into a carrier stream, propelled with a peristaltic pump. As the sample is transported towards the atomic absorption spectrophotometer, the sample disperses into the carrier stream. Fang (1995), in his review of flow injection AAS, notes the potential for increasing the normal analytical range by flow injection atomic absorption spectrometry (FI–AAS) by on-line dilution or preconcentration, and summarises geochemical applications for Bi, Se, Ca, Mg, Fe, Au, Mn, Hg, Ni, Ag and Sn. Lin and Hwang (1993) determined Au and Cu in ores, obtaining 1.8 and 1.0 ng ml^{-1}. Slurry sampling coupled with flow injection has recently been reported for the flame AAS determination of Cd, Zn and Mn in silicate-based materials (Lopez Garcia *et al.*, 1993).

ET–AAS has excellent sensitivity for many elements; its main constraint has been analysis times of the order of 2–3 min per determination arising from the need to use standard additions for calibration. Recently the United States Geological Survey (Kane, 1988) developed a simultaneous multi-element atomic absorption instrument with a continuum source ('SIMAAC'). The system employs a 300 W xenon arc lamp with a 20-channel echelle polychromator. Simultaneous capability may find favour with geochemical laboratories. Farah and Sneddon (1993) applied simultaneous multi-element AAS to determine Cu, Fe, Mn and Zn in sediments, using Smith–Hieftje background correction to obtain the best results.

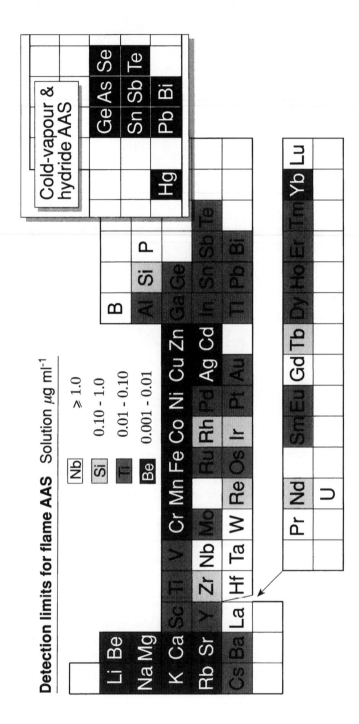

Figure 5.5　A periodic table showing 3s detection limit ranges for elements that can be determined by flame AAS (data from Cresser *et al.*, 1995, Fuller, 1979, Greenberg *et al.*, 1992, MEWAM, 1980; Whiteside, 1979). Additionally, those elements that can be determined by cold vapour and hydride techniques are shown in the inset.

Analysis of non-metals in aqueous solution

For geochemical and hydrological studies of aqueous solutions, quantification of non-metals, organic fractions or anions requires a range of analytical techniques which complement atomic spectroscopy. In certain situations projects may include the analysis of anions in streams to complete the ionic balance (Table 1.1) for input to numerical models or the study of the role of carbonate species in the aquifers of sedimentary rocks. The scope of this section is to review briefly the analytical methods available for the analysis of aqueous solutions such as precipitation, streams, soil solution, lakes and groundwater. Many of these measurements are important in environmental monitoring; some of the techniques described lend themselves readily to use in the field as well as the laboratory.

pH and ion-selective electrodes

pH pH is probably the most fundamental characteristic in water chemistry, and one of the most frequently made chemical measurements. Indicator papers are availabe but are unsuitable for pH measurement for characterisation of geological materials. Refer to MEWAM (1988) or Greenberg *et al.* (1992) for detailed methodology. The basic principle of the pH measurement is the determination of the activity of the hydrogen ions by potentiometric measurement, that is pH $= -\log_{10}[H^+]$. This parameter is simple to determine with inexpensive electrodes, buffers and a meter. The apparatus is shown in Fig. 5.6(a). Potentiometric measurements combine two 'half-cells' (a measuring electrode and a reference electrode) to form an electrochemical cell. pH is determined by comparison of the potential (emf) of the cell containing the test solution with the potential of the cell containing a buffer solution of known pH.

The electrode assembly stand routinely holds a glass electrode, a reference electrode and a temperature compensating device (thermistor probe). The H^+-sensitive device is usually a glass electrode, which on being immersed in a test solution develops a potential that varies linearly with pH. The Nernst equation defines the relationship between the measured potential E (volts) and pH as

$$E = E^\circ - \text{pH} \; [RT/F] = E^\circ - 0.59\text{pH at } 25^\circ C \qquad [5.2]$$

where R is the gas constant, T is the temperature (in K) and F is the Faraday constant (see Glossary).

pH-sensing glass electrode – universally used as the indicator electrode in the measurement of pH. The glass electrode is constructed with pH-sensitive glass in the form of a sealed glass bulb and contains HCl or a buffered chloride solution in contact with an internal silver reference electrode. The change in potential develops following ion exchange reactions between the outer surface of the glass and the experimental solution. To determine pH accurately in solutions with pH > 10, a low-sodium glass electrode should be used. A glass electrode should be stored in pH 4 buffer to maintain it in good condition. Research-grade glass electrodes, supplied by manufacturers, are adequate for survey and research (Davidson, 1987).

Reference electrode – provides a standard reference potential with respect to the hydrogen-sensitive glass electrode. A reference electrode consists of a metal in contact with a solution which is saturated with a sparingly soluble salt of the metal and an additional salt with a common anion. Common examples of the reference electrode are the calomel electrode $(Hg/Hg_2Cl_2(s), KCl(aq))$ and the silver:silver chloride electrode $(Ag/AgCl(s),$

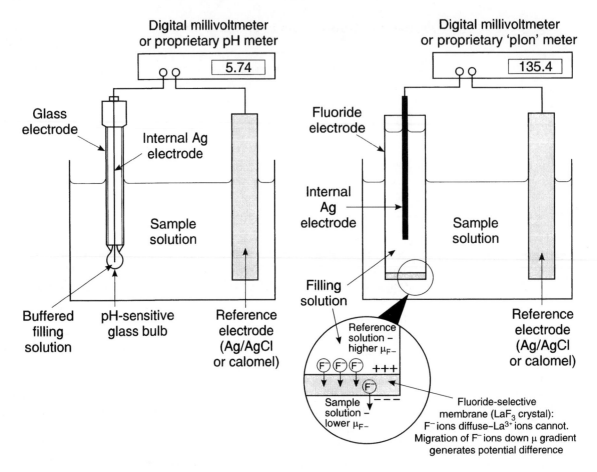

Figure 5.6   (a) Measurement of pH; (b) fluoride ion-selective electrode.

KCl(aq)). Contact with the test solution is through a ceramic, quartz or fibre frit located at the tip of the reference electrode. It is important to refill non-sealable electrodes with electrolyte to ensure a good flow of electrolyte through the liquid junction in the reference electrode in order to reduce pH errors. A reference electrode should be stored in concentrated KCl to maintain it in good working condition.

Combination electrode – incorporates glass and reference electrodes in a single probe. Combination electrodes are useful for determining soil pH of soil:water slurries (Grimshaw, 1989) or for field measurements used in conjunction with portable meters. In prime condition combination electrodes perform as well as separate glass and reference electrodes, although in the long term it is easier to maintain separate pH and reference electrodes.

Many pH meters are capable of reading pH or millivolts for ion-selective electrode measurement. For routine work pH values are accurate and reproducible to 0.1 pH units (Greenberg *et al.*, 1992). Routine procedures to set up the instrument involve checking the calibration of the instrument using two buffers, for example at pH 4 and 7, and adjusting the slope and intercept controls. Suspended matter or oil can interfere by blocking off the electrode surface. A calomel electrode contains toxic Hg and in the event of breakage should be disposed of safely. Low-resistance glass electrodes may be used for measuring pH in samples below $10°C$ for improved response times. The relationship of pH to electrode potential as expressed above in the Nernst equation is sensitive to changes in temperature, and most instruments contain a built-in temperature compensation (thermistor) probe.

Problems rarely occur with high-ionic-strength solutions, but in samples of low ionic strength (with an electrical conductivity $< 150\,\mu\text{S cm}^{-1}$) such as atmospheric precipitation or pure upland streams, pH can be subject to significant error resulting from the low conductance of the experimental solution and the significant variation in the liquid junction potentials. For measurement of low-ionic-strength solutions, the performance of the system may be checked using $5 \times 10^{-5}\text{M H}_2\text{SO}_4$ (pH 4.00). In addition, check the pH in stirred and quiescent solutions; the stirring shift for electrodes in good condition is < 0.02 pH units.

Ion-selective electrodes   Electrodes with specific sensitivity to other ions, such as F^-, Cl^-, have been developed based on membranes or solid-state crystals to separate the electrode internal reference solution from the test solution. The potential developed at the electrode is measured across a reference electrode, as for pH, commonly Ag/AgCl or the calomel reference electrode. Solid-state electrodes consist of a single crystal or a compacted disc of active material. For example, the solid-state fluoride-sensitive electrode employs a membrane of lanthanum fluoride (Fig. 5.6). The LaF_3 crystal separates the filling solution (0.001M F^-) from the test solution.

Fluoride is discharged from industrial processes such as brickworks, aluminium smelters or pottery factories. Atmospheric deposition may lead to direct increases of the pollutant in the soil or the increase in acidification may mineralise F-containing minerals. Fluoride is closely associated or complexed with Al which itself is an interferent in the determination of F^-. Acetate/citrate buffer is added to the water sample or soil/water extract to complex with the Al ions. Citrate itself causes a very slow electrode response; cyclohexamine diaminetetracetic acid is an alternative buffer to overcome this difficulty.

Solid-state electrodes are also available for Cu^{2+}, Ag^+, SO_2, Cl^-, CN^-, I^- and S^{2-}. Gas-sensing probes depend on gas diffusing through a membrane to modify the composition of the internal solution to change the potential. Gas-sensing electrodes are available for CO_2 and NH_3 determination. Plastic electrodes use a neutral matrix containing cavities of specific sizes. Electrodes sensitive to Ca^{2+} and NO_3^- perform well (Allen, 1989). Ion-selective electrodes have applications in flow analysers or for field measurements.

Conductivity

Electrical conductivity (usually quoted in units of $\mu\text{S cm}^{-1}$ – see Appendix D) provides a measure of the total dissolved salt content or total ionic strength of a solution. In geological studies, for example, conductivity may be used to monitor seasonal or daily fluctuations in aquifers, to compare cation/anion concentrations in solution, or to estimate sample size or dilution for analytical determinations. The conductivity is mainly dependent on the inorganic ions present as the relationship between concentration and species varies. However, for this parameter, absolute accuracy and precision may be of less importance than for other measurements. Hem (1982) provides a detailed discussion of the theory and practice of conductivity and conductance.

The conductivity probe contains inert electrodes with a fixed geometry. To avoid polarisation of the ions, the measurement uses an alternating current. Response varies with temperature of the solution. It is important to maintain reasonable control over the solution temperature even with temperature compensation in operation. The calibration of the instrument should be checked periodically with standard KCl solutions (1×10^{-3}M KCl $= 146.9\,\mu$S cm^{-1}). Interferences are not generally a problem, although high levels of suspended matter, or substances that coat the electrode surface such as oil or grease, will introduce bias.

Alkalinity

Alkalinity is a measure of dissolved carbonates, bicarbonates and hydroxides in stream water or groundwater:

$$\text{Alkalinity (mol kg}^{-1}) = 2[CO_3{}^{2-}] + [HCO_3{}^-] + [OH^-] - [H^+] \qquad [5.3]$$

where brackets denote concentrations of individual species. Alkalinity is expressed in units of mg l^{-1} CaCO$_3$ or occasionally mg l^{-1} HCO$_3{}^-$.

The equilibrium between carbonate (CO$_3{}^{2-}$) and bicarbonate (HCO$_3{}^-$) species in natural waters buffers pH and neutralises acidic inputs (Drever, 1988), and for most purposes alkalinity is synonymous with 'acid–neutralising capacity' or ANC (Kramer, 1982). Its determination is therefore simply a matter of titrating these naturally occurring bases in the water sample against acid of known strength. The end point, where neutralisation of the bases occurs and pH changes rapidly with small additions of acid, is close to the pH of 4.5, and methyl orange is used as indicator. Another end point, reflecting the contribution of carbonate (CO$_3{}^{2-}$) to total alkalinity, occurs at pH 8.3, with phenolphthalein as indicator . Using an autotitrator configured for potentiometric titration simplifies the measurement for coloured or turbid solutions. 'Gran plots' (Gran, 1950, 1952; Drever, 1988) linearise the titration curve to improve greatly the precision of the determination, allowing extrapolation of the equivalence (end) point to zero hydrogen ion concentration (Drever, 1988).

Because alkalinity varies with dissolved CO$_2$ (which is pressure and temperature sensitive), it is important that alkalinity of natural waters is measured in the field or with minimal delay.

Ion chromatography

Ion chromatography is a specific application of high–performance liquid chromatography (HPLC, see Chapter 16) that has been exploited commercially for the analysis of ionic or charged molecules. This field of chromatography includes ion exchange, ion pair and ion exclusion separation mechanisms. Ion chromatography has application in the geochemical analysis for the determination of anions, cations, transition metals and organics. MEWAM (1990) has recently produced a detailed review of theory and applications.

In practical terms, the liquid sample is injected via an injection valve into a carrier stream, termed the mobile phase or eluent, a fluid such as dilute NaOH which transports the solution through the chromatography column. As the mobile phase transports the sample through the column, ionic components compete for the exchange sites on the analytical separator column. Ions separate on the basis of size and ionic charge (Fig. 5.7), monovalent and smaller ions eluting more quickly. Fractions are registered in turn by an ion conductivity (Box 16.1), ultraviolet (UV) or electrochemical detector; components are identified from their retention time, the elapsed time between the injection and detection. The technique requires low-capacity ion exchange columns that are stable over a wide range of pH.

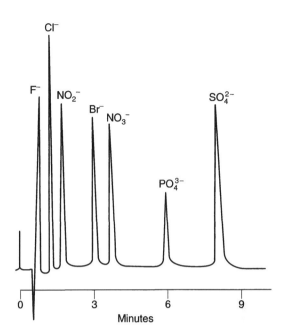

Figure 5.7 Ion chromatogram of a standard solution, showing the separation of anion species. Instrument Dionex 2010i, columns AG4A and AS4A, micromembrane suppressor, $NaHCO_3 + Na_2CO_3$ eluent.

Guard columns, commonly a shorter version of the analytical column, protect the more expensive analytical column from fouling.

Anion analysis Commercial columns for anion applications are based on sulphonated styrene divinyl benzene polymers bonded with a thin layer of ion-exchanging latex beads. This column type is supplied for anion applications with chemically suppressed conductivity detection for the analysis of major anions (Fig. 5.7). Alternatively silica-based columns are useful for indirect detection by UV detection (Cape, 1987). In carbonate/bicarbonate eluent systems with conductivity detection, the conversion of the ion to the hydrogen form through chemical suppression increases the conductance of the anionic species whilst at the same time reducing the background signal of the eluent. This form of ion chromatography has found widespread application owing to the associated improvements in sensitivity and detection limits. Interferences are not a major problem in water analysis. Because conductivity is a universal property of ionic species, interference may occur if species co-elute; that is, if species have identical retention times. Excessive concentration of one ion in the presence of trace levels of other analytes may present a problem in the analysis of brines or some groundwaters.

Colorimetric spectrophotometry

Colorimetric methods are based on the reaction of the solute with a chromogenic reagent to form a coloured compound or complex. Colorimetry has applications in geochemistry for the analysis of metals or non-metallic species in acid digests from rocks, soil, plant material and directly on aqueous solutions. Atomic spectrometry methods provide a measure of total concentration in solution, whereas colorimetric methods commonly allow analysis of speciation. For example, Driscol (1984) developed procedures to determine aluminium separately as total, labile and non-labile species.

The absorption of radiation by the substance in solution is the basis of spectrophotometry. The visible spectrum extends from approximately 400 to 850 nm; this relates to the absorption of energy by coloured solutions. The degree of absorption of monochromatic radiation by the sample contained in an optical cell or 'cuvette' is proportional to the number of molecules or ions in the path. The light transmitted is logarithmically related to the path length in the solution and the concentration of solute in solution. The **Beer–Lambert** relationship between **absorbance** and concentration (Equation 5.1) is linear over a range of concentrations. Deviations from linearity arise through chemical effects, solute/solvent interactions or through self-absorption as the concentration increases above absorbance values above about 0.7.

Manual colorimetric methods are appropriate for use by individuals with limited analytical facilities. These methods require a simple spectrophotometer and calibrated glassware (see Allen (1989) for manual procedures) allowing the operator to process up to 50 samples in a working day. In proprietary kit form, such methods are readily adapted to field analysis. Continuous or segmented flow colorimetric methods offer automation for up to 200 samples a day with the possibility to combine modules for simultaneous analysis, for example nitrate, ammonium, phosphate and silicate in waters (Greenberg *et al.*, 1992, Allen, 1989). More recently, flow injection (cf. AAS) analysis has emerged as an option to process analyses at the rate of 100 h^{-1}.

The analysis of nitrate is an important example of a colorimetric method for the analysis of groundwater and surface water. Colorimetry provides an alternative to ion chromatography with the potential for a higher throughput. Direct spectrophotometric methods involving either UV measurement or phenoldisulphonic acid are subject to interference from soluble organic matter. Routinely, either manual or automated, the method based on the reaction of nitrite with sulphanilamide and N-1-naphthyl ethylenediamine dihydrochloride compounds to form an azo dye is the most sensitive option. Reduction of nitrate to nitrite is usually achieved using copper–catalysed cadmium, either by batch processing or through a column, with the method quantifying nitrate plus nitrite.

Miscellaneous other determinations

Speciation of ferrous iron, Fe(II) or Fe^{2+}

Iron occurs in terrestrial rocks in both ferrous (Fe(II)) and ferric (Fe(III)) forms[1]. Because the Fe^{2+} and Fe^{3+} ions have different charge and ionic radii, they occupy different sites in silicate crystals. A complete analysis of an iron-bearing mineral should therefore report Fe(II) and Fe(III) separately. Ferrous iron in the digested rock solution may be determined

---

[1] In meteorites it may also occur as metallic Fe–Ni alloy. The natural occurrence of this oxidation state (Fe(0)) at the Earth's surface is rare, for example where FeO-bearing lava has interacted with carbonaceous sediment.

by titration against an oxidising agent, and the ferric iron concentration determined by subtraction from the total iron concentration. Two Fe(II) titration methods are available for silicate rocks and minerals:

(1) *Direct titration.* The sample is digested using hot concentrated H_2SO_4 + HF in a Pt crucible with a close-fitting lid to minimise atmospheric oxidation of Fe(II). After complexing excess HF with boric acid, the dissolved Fe(II) is titrated against potassium dichromate ($K_2Cr_2O_7$) using barium diphenylamine sulphonate as indicator.

(2) *Indirect titration.* The powdered sample, with a known excess of ammonium metavanadate (NH_4VO_3, an oxidising agent), is digested in cold 48% HF in a covered plastic beaker for a period of hours or days. When no dark particles remain, the excess HF is neutralised with boric acid and the excess metavanadate is titrated against ferrous ammonium sulphate, using barium diphenylamine sulphonate as indicator. This method (due to Wilson, 1955), by immediately oxidising Fe(II) to Fe(III) on dissolution, minimises atmospheric oxidation, but like direct titration is susceptible to interference from reduced sulphides present in the sample.

The ferrous iron analysis of a sample as collected may not represent the true FeO content of the material of interest (groundwater at depth, igneous melt, etc.) owing to oxidation.

Determining speciation is an important factor in understanding the environmental chemistry of arsenic (Box 5.2).

Box 5.2. Determining the speciation of arsenic in natural waters

Arsenic exists in the environment in two oxidation states, As(III) and As(V). Being the thermodynamically stable form in oxygenated environments As(V) predominates in most natural waters, but weathering of As-bearing sulphide ores releases As(III) which may therefore be abundant in mine effluents and in water courses polluted by them (as for example in the Carnon River incident in Cornwall in 1992, when contaminated mine waters burst out of the disused Wheal Jane tin mine – see Hunt and Howard, 1994). Arsenic is also readily metabolised by marine and freshwater organisms, giving rise to various alkylated species (such as dimethylarsenic) that may play an important role in its environmental chemistry. It follows that, for understanding the environmental distribution and fluxes of As, it is important to determine its speciation as well as the overall concentration.

Conventional hydride AAS methods of analysis are well suited for determining total arsenic (Fig. 5.5) but fail to distinguish between environmentally distinct As species. Hunt and Howard (1994) succeeded in using high-performance liquid chromatography (HPLC – see Chapter 16) to separate As(III) and As(V) prior to hydride–AAS analysis. A strong anion exchange column was used and the successive As(III) and As(V) fractions in the acidified eluent were reduced to arsine (AsH_3) by sodium boron hydride ($NaBH_4$) before passing sequentially into a heated silica glass atomiser cell for absorbance measurement.

Howard and Arbab-Zavar (1981) and Howard and Comber (1992) separated inorganic arsenic, monomethylarsenic and dimethylarsenic species for analysis using sequential hydride evolution. In this method, reduction using $NaBH_4$ produces respectively from these species the arsine derivatives AsH_3, $As(CH_3)H_2$ and $As(CH_3)_2H$, which are collected together in a cryogenic trap cooled by liquid nitrogen. Subsequent removal of the liquid nitrogen allows the trap to warm up slowly, releasing each of the arsine species in turn according to their relative volatility, enabling them to be determined sequentially using a conventional hydride AAS cell.

RG

Determination of volatile constituents of rock and mineral samples

Most rocks contain significant amounts of OH^-, CO_3^{2-} or other volatile constituents such as F^- and Cl^- bound in structural sites in their constituent minerals (Table 1.1), and the content of such components usually increases with hydrothermal alteration and weathering as hydrous minerals replace anhydrous ones.

Determination of these bound volatile constituents (which may provide a useful measure of the degree of alteration) usually involves removal by heating following preliminary drying of the sample at $110°C$ to remove adsorbed water. The H_2O and CO_2 evolved (see equations 1.1 and 1.2) can be measured in various ways:

- The weighed rock powder may be ignited to $1000–1200°C$ for 30 minutes, cooled, and the sample reweighed. This 'loss-on-ignition' (LOI) method does not discriminate between volatile species and is prone to interference from the oxidation of ferrous iron.
- Ignition of a known mass of sample to $1200°C$ in a stream of dried nitrogen, which passes through weighed absorption tubes containing a desiccant (e.g. magnesium perchlorate to absorb 'H_2O^+' – see p4 and a CO_2 absorber (e.g. 'soda asbestos'). The change in weight of these absorption tubes indicates the H_2O^+ and CO_2 evolved by the heated sample. Thermal conductivity sensors before and after the two absorption tubes provide an alternative to weighing. Not all hydrous minerals disproportionate even at $1200°C$.
- The weighed rock powder may be ignited at $1500°C$ by flash combustion in a tin capsule, and the evolved gases carried in a stream of carrier gas (e.g. He) through a gas chromatography column and measured by means of detector peak area, as described in Chapters 9 and 16.

Chapter 6

X-ray fluorescence spectrometry

Godfrey Fitton

X-ray fluorescence (XRF) spectrometry is one of the most widely used and versatile of all instrumental analytical techniques. An XRF spectrometer uses primary radiation from an X-ray tube to excite secondary X-ray emission from a sample. The radiation emerging from the sample includes the characteristic X-ray peaks of major and trace elements present in the sample. Dispersion of these secondary X-rays into a spectrum, usually by X-ray diffraction, allows identification of elements present in the sample. The height of each characteristic X-ray peak relates to the concentration of the corresponding element in the sample, allowing quantitative analysis of the sample for most elements (Fig. 6.1) in the concentration range of 1 ppm to 100%.

X-rays are electromagnetic radiation with wavelengths in the range 0.003–3 nm, more than 300 times shorter than the wavelength of visible radiation. They are produced when electrons jump between the inner, most tightly bound electron energy levels in atoms, those belonging to the K, L and M shells (Box 6.1). X-ray spectra are simple and easily interpreted. Their insensitivity to chemical bonding and valence effects allows the direct analysis of solid samples without the need for dissolution and other chemical pretreatment.

The layout of components in an XRF spectrometer is shown diagrammatically in Fig. 6.2. A sample, usually in the form of a solid disc, is irradiated with an intense beam of primary X-rays and emits secondary (**fluorescent**) X-rays. A small proportion of these is collimated into a parallel beam and dispersed into a spectrum through diffraction by a synthetic analysing crystal. The diffracted X-rays are further collimated and pass to an X-ray detector. The analysing crystal can rotate about an axis on its surface, and the detector and secondary collimator are mechanically coupled to the crystal so that they move in an arc around the rotation axis of the crystal. The mechanism through which the crystal, detector and secondary collimator are rotated, and which allows accurate measurement of the rotation angle, is called a goniometer. The detector and secondary collimator rotate at twice the angular rate of rotation of the crystal in order to keep the angle of incidence equal to the angle of reflection (θ). The XRF spectrometer produces a spectrum in which the intensity of the diffracted beam is related to the angle of reflection (for practical purposes it is customary to measure 2θ rather than θ). The angle θ is related to wavelength (λ) through the Bragg equation

$$n\lambda = 2d \sin \theta \qquad [6.1]$$

where n is an integer and d is the lattice spacing of the analysing crystal (see Box 6.2). Because long-wavelength X-rays are absorbed by air, it is usual for the XRF spectrometer chamber to be evacuated by a rotary pump (Appendix A). In cases where a vacuum cannot be used, as, for example, when analysing liquid samples, the spectrometer can be filled with helium which has a lower X-ray absorption coefficient than air.

The intensity of individual characteristic X-ray peaks may be measured by setting the goniometer to the appropriate 2θ angle and counting X-ray photons for a period long enough (typically 4–500 s) to achieve the required **precision** in the count rate. Count rates

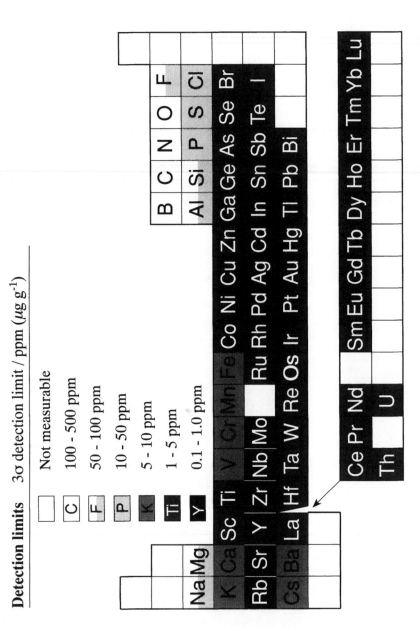

Figure 6.1   Periodic table showing elements (with approximate 3σ detection limits) that can be determined by XRF spectrometry in geological materials.

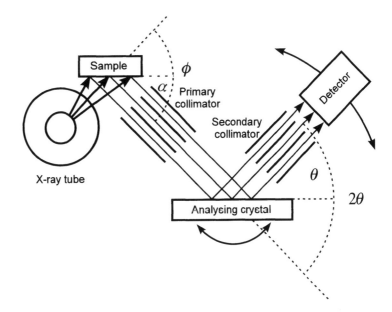

Figure 6.2   Schematic layout of a typical XRF spectrometer. The X-ray path is indicated by lines with arrows; bold lines indicate primary radiation, thinner lines indicate secondary (fluorescent) radiation. Varying the diffraction angle θ disperses the secondary radiation into an X-ray spectrum. The scattering angle ϕ, and the take-off angle α, are both fixed.

are generally recorded in units of thousands of counts per second (kc s^{-1}). Because the characteristic X-ray peaks are superimposed on a background of scattered radiation from the X-ray tube, it is also necessary to measure the intensity of this background radiation by counting at 2θ angles offset slightly from the peak. The net intensity (peak minus background count rates) can then be related to concentration by comparison with reference standards. The lower limit of detection (Box 4.4) for an element is determined by the precision with which the appropriate background intensity can be measured.

X-ray excitation

The primary energy source in XRF analysis is an X-ray tube designed specifically for the purpose. The principal components are a tungsten filament and metal anode contained within a glass tube under a high vacuum. The filament is heated by means of an electric current and produces a cloud of electrons. These are accelerated away from the filament, through the application of a high negative potential (the tube voltage, typically 40–100 kV), and strike the anode, which is earthed. The rapid deceleration of these electrons by interaction with the anode atoms releases energy as X-rays with a distribution of energy (and therefore wavelength) forming a continuum with an upper limit, expressed in keV, numerically equal to the applied tube voltage in kV. Electrons accelerated by a potential dif-

Box 6.1 An introduction to X-ray spectra

X-rays are electromagnetic waves with much shorter **wavelengths** (λ) than visible light, in the range 0.003–3 nm. X-rays consist of high-energy **photons**, with photon energies (E_q) in the range 500–500,000 **eV**. These properties are related through Equation 4.1.2 (Box 4.1):

$$E_q = hc/\lambda \qquad [6.1.1]$$

where h is Planck's constant and c is the speed of light (for values see the Glossary). Owing to the high photon energy, X-rays fall in the category of ionising radiation and the design of analytical equipment that generates X-rays is therefore closely regulated; appropriate radiation precautions must always be observed by analysts using such equipment.

X-rays are the most energetic form of radiation originating in the electron 'shells' of atoms; only γ-rays have higher photon energies, and they are produced by atomic nuclei. X-rays are produced by electron transitions into vacancies created in the deepest, most tightly bound electron energy levels in the atom, the so-called **K, L and M shells**. Compared with the complexity of valence electron energy levels, the 'inner' electron shells have a very simple structure:

Shell	Value of n[1]	Orbitals included	Number of discrete energy levels[2]
K shell	1	1s	1
L shell	2	2s, 2p	3
M shell	3	3s, 3p, 3d	5
N shell	4	4s, 4p, 4d, 4f	7

[1] n is the principal quantum number (for explanation, see Gill, 1995, Chapter 5).
[2] The number of levels is known *empirically* from the number of K, L and M **absorption edges** (see Box 6.5, Fig. 6.5.1). *Theoretical* explanation requires consideration of the number of ways the l (orbital angular momentum) and s (spin) quantum numbers can be combined to form the total angular momentum quantum number j and is beyond the scope of this book.

X-ray spectra, the product of transitions between these energy levels, thus consist of fewer lines (and experience fewer potential line overlaps) than atomic spectra in the UV/visible wavelength range; hence the attractiveness of X-ray spectrometry for elemental analysis. Moreover, as most X-rays penetrate materials to a significant depth, X-ray methods allow the analysis of solid samples directly.

Figure 6.1.1 shows diagrammatically a part of the **characteristic** X-ray spectrum of iron. Each peak represents a specific electron transition, as indicated (in simplified form) in the inset diagram. The wavelength of the peak is related to the energy difference ΔE between the electron's initial and destination energy levels:

$$\Delta E = E_q = hc/\lambda \qquad [6.1.2]$$

Peaks (more commonly known as 'lines') are conventionally labelled according to the identity of the *destination* shell. Unlike visible/UV spectra, the general form of the X-ray spectrum (in terms of K, L and M lines) is common to all elements, though the wavelength λ of each line varies systematically with the element's **atomic number**, Z. The relationship is a simple one, given by the Moseley equation:

$$1/\lambda = k(Z - k')^2 \qquad [6.1.3]$$

where k and k' are constants depending on the X-ray line transition being considered. This simple relationship makes peak identification straightforward. Figure 6.1.1 illustrates how the Kα line migrates with change of Z.

Emission spectra

To excite X-ray emission, it is first necessary to create a vacancy in an inner electron shell (for most analytical applications, the K or L shell) by ejecting one of its electrons out of the atom; the cascading of one or more electrons from higher levels to fill this vacancy is what causes X-ray emission. The energy source deployed to ionise the inner shell may be either of the following:

ference of 50 kV, for example, acquire a kinetic energy of 50 keV, and X-rays excited by such electrons cannot have a photon energy exceeding this value.

The impact of high–energy electrons can also ionise the anode atoms by ejecting electrons from their inner shells. The subsequent replacement of the ejected electrons by electrons from the outer shells generates X-rays with wavelengths characteristic of the anode material. The X-ray spectrum generated thus consists of continuous and discontinuous (or characteristic) components (Fig. 6.3). The radiation passes out of the tube via a thin window, usually made of beryllium because its transparency to X-rays extends to long wavelengths. An X-ray tube is analogous in its component parts and operating principles to an electron microprobe (Chapter 14). Both use electrons emitted by a heated filament, and accelerated

(1) *A high-energy electron beam* ('primary excitation', used in X-ray tubes and in the electron microprobe – Chapter 14). Such electrons interact with target atoms in two ways:

(a) They collide with, and eject, electrons in the atom's electron shells. In filling the vacancies thus created the target atoms emit sharp **characteristic** X-ray lines whose wavelengths are specific to the element concerned.

(b) Electrons in the beam may undergo deceleration under the influence of the electrostatic field of the nucleus. The resulting change in the electron's own kinetic energy also leads to X-ray emission, but as such interactions do not involve quantised energy levels they generate an *X-ray continuum* rather than sharp peaks. See **bremsstrahlung**.

Thus *primary* X-ray spectra consist of sharp **characteristic** lines superimposed on a smooth continuum (see Fig. 6.3).

(2) *A beam of X-rays* of sufficiently high photon energy to ionise the relevant shells ('secondary excitation' or X-ray **fluorescence**).

Fluorescent X-rays generated in this way consist in principle of characteristic lines without the continuum, though some continuum X-rays are unavoidably carried forward from the X-ray tube spectrum due to scattering from the specimen (Fig. 6.5).

RG

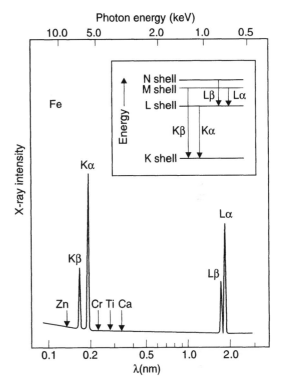

Figure 6.1.1 The X-ray spectrum of iron (simplified). The inset shows the electron transitions involved. The arrows in the main diagram show how the Kα wavelength shifts from element to element. From Gill (1995).

through the application of a high negative voltage, to generate characteristic X-ray spectra from a target. In the case of the X-ray tube the target is a metal anode whereas in the electron microprobe it is the specimen to be analysed.

In XRF analysis, primary radiation from the X-ray tube is used to irradiate a sample (Fig. 6.2). This has the effect of ionising the sample by displacing electrons from its component atoms, since X-ray photons have energies comparable with the energies binding electrons in the inner shells of the atoms concerned (Box 6.1). This process can only occur with photons more energetic than any given binding energy, but is most efficient for photons with energies only a little higher. X-ray tubes emit radiation with a range of wavelength and energy (Fig. 6.3) encompassing the K- or L-shell binding energies of all but the

Box 6.2 X-ray diffraction

X-rays, being electromagnetic waves, are susceptible to **diffraction** (cf. Box 4.3). Having wavelengths similar to the inter-atomic distances in crystals, they interact with a crystal lattice in the same way that light interacts with a conventional diffraction grating (or with the tracks of a compact disc): they are dispersed in different directions according to wavelength.

There is, however, one key difference: whereas the conventional diffraction grating is a pattern repeated in *one* dimension, a crystal is a *three-dimensional* repeat structure. Notwithstanding this additional complexity, W.H. and W.L. Bragg showed that the diffraction of X-rays by crystals is described by a remarkably simple equation:

$$n\lambda = 2d \sin\theta \qquad [6.2.1]$$

where the symbols are as shown in Fig. 6.2.1. Incoming rays are scattered by atoms in successive planes of atoms. Scattered rays for which the Bragg equation is *satisfied* will experience *constructive* interference, and will form a diffracted beam at an angle θ equal to the angle of incidence of the incoming rays. In these circumstances, the crystal planes appear to act as a selective mirror, and diffracted beams are commonly referred to as 'reflections'.

Rays that do not satisfy the Bragg equation undergo destructive interference and do not form refections.

Uses

Equation 6.2.1 has two applications in Earth science:

(1) *In X-ray crystallography* If the X-ray beam is **monochromatic** and the wavelength λ is known, the beam will undergo diffraction by the crystal (or cluster of crystals) to form a diffraction pattern of sharp reflections. Measurements of the various θ angles can be used to determine the inter-planar spacings d characteristic of the diffracting crystal. X-ray diffraction provides a powerful technique for mineral identification.

(2) *In X-ray spectrometry* If a large single crystal is available, cut parallel to a particular set of planes whose spacing d is known, the crystal can be used to separate various wavelength components of a polychromatic X-ray beam, whose wavelengths (λ) can be calculated from measured θ values. This is the basis of the 'wavelength-dispersive' X-ray spectrometer.

RG

Figure 6.2.1 Diffraction of X-rays by a crystal lattice.

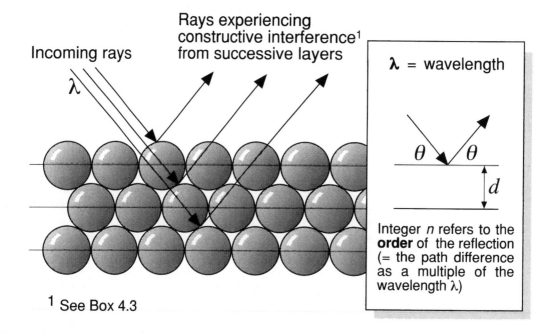

Rays experiencing constructive interference[1] from successive layers

Incoming rays

λ

λ = wavelength

θ θ

d

Integer n refers to the **order** of the reflection (= the path difference as a multiple of the wavelength λ)

[1] See Box 4.3

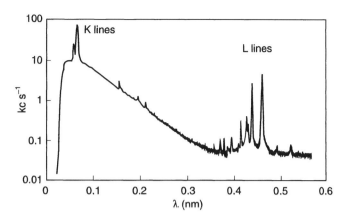

Figure 6.3   Part of the X-ray spectrum from a rhodium–anode X-ray tube showing the continuous and characteristic components. Discrete peaks (lines) with wavelengths characteristic of Rh are seen superimposed
on a broad continuum. The characteristic peaks fall into two groups (K and L) produced by electrons
jumping into the K and L electron shells respectively. Tiny Cu, Fe, Mn and Cr peaks, with wavelengths between 0.15 and 0.22 nm, are generated from metal components within the spectrometer.
         The lower-wavelength limit of the continuous radiation corresponds to the voltage (50 kV) applied
to the X-ray tube. A break in the continuous spectrum at 0.38 nm is due to a change of analysing crystal
(from LiF200 to Ge111) at this point. The spectrum was generated by scattering primary tube radiation off a boric acid sample in a Philips PW1480 XRF spectrometer.

lightest atoms. An electron vacancy is rapidly filled by an electron transferring from an outer
shell. The energy released, equal to the difference in binding energy of electrons in the two
levels, is emitted as a secondary, or fluorescent, X-ray photon with a discrete energy and
wavelength, characteristic of the individual element. This photon will then either escape
from the atom to form part of the characteristic spectrum, or it may displace an outer electron and be absorbed. Absorption of photons within the atom (known as the Auger effect)
is most pronounced for smaller atoms, and severely limits the yield of secondary X-rays
from the lighter elements.

By convention, lines in the X-ray spectrum are named after the sites of the respective
electron vacancies. Thus, electrons jumping from outer shells into the K and L shells give
rise to K and L lines respectively. The two K-shell electrons have the same energy but this
is not true for the electrons in the other shells. The L shell can accommodate eight electrons
in three energy levels, the M shell 18 electrons in five energy levels, and the N shell 32 electrons in seven energy levels. A set of selection rules limits the number of possible electron
transitions and therefore the number of emission lines in the X-ray spectrum. Figure 6.4
shows the principal K and L lines, with their names and the electron transitions involved.
These are the lines of most interest in XRF analysis. The conventional naming of lines is
far from systematic with respect to electron transitions, but instead reflects the relative
intensity. Thus α lines are more intense than β lines which are more intense than γ lines,
and K lines are generally more intense than the corresponding L lines.

The choice of anode material in the X-ray tube is critical to the usefulness of XRF spectrometry as an analytical technique. The most intense continuous radiation is obtained
from anode materials with a high atomic number, and these are generally chosen in preference to lighter materials. However, the excitation of secondary radiation from a sample is
most efficient using primary (tube) radiation of only slightly shorter wavelength than that
of the secondary peak. There is, therefore, an advantage in using anodes of low atomic num-

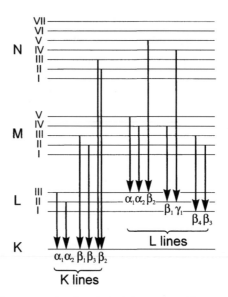

Figure 6.4   An electron energy diagram showing the principal electron transitions responsible for K and L lines in the X-ray spectrum.

ber for the analysis of light elements so that the characteristic K radiation from the anode can contribute to the excitation of secondary radiation from the sample. Several metallic elements (W, Mo, Rh, Au, Ag, Cr and Sc) have been used as anode materials for various XRF applications. Of these, W, Cr and Rh are most commonly used in geological applications. Some laboratories routinely use a W-anode X-ray tube for the analysis of heavy elements with short-wavelength peaks, and a Cr-anode tube for the analysis of lighter elements. However, many laboratories use a Rh-anode tube as a good compromise for all applications. The moderately high atomic number of Rh gives intense continuous radiation and the Rh K lines ($\lambda \approx 0.06$ nm) are well placed to excite K-line emission from the geologically useful elements Rb, Sr, Y, Zr, Nb and Mo ($\lambda \approx 0.07$–0.09 nm). The long wavelengths of the Rh L lines contribute to the excitation of lighter elements (from Cl downwards). Furthermore, the characteristic spectral lines of Rh do not overlap seriously with those of any of the elements routinely analysed in geochemical work.

The conversion of electrical energy into X-rays is a very inefficient process as only about 1% of the total applied power is released as X-rays. Most of the rest is released as heat which has to be dissipated by cooling the anode. This is usually achieved by welding the anode to a water-cooled copper block. Some of the electrons are scattered from the anode and absorbed by the tube window which can become very hot as a consequence. Beryllium is not a very good conductor of heat and thermal stresses can lead to fracture if the window is too thin. This constrains the window thickness to 150 μm or more and hence limits the output of long-wavelength X-rays, with consequent limitations on the application of XRF in the analysis of light elements. Fluorine is the lightest element routinely analysed in geological applications, although carbon is sometimes determined by XRF spectrometry in the steel industry, and the thickness of boron films on silicon semiconductor wafers is routinely measured by XRF spectrometry.

The accelerating voltage (kV) applied to the X-ray tube, and the tube current (mA), can be varied to optimise the X-ray ouput for individual applications. Short-wavelength spectral lines ($\lambda < 0.13$ nm; e.g. K lines of Ga to I, and L lines of Pb, Th and U) are excited by the short-wavelength (high-energy) parts of the primary radiation. Since the intensity of this part of the spectrum is controlled by the voltage applied to the X-ray tube, secondary emission can be optimised by operating the X-ray tube at maximum voltage (typically 80–100 kV). Secondary emission from light elements ($\lambda > 0.6$ nm; e.g. K lines of F to P) is governed by the intensity of long-wavelength tube radiation, which is controlled more by the flux of electrons through the X-ray tube than by their energy. This flux can be increased by increasing the small voltage applied to heat the tube filament, with a consequent increase in the current flowing through the tube. Since the total power that an X-ray tube can consume is limited to about 3 kW, an increase in tube current must be accompanied by a reduction in tube voltage. Thus, for analysis of light elements it is usual to operate the X-ray tube at maximum tube current (typically 60–75 mA) and lower applied voltage (40 kV). The high voltages required for the excitation of short-wavelength secondary radiation (80–100 kV) must be accompanied by a lower tube current (30 mA). Elements with spectral lines between these two wavelength extremes (e.g. K lines of S to Zn, L lines of Ba and the rare earth elements) are generally analysed with intermediate tube voltage and current (typically 50 kV, 50 mA). Tube voltage and tube current relate to the flow of electrons through the vacuum of the X-ray tube and should not be confused with the current flowing through the filament.

In an XRF spectrometer the sample is placed close to the tube window in order to irradiate it with as intense a beam of X-rays as possible. The flat surface of the sample is irradiated over an elliptical area roughly 30 mm in maximum dimension and the sample is usually spun slowly about a vertical axis in order that its surface should be irradiated uniformly. Most modern spectrometers take disc-shaped samples of 40 mm diameter, although smaller samples can be analysed by inserting a mask in the X-ray path between the sample and the primary collimator. Samples can be formed by pressing powder into a die under pressure or by casting a molten mixture of sample and lithium borate flux into a suitable mould. Metal samples can be machined into a disc. Particulate matter filtered from air or water may be analysed directly by irradiating the filter.

Dispersion of the X-ray spectrum

The radiation entering the spectrometer consists of characteristic fluorescent radiation from the sample, superimposed on primary tube radiation scattered by the sample. Figure 6.5 shows part of the X-ray spectrum produced by irradiating a powdered basalt sample with X-rays from a Rh-anode tube. The continuous part of the spectrum is similar in shape to that of the primary X-rays (Fig. 6.3) except that it has been truncated through absorption by the sample. Primary Rh K and L lines are still identifiable but have been joined by a large number of secondary lines from the sample. The intensity of these secondary lines can be measured and used to provide quantitative information on the composition of the sample.

In order for radiation from the sample to be dispersed into a spectrum it must be formed into a parallel beam. A collimator consisting of a stack of thin, regularly spaced, parallel metal plates is used for this purpose because X-rays cannot be focused using optical lenses.

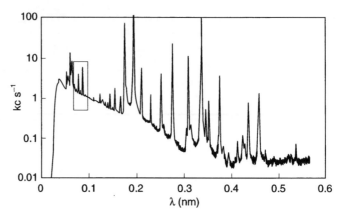

Figure 6.5   Part of the X-ray spectrum from a sample of basalt (USGS reference standard BHVO-1) irradiated with primary radiation from a Rh-anode X-ray tube (conditions as in Fig. 6.3). The Rh K and L lines can still be seen but are accompanied by a large number of lines from elements in the sample. The box encloses the segment of the X-ray spectrum illustrated (for a different sample) in Fig. 6.6(b).

The collimator allows only those rays emitted in a direction parallel to the plates to be transmitted, all other rays being absorbed. The X-ray beam is then scattered from an analysing crystal and a secondary collimator selects a parallel beam of scattered X-rays to pass to the detectors. The purpose of the secondary collimator is to ensure that only those rays scattered from the crystal at an angle (θ) equal to the angle of incidence are allowed to reach the detectors. These geometrical constraints are required in order that the secondary X-ray beam can be diffracted (see Box 6.2). Only rays with a wavelength (λ) satisfying the Bragg equation ($n\lambda = 2d \sin \theta$) are strongly reflected from the crystal into the detector. Varying the angle θ while maintaining this geometry allows the spectrometer to record the intensity of radiation as a function of wavelength, and thus produce spectra like those in Figs 6.3 and 6.5. The angular (and therefore wavelength) resolution of the spectrometer is partly determined by the spacing of plates in the primary collimator. Widely spaced plates allow a high-intensity beam of X-rays to reach the crystal, but with low angular resolution in the spectrum. More closely spaced plates improve angular resolution but at the cost of intensity. High angular resolution is necessary when X-ray spectral lines of different elements are close together, as is usually the case in the short-wavelength part of the spectrum. XRF spectrometers are usually equipped with both 'coarse' and 'fine' primary collimators, mounted side by side on a rotating turret so either can be selected during operation.

The construction of XRF spectrometers (Fig. 6.2) restricts the effective range of 2θ to a maximum of about $145°$ of arc. No single analysing crystal can cover the whole X-ray wavelength range with this range in 2θ so several different crystals must be used. A large number of crystals are commercially available for a wide range of applications and, in recent years, several synthetic multi-layer pseudocrystals with large inter-layer spacings (d) have become available for the dispersion of long-wavelength radiation. Pseudocrystals are produced by depositing alternate layers of heavy (e.g. W) and light (e.g. Si) elements onto a silicon or glass substrate. The parameters of the crystals most commonly used in geological applications are given in Table 6.1. Crystals used in XRF spectrometers with the geometry shown in Fig. 6.2 are generally flat, unlike those used in the electron microprobe (Chapter 14) which are curved. Curved crystals are of limited use in XRF spectrometry because the secondary X-rays originate on a surface rather than from a point source. However, transversely

Table 6.1   Parameters of analysing crystals and pseudocrystals used in geological applications

Crystal	Reflection plane	$2d$ spacing (nm)	Routine uses
Lithium fluoride (LiF)	220	0.2848	V to I (K); Ce to U (L)*
Lithium fluoride (LiF)	200	0.4028	K to I (K); Sn to U (L)
Germanium (Ge)	111	0.6532	P to Cl (K)
Pentaerythritol (PE)	002	0.8742	Al and Si (K)
Thallium acid phthalate (TlAP)	001	2.575	F to Mg (K)
Pseudocrystals[†]	–	5 to 20	B to Mg (K)

*Gives better wavelength resolution than LiF200 but with poorer reflectivity.
†Also known as MLDEs (Multi-Layer Dispersing Elements).

curved crystals (curved along an axis normal to the crystal rotation axis) can improve count rates at long wavelengths by focusing the diffracted X-rays onto the secondary collimator.

Not all XRF spectrometers use analysing crystals to disperse the secondary X-ray beam into a spectrum. Energy-dispersive (ED) X-ray detectors (Box 7.1) are suitable for applications where low detection limits and good energy resolution are not essential. The principal advantages of energy-dispersive spectrometry are that the entire X-ray spectrum is measured simultaneously, the sample is not constrained to be a flat disc, and it is not necessary to use collimated X-ray beams. This is useful in the analysis of valuable artefacts, and ED XRF spectrometry is widely used by museums and in archaeological research.

X-ray detection

The diffracted X-ray beam in an XRF spectrometer is directed, via the secondary collimator (Fig. 6.2), into one or more X-ray detectors. Most spectrometers employ two detectors: a gas flow proportional detector and a scintillation detector (Box 6.3). Gas flow proportional detectors (usually referred to as 'flow counters') are used for measuring intensities of X-rays with wavelengths longer than about 0.15 nm (K spectra of elements lighter than Zn). Because of their sensitivity to long wavelengths, flow counters are mounted within the spectrometer vacuum chamber, and X-rays pass into the detector gas (Ar–CH_4 mixture) through a thin (typically 1 or 2 µm) **Mylar**® window. The window is coated with aluminium to make it electrically conducting.

Short-wavelength X-rays are measured using a scintillation detector which is usually mounted outside the spectrometer vacuum chamber. These X-rays pass through a Mylar exit window on the back of the flow counter and out of the vacuum chamber via another Mylar window. The scintillation detector is normally equipped with its own auxiliary collimator and this, together with its greater distance from the analysing crystal, gives the detector a better angular resolution than the flow counter. Scintillation detectors are used alone for measuring radiation at the short-wavelength end of the X-ray spectrum, especially where high angular resolution is required (e.g. K spectra of elements from Nb to I, L spectra of Th and U). Radiation with intermediate wavelengths (e.g. the K radiation of elements

Box 6.3 X-ray detectors

X-ray detectors are devices that convert the quantum energy of an X-ray photon (or other ionising radiation) into an electrical pulse amenable to digital processing. Three kinds are in common use:

(1) gas-filled detector;
(2) scintillation detector;
(3) semiconductor detector.

The first two, widely used in X-ray fluorescence analysis, are described in this box and illustrated in Figs 6.3.1 and 6.3.2. The semiconductor detector is dealt with in Box 7.1.

Gas proportional detector

Figure 6.3.1 shows the construction of a typical gas-filled detector. An X-ray photon passing through the thin 'window' in the wall of the detector collides with, and ionises, numerous gas molecules. Electrons liberated by ionisation are strongly accelerated towards the central wire, which is held at a high positive potential (typically 1.0–1.5 kV) relative to the earthed casing; they in turn collide with and ionise many more gas molecules (a process called *gas amplification*), causing an *avalanche* of electrons to converge on the anode wire. The charge col-

lected on the anode in the short interval of time following the initial ionisation is transformed into a voltage pulse that can be amplified and registered in a computer.

The greater the quantum energy of the incoming photon, the more primary ionisations it causes in the gas and the more free electrons are thereby produced. With a stable operating voltage, the detector therefore generates pulses whose amplitudes are related to the quantum energies of the X-ray photons producing them; hence the widely used term *proportional counter*. This proportionality, though crude in comparison with semiconductor detectors (Box 7.1), is analytically useful as it allows certain categories of interfering radiation to be filtered out by pulse-height discrimination (see text). When higher voltages are applied ($\sim$2.0 kV) the gas amplification is sufficiently great (10^8–$10^9\times$) to saturate the detector, producing large pulses with amplitudes independent of quantum energy. This *Geiger regime* finds little application in modern X-ray analysis.

Figure 6.3.1 shows a *sealed proportional detector* used for the relatively penetrating, short-wavelength X-rays from heavier elements such as Fe, Zn and Sr. The window is made of 100 μm thick beryllium foil; the inert gas is typically xenon, whose high atomic number ensures efficient absorption of X-ray energy. The

Figure 6.3.1 Gas proportional X-ray detector.

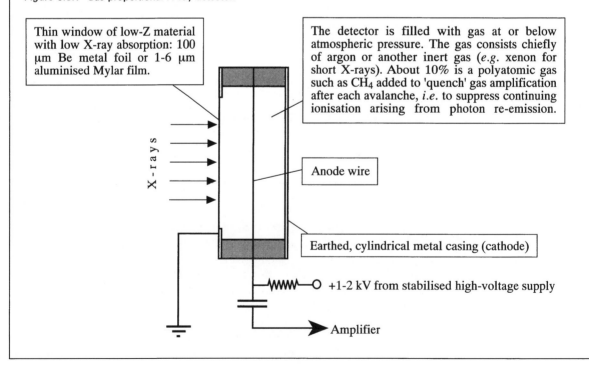

Thin window of low-Z material with low X-ray absorption: 100 μm Be metal foil or 1-6 μm aluminised Mylar film.

The detector is filled with gas at or below atmospheric pressure. The gas consists chiefly of argon or another inert gas (*e.g.* xenon for short X-rays). About 10% is a polyatomic gas such as CH_4 added to 'quench' gas amplification after each avalanche, *i.e.* to suppress continuing ionisation arising from photon re-emission.

X - r a y s

Anode wire

Earthed, cylindrical metal casing (cathode)

+1-2 kV from stabilised high-voltage supply

Amplifier

longer-wavelength spectra of lighter elements such as Nd and Si, however, are too attenuated by the thick window of a sealed detector. Thinner windows made from 6 μm or 1 μm thick sheets of the polyethylene derivative **Mylar** are used for such elements. Because these thinner windows are gas permeable, however, the detector design is modified to feed a slow stream of gas – usually an Ar–CH_4 mixture – through the detector during use. Such *flow counters* are mounted within the spectrometer vacuum chamber to minimise X-ray attenuation. The flow counter can be used to detect X-ray wavelengths longer than about 0.15 nm (*k* spectra of elements lighter than Zn).

Scintillation detector

Crystals such as sodium iodide, when doped with a suitable impurity (e.g. thallium), respond to ionising radiation by emitting tiny flashes of visible light (scintillations). A scintillation detector consists of three elements (Fig. 6.3.2):

(1) A *phosphor* (scintillator crystal) that converts X-ray photons into light pulses.

(2) A *photocathode* of antimony–caesium alloy that reacts to the light pulses by emitting **photoelectrons**.

(3) An *electron multiplier* that amplifies each photoelectron to form a measurable burst of electrical charge

(typically 10^5–10^6 electrons) collected on an anode. This is accomplished by secondary electron emission at a series of 'dynodes' held at progressively higher electric potential V (Fig. 6.3.2).

The photocathode and electron multiplier are combined in a vacuum tube called a *photomultiplier*.

Because the output signal from each of these stages is proportional in magnitude (light intensity, number of photoelectrons and charge collected respectively) to the corresponding input signal, a beam of X-ray photons will cause a scintillation detector, like a gas proportional detector, to generate a series of electrical pulses whose individual amplitudes are proportional to the quantum energies of the photons that generated them. The energy resolution is poorer than for gas-filled detectors, but the scintillation detector can handle higher count rates (X-ray intensities). It can be used for radiation in the wavelength range 0.02 (W Kα) to 0.15 nm (Zn Kα).

Certain organic liquids also scintillate in response to ionising radiation. ^{14}C may, for example, be analysed by counting its radioactivity in a very sensitive scintillation counter in which the sample is actually immersed in the scintillating liquid.

X-rays can also be detected by photographic film, but this method is restricted today to X-ray diffraction studies of crystal structure and is not relevant to chemical analysis.

RG

Figure 6.3.2 Construction of a scintillation detector.

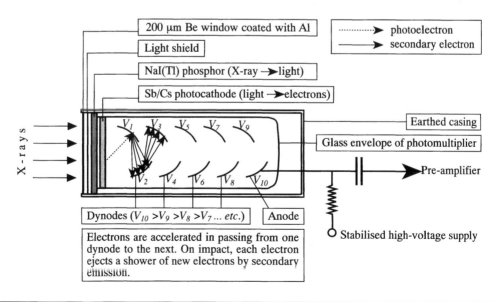

from Zn to Zr; L radiation of Ba and the rare earth elements) is usually measured using the two detectors in tandem.

Flow and scintillation detectors convert X-ray photons into electrical pulses. Both are *proportional counters* in the sense that the electrical pulses have amplitudes proportional to the energies of the X-ray photons. This allows the detector signals to be filtered electronically through a process known as *pulse-height selection*, the principal use of which is to filter out high-order reflections ($n > 1$ in the Bragg equation, Box 6.2) from the analysing crystal. Radiation entering the detectors at an angle 2θ to the undiffracted secondary beam will comprise all wavelengths satisfying the relationship $n\lambda = 2d \sin \theta$. Thus the radiation will have discrete wavelengths of λ ($n = 1$), $\lambda/2$ ($n = 2$), $\lambda/3$ ($n = 3$), etc., and the corresponding X-ray photons will have energies of E_q, $2E_q$, $3E_q$, etc. Filtering the detector pulses through a narrow energy window allows the selection of only pulses corresponding to photons of a particular energy and hence only one value of n in the Bragg equation (usually $n = 1$, but occasionally $n = 2$). Second-order ($n = 2$) reflections will give better spectral resolution than first-order reflections of the same emission line, as can be seen by differentiating the Bragg equation:

$$\frac{d\theta}{d\lambda} = \frac{n}{2d} \frac{1}{\cos \theta} \qquad\qquad [6.2].$$

High wavelength-resolution is favoured by high values of n and low values of d. However, since the increase in resolution given by second-order reflections is obtained at the cost of intensity, these are rarely used in routine XRF analysis. Pulse-height selection also reduces spectral background intensities by filtering out electronic 'noise' from the X-ray detectors.

Flow counters filled with argon produce electrical pulses with two discrete amplitudes when detecting X-ray photons with energies higher than the K absorption edge (Box 6.5) of argon. One set of pulses represents the energy of the photons being detected while the other is produced by secondary fluorescence of the argon. The latter is called an **escape peak**. Since both sets of pulses usually relate to the same incident photons, the energy window used in pulse-height selection is generally set wide enough to allow both sets of pulses to be counted. Occasionally, the escape peak is produced partly or wholly by an overlapping high-order reflection from an X-ray emission line of a different element. In such instances it is necessary to set the energy window to exclude the escape peak. A good example of this effect is the overlap of second-order Ca Kβ on P Kα.

The filtered output from the two detectors is amplified and individual pulses are counted electronically and converted into a count rate, usually expressed as thousands of counts per second (kc s^{-1}). In modern XRF spectrometers, this information is stored digitally and then passed to an on-line computer for further processing. If the spectrometer has been scanned to measure count rates as a continuous function of the 2θ angle, the resulting information can be displayed as a spectrum as in Fig. 6.5, for example. More usually the spectrometer will be driven to predetermined 2θ angles to measure count rates from individual emission lines and adjacent background positions. The resulting information will then be used to determine the concentrations of the respective elements.

Automation

Most of the recent technological advances in XRF spectrometry have been connected with the automation of the spectrometer and the computer processing of intensity data. Analytical programs specifying instrument conditions for each element to be determined are written by the operator and stored in an on-line computer. Samples are loaded automatically into a sample chamber through an airlock to avoid having to evacuate the spectrometer after each sample change. The applied tube voltage and current, the primary collimator, analysing crystal, 2θ angle, background positions, and detectors can all be specified and changed automatically during operation. Small electric motors or pneumatic systems are used to operate airlocks and change crystals and collimators. The goniometer angle 2θ is usually varied by an optically encoded stepping motor which, in the most modern instruments, controls its position to an accuracy of $\pm 0.0025°$. Other instrumental parameters, such as X-ray tube voltage and current, pulse-height window, detector selection, and counting times, are controlled electronically.

Quantitative analysis

The output from the spectrometer consists of peak and background count rates for an X-ray emission line of each of the elements whose concentration is to be determined. The choice of suitable emission lines is governed by several factors, the most important of which are relative intensity and freedom from interference by lines of other elements. Where possible, K lines are chosen in preference to L lines, and α lines in preference to β lines. The ability of the primary radiation from the X-ray tube to eject electrons from the K shells of atoms limits the use of K lines in XRF spectrometry. The intensity of available primary radiation falls off sharply at short wavelengths (Fig. 6.2) with a consequent reduction in the efficiency of secondary K-line excitation for heavy elements. For most applications, iodine is the heaviest element for which the $K\alpha$ line can be used in XRF spectrometry. L lines must be used for heavier elements.

The net (peak minus background) intensity of the chosen emission line governs the precision of the analysis (Box 6.4). The net intensity of a characteristic emission line is governed by the number of electron jumps per second which can be induced by the primary X-ray tube radiation. This in turn relates to the number of atoms of the particular element present in the sample and, for samples of equal mass, to the concentration of that element. It follows, therefore, that the net intensity I of an emission line should be proportional to the concentration C of the element producing it, such that

$$I = kC \qquad\qquad [6.3]$$

where k is a proportionality constant determined by relating intensity to concentration for one or more reference standards. XRF spectrometry, like most instrumental analytical techniques, is not an absolute method but relies on the availability of materials of known composition. Such materials may be produced synthetically or analysed by other methods. Reference standards covering most of the compositional range of geological materials and

---

Box 6.4 The normal distribution III: random error and detection limit in radiation measurements

A measurement of X-ray or γ-ray intensity is made by counting the arrivals of individual photons, as registered by the detector, during a predetermined interval of time[1]. Because photon arrivals are randomly distributed in time (recall the irregular series of 'clicks' emitted by a Geiger counter), repeated measurements of any constant γ-ray or X-ray intensity are subject to random error. Accurate determination of these random errors is important for two reasons:

(1) the random error associated with the measurement of a peak intensity contributes to the *precision* of the analysis;

(2) the random error associated with background measurement determines the size of the smallest peak that can be distinguished from the background, and therefore determines the *detection limit*. (cf. Box 4.4) for the analysis.

Repeated counts of a constant radiation intensity using the same counting interval conform to the **Poisson distribution**, a statistical distribution describing the probability of events that occur randomly in time. For count rates sufficiently large to be acceptable for analysis, however, the Poisson distribution approximates to a normal distribution (Box 1.1). Both distributions share a convenient property: if the mean of several replicate intensity measurements is n counts, the standard deviation among the individual readings will be found to equal $\sqrt{n}$. An approximate value of the standard deviation can therefore be estimated from an individual reading n_i as $\sqrt{n_i}$. For example, if an X-ray or γ-ray signal is counted for 10 seconds and the count obtained is 3129 counts, the standard deviation associated with that count will be approximately $\sqrt{3129} = 56$ counts $= 1.8\%$ relative. The predictability of the random error associated with radiation measurements has led to XRF and INAA being widely used for the certification of reference materials.

This property of radiation measurements also simplifies the estimation of detection limits for XRF and neutron activation analysis. From a measured background count of n_B the standard deviation on the background can be estimated as $\sqrt{n_B}$, and the $3s$ detection limit (Box 4.5) will be equal to $3\sqrt{n_B}$ divided by the sensitivity. For example, if the background count were found to be 5821, $3s_{background}$ would be $3 \times \sqrt{5821} = 228$ counts. If the sensitivity for the element in question were 175 counts per ppm, the detection limit would be $228/175$ ppm $= 1.3$ ppm. Note that it is the *total background count* that must be used in this calculation, not the count *rate*.

If the first experiment were repeated with a counting time of 100 seconds, yielding a count of say 30,889, the standard deviation would be 176 counts $= 0.6\%$ relative; increasing the total count has *increased* the standard deviation in counts, but has *reduced* the relative error in percentage terms. It follows that both precision and detection limit are improved if the total count recorded is increased (e.g. by extending counting time).

RG

---

[1]   With or without extension to allow for **dead time** – see Chapter 14.

---

most of the chemical elements are now readily available (see e.g. Govindaraju, 1994) and are widely used in XRF analysis.

In practice, the simple relationship expressed in Equation 6.3 rarely applies. Several sources of systematic error inherent to XRF analysis conspire to disturb this relationship, and these must be overcome in order to achieve fully quantitative analysis. The most important of these are detector dead time, line overlap, X-ray absorption and enhancement effects, and particle size effects.

Dead time

After an X-ray detector has responded to a photon, a finite time (the **dead time**) must elapse before the detector can respond to another photon. In the case of a flow detector, the dead time is the time required for positive Ar ions to decay and is in the order of 200 nanoseconds. In a scintillation detector, the time required for an electron in an excited state to return to the ground state is around 100 nanoseconds. The effect of dead time, though negligible at low count rates, becomes significant at high count rates and results in measured rates significantly lower than the rate at which photons enter the detectors. Fortunately the effect is predictable if the detector dead time t_D is known, and is therefore correctable. The measured intensity I relates to the true intensity I^* through

$$I^* = \frac{I}{1 - It_D} \qquad [6.4]$$

Modern XRF spectrometers contain electronic circuitry designed to overcome this problem by firstly imposing an artificial dead time of around 300 nanoseconds, longer than the dead time of either detector, and then generating extra pulses to replace those lost. The result is that X-ray detectors can now produce a linear response at count rates of up to 2000 kc s^{-1}.

Line overlap

An XRF spectrometer resolves X-ray emission lines as intensity peaks with an angular width of, at best, about 0.4° 2θ. At a 2θ angle of 20° and with a LiF200 analysing crystal, this is equivalent to a wavelength resolution of 0.0014 nm (Equation 6.2). Spectral lines with wavelength differences of less than this will not, therefore, be completely resolved. Resolution is limited by the geometry of the XRF spectrometer, with secondary radiation emitted from a large sample area and dispersed by a flat analysing crystal. The electron microprobe (Chapter 14), with a point source of radiation and curved crystals, can achieve much better wavelength resolution. Overlap of lines from different elements is a problem frequently encountered in XRF spectrometry.

The problem of line overlap is illustrated in Fig. 6.6 which shows the part of the X-ray spectrum that includes the K lines of Nb to Rb. Figure 6.6(a) consists of superimposed spectra from two synthetic standards of granitic composition. One of these has been spiked with a few hundred parts per million (ppm) of Rb, Sr and Nb, and the other with similar concentrations of Y and Zr. It is clear from these spectra that the Kβ lines of Rb, Sr and Y overlap with, respectively, the Kα lines of Y, Zr and Nb to a greater or lesser degree. The overlap of Rb Kβ on Y Kα is almost complete, while that of Y Kβ on Nb Kα is only very slight. Figure 6.6 (b) shows the same part of the X-ray spectrum from a sample of granite (USGS reference standard G2). The recorded intensities of the Y, Zr and Nb Kα lines clearly include a component from the interfering elements, and this effect must be removed before these intensities can be converted into concentrations.

The simplest way to remove the effects of line overlap is to calculate overlap factors by analysing synthetic standards containing only the interfering element. These factors are expressed in apparent concentration per unit concentration of the interfering element. The overlap factor for Sr on Zr, for example, is approximately 1 ppm Zr for every 10 ppm of Sr. It is better to express factors in concentration than intensity units because an absorption edge between the two wavelengths would cause an overlap factor expressed in intensity units to vary with the composition of the material being analysed. It is also possible to calculate an overlap factor by minimising residuals during regression of intensity against concentration for reference standards.

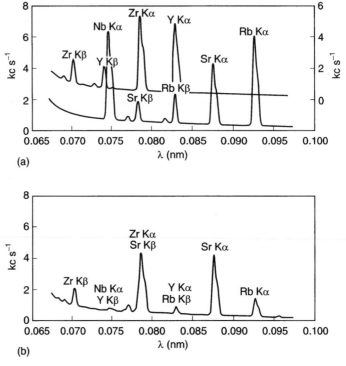

Figure 6.6 (a) Superimposed X-ray emission spectra from two synthetic standards of granitic composition showing overlap of Rb, Sr and Y Kβ lines on, respectively, Y, Zr and Nb Kα lines. The small unlabelled peaks represent Kβ₂ lines. The baseline of the upper trace has been deliberately elevated for clarity (see right-hand scale). (b) An X-ray emission spectrum, over the same wavelength range, from a sample of granite (USGS reference standard G2).

Absorption and enhancement effects

These effects account for the most serious problems encountered in XRF spectrometry and their removal is an essential prerequisite for successful XRF analysis. Secondary X-ray photons emitted by a sample must be able to escape from the sample before they can be dispersed and counted. Some photons, however, will be absorbed by atoms in the sample itself and thereby stimulate the emission of tertiary fluorescent X-rays. The proportion of X-rays absorbed will depend on their wavelength and the composition of the sample (Box 6.5). Such effects will lead to systematic errors in analysis when the samples and calibration standards differ significantly in composition. Enhancement effects due to the stimulation of tertiary X-ray emission are generally a less serious problem, but can become important during the determination of the concentration of an element in the presence of large and variable concentrations of a slightly heavier element. A full theoretical treatment of absorption and enhancement is beyond the scope of this book and the interested reader is referred instead to a more specialised textbook (e.g. Tertian and Claisse, 1994). Some explanation of absorption effects, however, is essential to an understanding of XRF spectrometry.

The intensity of secondary radiation emitted by a sample is influenced by the degree to which both primary and secondary radiation is absorbed. Because the primary radiation will necessarily have a shorter effective wavelength than the secondary radiation, it will be less strongly absorbed and it is convenient to consider only the absorption of secondary X-rays. Consider, therefore, a sample which is uniformly irradiated with primary X-rays and emitting characteristic radiation of net intensity I_0 per unit thickness from an element with a concentration C. Since the mass per unit thickness is proportional to the density ρ, Equation 6.3 gives

$$I_0 = k'\rho C \tag{6.5}$$

where k' is a proportionality constant. If the sample has a thickness x, the total radiation emitted, before absorption, will be equal to $I_0 x$, and a thin layer of material of thickness $\mathrm{d}x$, located at the back of the sample, will therefore emit radiation with intensity $I_0\,\mathrm{d}x$. This radiation will reach the surface of the sample with intensity $\mathrm{d}I$. Substituting $\mathrm{d}I$ and $I_0\,\mathrm{d}x$ for I and I_0, respectively, in Equation 6.5.1 (Box 6.5) gives

$$\frac{\mathrm{d}I}{I_0\,\mathrm{d}x} = e^{-\mu\rho x} \tag{6.6}$$

The *total* intensity I of secondary radiation reaching the surface of the sample can be found by integrating Equation 6.6. Thus

$$\frac{I}{I_0} = \frac{1 - e^{-\mu\rho x}}{\mu\rho} \tag{6.7}$$

Combining Equations 6.5 and 6.7 gives

$$I = \frac{k'C(1 - e^{-\mu\rho x})}{\mu} \tag{6.8}$$

where I is the intensity of an emission line from an element present in concentration C, emitted by a sample of thickness x, μ is the total mass absorption coefficient (Equation 6.5.2) of the sample at the wavelength of the emission line, and k' is a proportionality constant. Figure 6.7 shows how I/I_0 (Equation 6.7) varies with sample thickness for long-, intermediate- and short-wavelength radiation (Na, Ti and Nb $K\alpha$ respectively) emitted

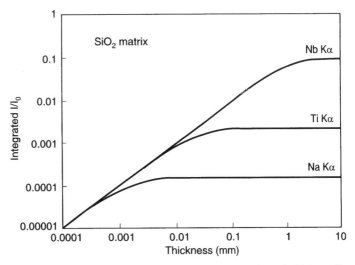

Figure 6.7   Intensity I of emitted Na, Ti and Nb $K\alpha$ radiation as a function of sample thickness (from Equation 6.7). I_0 is the intensity emitted, before absorption, by a sample of unit thickness (1 cm in this case).

Box 6.5 Absorption of X-rays

X-rays, like all electromagnetic radiation, are attenuated by matter through which they pass. The intensity I of a beam of X-rays is reduced by an amount dI proportional to the thickness of the material dx, so that

$$dI/I = -\mu_{lin}\, dx$$

The proportionality constant μ_{lin} in this equation is called the **linear absorption coefficient**. Expressed in mass units, this equation becomes

$$dI/I = -\mu\rho\, dx$$

where μ is the **mass absorption coefficient** (usually expressed in units of $cm^2\ g^{-1}$) and ρ is the density of the material (in $g\ cm^{-3}$). Integrating this equation gives

$$I/I_0 = e^{-\mu\rho x} \qquad [6.5.1]$$

where I_0 is the initial intensity of the X-ray beam and I the intensity after passing through material of thickness x. Equation 6.5.1 is an expression of the well-known **Beer–Lambert law**. The fraction I_0-I of the X-ray intensity not transmitted in the direction of the incident beam is lost mostly in two ways:

(1) Absorption of photons in expelling electrons to generate secondary X-rays (Box 6.1).
(2) Random scattering of incident photons through collisions with atoms (Box 6.6).

The mass absorption coefficient μ increases with wavelength λ and atomic number Z such that μ is approximately proportional to the cube of λ and the fourth power of Z. This continuous variation is interrupted, however, at several discontinuities. Figure 6.5.1 illustrates this using the molybdenum absorption spectrum. On this dia-

gram, μ is seen to increase with λ up to 0.062 nm when it falls abruptly. It then rises again until, at 0.43 nm, it falls again, this time in three steps (at 0.43, 0.47 and 0.49 nm). These discontinuities, referred to as **absorption edges**, correspond with the **binding energies** of electrons in the K and L shells respectively. K-shell electrons have one energy level and L-shell electrons have three (Box 6.1; Fig. 6.4). The K and L X-ray emission lines have wavelengths slightly longer than their respective absorption edges (Fig. 6.5.1) since the absorption edges represent ejection of electrons from K and L shells right out of the atom, whereas the emission lines corre-

Figure 6.5.1 Part of the X-ray absorption spectrum of molybdenum showing the absorption jumps due to the binding energies of K- and L-shell electrons (K and L absorption edges). The vertical bars represent the Mo K and L emission lines.

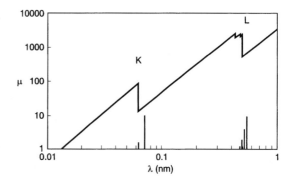

from a matrix composed essentially of silica (chosen to represent the lightest common material of geological interest). The intensity at each wavelength initially increases with thickness but then levels off as radiation from deeper levels within the sample can no longer reach the surface. The point at which the intensity levels off varies with wavelength from about 7 μm for Na $K\alpha$ to about 4 mm for Nb $K\alpha$ and the sample is then effectively infinitely thick with respect to radiation of the respective wavelength. An infinitely thick sample can conveniently be defined as one thick enough to absorb >99% of radiation passing through it. This condition is satisfied in samples with $\mu\rho x > 4.6$ from Equation 6.5.1; $-\ln 0.01 = 4.6$) and leads to a useful simplification of Equation 6.8. For infinitely thick samples, $\exp(-\mu\rho x)$ becomes very small and Equation 6.8 becomes

$$I = k'C/\mu \qquad [6.9]$$

To eliminate absorption effects, therefore, it is only necessary to multiply the measured intensity by the appropriate mass absorption coefficient. A more rigorous treatment of absorption would allow for the fact that the primary tube radiation is also absorbed by the sample which is not, therefore, irradiated uniformly. However, secondary X-rays are produced by the absorption of primary radiation of slightly shorter wavelength than the result-

spond to transitions to those shells from higher energy levels *within the atom* (therefore involving smaller energy differences). Absorption edges exist because a photon with a slightly higher energy than the binding energy of an electron can eject that electron and be absorbed, whereas a photon with a slightly lower energy cannot.

Mass absorption coefficients (μ) have been determined experimentally over a wide range of Z and λ, and their values can be estimated by interpolation in cases where direct measurement is not possible. Tables are available giving values of μ for each element (as absorber) over the whole range of atomic emission wavelengths used in XRF analysis. Composite values of μ for compounds or mixtures at any wavelength can be obtained by calculating averages of individual values of μ weighted to the mass fraction of each element present. Thus the composite value of μ at some specified wavelength is given by

$$\mu = \sum_{i=1}^{n} C_i \mu_i \qquad [6.5.2]$$

where C_i are the weight fractions of elements i, and μ_i their corresponding μ values at the wavelength in question. For example, the μ value appropriate for the absorption of Ti Kα radiation by SiO_2 is given by

μ = mass fraction of Si $\times$ μ(Ti Kα by Si) +

mass fraction of O $\times$ μ(Ti Kα by O)

= $0.467 \times 328 + 0.533 \times 63 = 186.8$ cm^2 g^{-1}.

Values of μ for more complex materials may, of course, be calculated by adding more terms to the equation.

The absorption of X-rays of various wavelengths by SiO_2 is illustrated in Fig. 6.5.2. The curves, covering most of the range of wavelengths used in geological applications of XRF spectrometry, were derived from Equation 6.5.1, using appropriate values of μ. A 0.01mm layer of SiO_2 will completely absorb Na Kα but will hardly affect Nb Kα at all. The same layer will attenuate Ti Kα by about 40%. Variations in μ between materials and at different wavelengths is of considerable importance in XRF spectrometry.

JGF

Figure 6.5.2 Fraction (I/I_0) of incident Na, Ti and Nb Kα radiation transmitted by a layer of SiO_2 of variable thickness. The curves are derived from Equation 6.5.1.

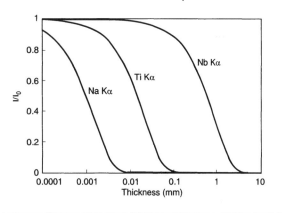

ing emission line, and therefore μ (primary) $<$ μ (secondary). In the absence of a major element absorption edge between the primary and secondary wavelengths, μ (primary) will be proportional to μ (secondary) and Equation 6.9 will still apply.

Infinitely thick samples are used wherever possible in XRF analysis because of the relative ease with which absorption effects can be reduced or eliminated in such samples. Since secondary X-rays leave the sample at an angle α to the surface of the sample (the take-off angle) in XRF spectrometers (Fig. 6.2), the critical thickness x_c for an infinitely thick sample will be given by

$$x_c = \frac{4.6 \sin \alpha}{\mu \rho} \qquad [6.10]$$

With $\alpha \cong 45°$ (typical of most spectrometers; Fig. 6.2) a 40 mm diameter pressed-powder disc with a mass of 8 g will be infinitely thick for K radiation of all elements up to Nb ($Z = 41$) in most geological materials. Thicker samples will be needed for analyses involving shorter wavelengths or unusually light materials.

Figure 6.7 shows that intensity varies linearly with sample thickness in thin samples. A useful definition of a thin sample in this context is one in which $\mu\rho x < 0.1$. For such samples, the term $1 - \exp(-\mu\rho x) \cong \mu\rho x$ and so Equation 6.8 simplifies to

$$I = k'C\rho x \qquad\qquad [6.11]$$

Intensity is simply proportional to the mass (ρx) of the sample and the concentration of the element being measured. Absorption effects are negligible for thin samples. This allows the analysis of thin films of particulate matter collected on filter paper, where measurements of the total mass of an element are required. Analysis of particulate matter by XRF spectrometry is applied extensively in environmental and oceanographic studies. XRF analysis of thin films is also widely used in the semiconductor industry for measuring the thickness of layers of material deposited on silicon wafers.

Most practical applications of XRF spectrometry assume either infinitely thin or infinitely thick samples, since both ideals allow absorption effects to be overcome. The correction of absorption effects in infinitely thick samples, though theoretically straightforward, still poses the problem of determining values for the bulk mass absorption coefficient at each analyte emission wavelength. This problem is most acute for major element analysis, particularly for complex, chemically variable matrices such as silicate rocks. As noted in Box 6.5, bulk values of μ can be calculated from the mass fractions of each element (in practice, each major element) present in the sample, yet these mass fractions are the very unknowns we are seeking to determine. **Iterative** calculations are therefore frequently used, in which an initial estimate of the composition, derived from uncorrected intensities, is used to calculate crude values of the bulk mass absorption coefficient μ and hence refine the estimate of composition. This leads to cyclic calculations in which successive estimates converge on a corrected analysis that is taken to be the best approximation to the true composition. Iterative absorption corrections generally use composite 'influence' or 'alpha' factors which take account of enhancement (treated mathematically as negative absorption) and primary absorption, in addition to secondary absorption effects. Alpha factors vary with the type of material being analysed and are calculated for each application. Their use requires the provision of a dedicated computer as part of the analytical system, and is now standard practice in XRF analysis.

Trace element X-ray intensities are often corrected using mass absorption coefficients based on predetermined major element compositions. Another method widely used for trace element absorption corrections exploits the properties of primary X-rays scattered from the sample. When monochromatic X-rays, such as a characteristic X-ray tube line, are scattered, the scattered radiation contains two wavelength peaks. One of these is unmodified and due to coherent scattering, while the other has a longer wavelength due to incoherent or Compton scattering (Box 6.6). The intensity $I(Comp)$ of the Compton peak is closely related to the mass absorption coefficient μ_λ of the scattering medium at the wavelength λ of the monochromatic radiation such that

$$I(Comp) \propto 1/\mu_\lambda \qquad\qquad [6.12]$$

This relationship is illustrated in Fig. 6.6.2. Mass absorption coefficients at different wavelengths are proportional to each other between absorption edges (Fig. 6.5.1) so that

$$\mu \propto \mu_\lambda \qquad\qquad [6.13]$$

where μ is the mass absorption coefficient at the wavelength of an emission line of an element of interest. Combining Equations 6.9, 6.12 and 6.13 gives

$$\frac{I}{I(Comp)} = k''C \qquad\qquad [6.14]$$

Box 6.6 Scattering of X-rays

When an incident beam of monochromatic X-rays is scattered by, for example, a sample in an XRF spectrometer (Fig. 6.2), the scattered radiation is found to consist of two components:

(1) unmodified radiation with the same wavelength as the incident beam; and
(2) radiation with a longer wavelength than the incident beam.

The first type of scattered radiation is due to coherent, or **Rayleigh scattering** resulting from elastic collisions between X-ray photons and atomic electrons, in which the photon changes direction but does not lose energy. The second type was first studied by A.H. Compton and is referred to as incoherent or **Compton scattering**. Incoherent scattering is considered to be the result of the inelastic interaction between X-ray photons and loosely bound electrons. Electron recoil results in the scattered photon losing energy and therefore taking on a longer wavelength. The wavelength difference between incoherent and coherent scattered radiation $(\lambda' - \lambda)$ is independent of the nature of the scattering medium and depends only on the angle (ϕ) through which the radiation is scattered. This relationship is expressed in the equation, derived by Compton

$$\lambda' - \lambda = \frac{h}{m_0 c}(1 - \cos \phi) \qquad [6.6.1]$$

where h is Planck's constant, m_e is the mass of an electron at rest, and c is the velocity of light. The constant $h/m_e c$ has the dimension of length and a value of 0.002426 nm. Since most XRF spectrometers (Fig. 6.2) have a scattering angle (ϕ) close to 90° (cos ϕ = 0), the wavelength difference observed is about 0.0024 nm. The effects of coherent and incoherent scattering of Rh Kα and Rh Kβ is shown in Fig. 6.6.1. These spectra were obtained by scattering the radiation from a Rh-anode X-ray tube off a range of materials in an XRF spectrometer. Each emission line is represented by two peaks: a sharp coherent peak and a broad incoherent (Compton) peak. The wavelength difference between the two peaks is independent of the scattering medium but their intensities are strongly influenced by the mean atomic number $\bar{Z}$ of the medium. Materials with a low average atomic number (e.g. boric acid, H_3BO_3) produce very efficient Compton scatter, while materials of moderately high atomic number (e.g. Cu) are very inefficient. Geological materials are represented on this diagram by silica and a sample of basalt (BHVO-1).

The clear relationship between scattering efficiency and atomic number (Z) apparent in Fig. 6.6.1 can be exploited to estimate the mass absorption coefficient (μ) at the wavelength of an X-ray tube line. The strong dependence of μ on Z noted in Box 6.4 suggests that scattering efficiency and μ should have an inverse rela-

tionship. This has been confirmed experimentally, and the intensities of both coherent and incoherent scattered tube lines are found to vary with $1/\mu$. Figure 6.6.2 shows an excellent linear correlation, with a near-zero intercept, between $1/\mu$ and the intensity of the incoherent (Compton) Rh Kα scatter peak. Thus

$$I(Comp) \propto 1/\mu(\text{Rh K}\alpha) \qquad [6.6.2]$$

The intensity of coherent scattered Rh Kα radiation correlates equally well with $1/\mu$, but with a non-zero intercept.

JGF

Figure 6.6.1 Coherent (Rayleigh) and incoherent (Compton) scattering of Rh Kα and Rh Kβ. The sharp coherent peaks have the same wavelength as the incident radiation whereas the broad Compton peaks have longer wavelength and lower energy. The radiation was dispersed into a spectrum using a LiF220 analysing crystal and a 'fine' primary collimator.

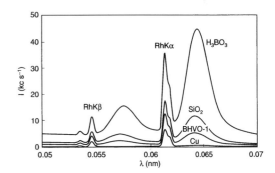

Figure 6.6.2 The relationship between the intensity of incoherently scattered Rh Kα primary radiation (I Comp) and $1/\mu$ for a wide range of materials. The data points range from copper (lowest intensity) to silica (highest intensity) with natural rock samples (ironstone to granite) between. The count rates are higher than those in Fig. 6.6.1 because they were measured with a LiF200 analysing crystal and a 'coarse' primary collimator.

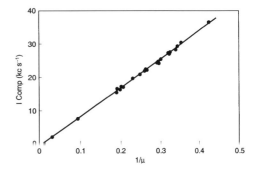

where k'' is a proportionality constant. Absorption effects can therefore be effectively eliminated by dividing the measured intensity I by the intensity $I(Comp)$ of the Compton scatter peak from one of the tube lines. This way of dealing with absorption effects was first described by Reynolds (1963) and is often referred to as the Reynolds method. The method is only applicable in cases where there is no major element absorption edge between the wavelengths of the tube line and the element of interest. It is widely used in geological applications for the determination of concentrations of trace elements with K-line (e.g. Rb, Sr, Y, Zr, Nb) or L-line (e.g. Pb, Th, U) wavelengths shorter than the iron absorption edge, since iron is the heaviest of the common major elements. Enhancement effects are negligible in such situations because elements heavier than iron are generally present only in low concentrations. The application of the Reynolds method to the determination of Sr concentrations in igneous rocks is shown in Fig. 6.8. This figure shows the variation of (a) the intensity of the Sr Kα line and (b) the intensity ratio I(Sr Kα)/I(Rh Kα Comp) with the concentration of Sr in a set of reference standards. The uncorrected data are clearly affected by absorption differences between standards because basic rocks (high Fe and Ca contents) give lower count rates than acid rocks (low Fe and Ca contents) for the same content of Sr.

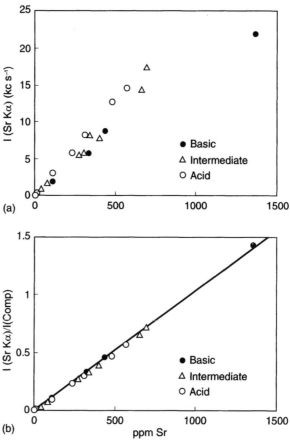

Figure 6.8   The use of the Reynolds method for eliminating absorption effects during the determination of Sr in igneous rocks. The data in diagram (a) are clearly affected by absorption differences, but these are effectively eliminated (b) by dividing the intensity of Sr Kα by the intensity of the Rh Kα Compton scatter peak.

In principle, the intensity of any scattered tube radiation can be used to eliminate absorption effects, as was proposed by Anderman and Kemp (1958). These authors observed that the intensities of fluorescent radiation and scattered tube radiation at a wavelength near the fluorescent line will be similarly absorbed by the sample. The ratio of the two intensities (net peak/background) should therefore be insensitive to absorption effects. A limitation of this method is that the background intensity must be measured with the same precision as the peak intensity.

A more radical solution to the absorption problem is to reduce the effect at source by mixing the sample with a substance with a high mass absorption coefficient (a so-called 'heavy absorber'). Lanthanum oxide (La_2O_3) is often used as a heavy absorber in geological applications of XRF spectrometry (e.g. Norrish and Hutton, 1969). It is customary to add the absorber while fusing the sample with a suitable flux (usually a lithium borate) to make a glass disc. Adding a heavy absorber in a fixed ratio to the sample mass raises the overall mass absorption coefficients thereby reducing the proportional differences caused by absorption variations between samples. With absorber:sample ratios of around 1:1, linear calibrations can be obtained over a very wide range of compositions although X-ray intensities are considerably reduced. The use of a heavy absorber can only *reduce* the effects of absorption, and their *elimination* still requires mathematical corrections in addition. The heavy-absorber method is limited to the determination of major element concentrations. It is not applicable to trace element analysis owing to the attenuation of the low X-ray intensities by the absorber and their dilution by the addition of flux.

Particle-size effects

All the methods, outlined in the previous section, for eliminating absorption effects make the fundamental assumption that the sample is homogeneous and that fluorescent radiation is absorbed by material with the composition of the sample as a whole. For short-wavelength radiation, emerging from a considerable depth within the sample, this assumption is valid. At longer wavelengths, however, radiation reaching the surface has travelled only from a shallow depth below the surface. For example, Ti Kα radiation emitted by a sample composed essentially of silica will only be able to escape from a surface layer less than 100 μm thick, and Na Kα from a layer only a few micrometres thick (Figs 6.5.2 and 6.7). Most materials analysed in geological applications are more absorbing than silica and so their critical thicknesses (Equation 6.10) will generally be significantly less in practice. Since most grinding equipment produces powders with a mean particle size of a few tens of micrometres (Box 2.2), this powder will be heterogeneous on the scale of penetration of long-wavelength radiation. A large proportion of the long-wavelength radiation emitted by a typical powder sample will be absorbed only by the mineral phase in which it originates, and absorption corrections based on the *bulk* composition of the sample will be ineffective as a consequence. These particle size effects limit the usefulness of XRF spectrometry in the analysis of powders for light elements and, for this reason, rock samples should be ground as finely as possible for all XRF analytical work.

The only way to eliminate particle size effects completely is to homogenise the sample totally by fusion with a flux to form a glass. Lithium borates (metaborate, tetraborate, or mixtures of the two) are most commonly used for this purpose because of their low average atomic number and therefore low mass absorption coefficients. A powdered sample is mixed with flux (generally in a sample:flux ratio of between 1:5 and 1:10) and the mixture fused, either in a muffle furnace or over a high-temperature gas burner, and cast into a suitable mould to form a glass disc or bead. A heavy absorber may be included at this stage if required. Dilution of the sample with light flux raises X-ray background levels by increasing the scattering efficiency, and reduces the net intensity of fluorescent radiation. Consequently fused glass beads are only routinely used in the determination of the major elements and some of the more abundant light trace elements (e.g. Cr, Ni).

Box 6.7 Ultra-trace element analysis by total reflection X-ray fluorescence analysis (TXRF)

This novel X-ray fluorescence technique exploits two peculiarities of X-ray optics that arise when the beam is directed onto a polished solid surface *at a very low angle of incidence* (typically 0.1°)[1]:

(1) The beam is totally reflected[2] with negligible penetration into the solid.

(2) The incident and reflected rays undergo mutual **interference**, generating a standing wave in a zone (several 100 nm thick) immediately above the solid surface. Any evaporated residue present in this zone undergoes efficient excitation by the combined intensity of the incident and reflected beams, providing high trace element sensitivity.

Equipment

Figure 6.7.1 shows a typical TXRF spectrometer. A 'line-focus' X-ray tube (in which X-rays are generated at the tip of a knife-edge-shaped target instead of a flat one) produces a planar X-ray beam of extremely low divergence (∼0.01°). It is diffracted by a crystal[3] monochromator to filter out wavelengths other than the selected target peak (e.g. X-ray continuum), and then directed at grazing incidence onto the polished surface of the sample support, which can be finely adjusted in elevation and tilt. An evaporated film or fine particulate sample mounted on the support surface fluoresces, and the characteristic radiation is detected by a Si(Li) or Ge energy-dispersive

Figure 6.7.1 Layout of a TXRF spectrometer (after Klockenkämper *et al.*, 1992). The magnitude of the angle of incidence ϕ has been exaggerated.

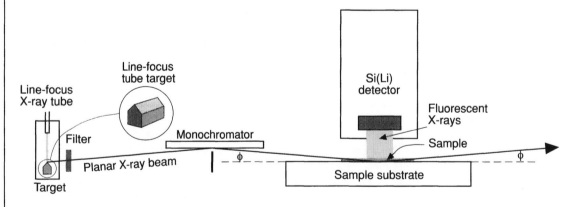

[1]   The angle of incidence of the primary beam in conventional XRF analysis is ∼ **45°**.
[2]   Refractive indices of solids at X-ray wavelengths are very slightly *less* than unity. At very low angles of incidence, therefore, an X-ray beam experiences total *external* reflection, analogous to the total *internal* reflection observed in light optics when RI > 1.0000.
[3]   A manufactured multi-layer dispersion element (Chapter 14) may be used in place of a crystal.

Applications and developments

XRF spectrometry is used for the bulk chemical analysis of materials in an enormous variety of applications in industry and in academic research. In its geological applications it is mostly used for determining the major and trace element composition of rock and sediment samples, and XRF spectrometry is the technique most widely used for this purpose. The long-term stability of X-ray spectrometers is such that very long count times can now be used to measure very low concentrations (< 1 ppm) of some trace elements (e.g. Mo, Nb) with an accuracy of ± 0.1 ppm (Fig. 6.1). In addition to bulk analysis, XRF spectrometry is increasingly used in environmental studies. Thin films, with their freedom from absorp-

detector (see Box 7.1) mounted about 1 mm vertically above the sample surface.

Performance

TXRF offers two remarkable advantages over conventional XRF:

(1) Because fluorescence occurs in a layer less than 1 μm thick, matrix absorption in the specimen is negligible and therefore matrix corrections are unnecessary.

(2) The combination of a monochromatic primary beam (without continuum) and negligible penetration of X-rays into the substrate results in an extremely low background signal, and therefore exceptionally low detection limits are attainable (less than 10 pg of analyte).

Figure 6.7.2 shows the TXRF spectrum of ultra-trace elements in a droplet of rain water, with Ga (stippled peak) added as an **internal standard** (25 ng ml^{-1}), before evaporation onto a polished quartz glass substrate. The numbers below each element symbol indicate the concentrations in the raindrop of the elements (in ng ml^{-1} = ppb).

TXRF has many applications as a microanalysis technique (only a few microlitres of solution are required), a trace element technique (detection limits as low as 0.2 pg have been reported) and surface analysis technique (e.g. for semiconductor wafers). It is particularly suited to the analysis of microscopic samples of artefacts (e.g. provenance of oil paint), of biological tissue, and of natural waters. It has a non-destructive multi-element capability of up to 80 elements.

RG

Figure 6.7.2 TXRF spectrum of a droplet of rain water (reproduced from Klockenkämper and von Bohlen, *Journal of Anal. Atomic Spectrometry* **7**, p273–9, 1992, by kind permission of the Royal Society of Chemistry). Concentration values (in the raindrop before evaporation) are given in ng ml^{-1}.

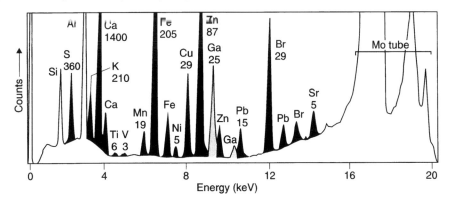

Box 6.8 Spatially resolved trace element analysis by synchrotron X-ray fluorescence (SXRF)

Spatially resolved trace element analysis using X-rays has until recently proved to be an unattainable goal. X-ray fluorescence analysis, though capable of trace element analysis at the ppm level, lacks the potential for spatially resolved microbeam analysis as X-rays do not lend themselves to optical focusing into a convergent beam. The electron microprobe (Chapter 14), on the other hand, offers spatially resolved analysis but detection limits in most cases exceed 100 ppm (Fig. 14.9) owing to the high X-ray continuum background. One avenue that offers a way round these restrictions is to exploit the super-intense X-ray beams available from the synchrotron storage rings designed for high-energy physics experiments.

A synchrotron is a narrow near-circular tube up to 1 km in circumference, in which bursts of high-energy electrons or positrons (travelling at speeds close to that of light) are 'stored' by being cycled repeatedly round the tube (under high vacuum) under the control of synchronised electromagnetic fields. A synchrotron incorporates

Figure 6.8.1 Geometry of an X-ray fluorescence microprobe (reproduced with permission from Smith and Rivers, Fig. 5.12 *Microprobe Techniques in the Earth Sciences*, P. J. Potts *et al* (eds), 1985, Chapman & Hall.).

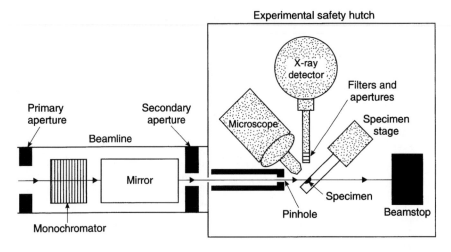

tion effects, make them an ideal medium for the XRF analysis of particulate matter filtered from water or air.

Two exciting new developments have extended considerably the analytical capabilities of XRF spectrometry. One of these (TXRF; Box 6.7) allows the analysis of thin films of material with extreme sensitivity, and is currently used extensively in the determination of impurities on the surfaces of silicon wafers. The other (SXRF; Box 6.8) uses a finely collimated beam of high-intensity synchrotron X-rays to achieve microbeam analysis down to the ppm concentration level. Both techniques are currently limited by cost to very specialised applications.

numerous 'bending magnets' that transform a straight electron trajectory into an inward-curved arc, and as each burst of particles experiences inward acceleration under the influence of a bending magnet it radiates an intense pulse of X-rays as a narrow fan tangential to the ring. Successive bursts of electrons passing round the ring generate a pulsed beam. An X-ray fluorescence experiment set up along this 'beam line' can exploit the following characteristics of such beams:

- Depending on the technological refinements used, the beam has a **brilliance** 10^4–10^{12} times higher than a conventional X-ray tube (Smith and Rivers, 1995).

- It is confined within a fan of very low divergence and, given the high photon flux, can be collimated to a microbeam a few micrometres in diameter using a simple pinhole.

- The beam has a continuum energy spectrum (similar in origin to **bremsstrahlung**), but a recent development (the 'undulator') allows spectral 'tuning' to select a narrow band of continuum just on the high-energy side of a target analyte's absorption edge, giving optimum fluorescence efficiency for that element.

- The beam is highly polarised, and by careful choice of geometry it is possible to eliminate virtually all background from **Rayleigh** and **Compton scattering**. Trace element peaks consequently have much higher peak/background ratios than in the electron microprobe and thus lower detection limits.

These characteristics enable synchrotron XRF to achieve microbeam analysis of trace analytes at the ppm level with a spatial resolution of a few micrometres. An SXRF spectrometer layout is shown in Fig. 6.8.1. The finely collimated beam impacts on the specimen at an angle of 45°. Fluorescent X-rays from the sample are analysed by an energy-dispersive spectrometer (see Box 7.1), positioned in the plane of the synchrotron ring (where scattering of the polarised beam is least). The specimen stage can be computer driven in x, y and z movements, and the user views the specimen through an optical microscope. Analyses are usually calibrated using a major element internal standard whose concentration has been determined by electron microprobe analysis (Chapter 14). Further details of the technique are given by Smith and Rivers (1995).

Unlike the ion microprobe (SIMS, Chapter 15) SXRF is not capable of resolving isotopic compositions. SXRF time on a major high-energy physics facility such as Daresbury in the UK or Cornell University in the USA has to be competed for on a peer-review basis. Perhaps the most powerful aspect of SXRF is the ability to use the high brilliance of the X-ray source for closely related, spatially resolved studies of chemical bonding, surface analysis and crystal structure in conjunction with chemical analysis.

RG

Chapter 7

Neutron activation analysis

Susan J. Parry

Introduction

In neutron activation analysis (NAA) a sample is irradiated in the neutron flux inside a nuclear reactor. Nuclei of a stable analyte isotope are transformed by neutron capture into a heavier, radioactive isotope that undergoes β-decay with the emission of a characteristic γ-ray spectrum. The analyte can be identified in this spectrum from the energy of the γ-ray peaks, and its concentration measured from the peak integral.

NAA is a powerful multi-element technique for determining a number of geochemically important trace elements. It is particularly applicable to geochemical problems because samples are analysed in solid form. It therefore has advantages where complete dissolution of the sample is difficult, or where elements are likely to be selectively lost during dissolution prior to analysis.

A nuclear reactor is the most useful source of neutrons for activation analysis. Irradiation services are commercially available to analysts at research reactors (usually operated by government establishments or universities) in most industrialised countries. Analysts do not have to visit the site of a reactor to carry out their measurements; samples may be sent for irradiation at the nearest reactor facility and returned (with appropriate handling procedures for radioactive materials) to the analyst's own laboratory for counting. Spectrometry equipment in the form of high-resolution semiconductor γ-ray detectors has wide application in the measurement of environmental radioactivity.

Relevant publications on the technique include books by Alfassi (1989), Amiel (1981a), Ehmann and Vance (1991) and Parry (1991). For geochemical analysis and environmental studies in particular, the reader is referred to Das *et al.* (1989a,b) and Tolgyessy and Klehr (1987).

Theory

Activation analysis is based on the fact that nuclei of a particular isotope exposed to a neutron flux are liable to absorb thermal and slow neutrons, a nuclear reaction known as 'neutron capture'. For example,

$$^{59}\text{Co} + \text{n} \rightarrow {}^{60}\text{Co} + \gamma \qquad\qquad [7.1]$$

This reaction, which nuclear physicists represent in the shorthand form $^{59}\text{Co}(\text{n},\gamma)^{60}\text{Co}$, is shown graphically in Fig. 7.1(a). The γ-ray photon emitted during this reaction is called a 'prompt gamma'. For many stable nuclides (such as ^{60}Ni and ^{61}Ni in Fig. 7.1a) the new nucleus generated by neutron capture is also stable. The majority of elements, however, have at least one isotope (e.g. ^{59}Co) that produces a *radioactive* product (^{60}Co), and it is these isotopes that are utilised in NAA. The term 'activation' refers to the generation of such a radioisotope by exposure to nuclear particles.

The radioactive isotope ^{60}Co decays to ^{60}Ni by emitting a β-particle (plus an anti-neutrino), a reaction symbolised as ^{60}Co$(,\beta+\bar{\nu})^{60}$Ni. The daughter nucleus (in this case ^{60}Ni) is not usually in its ground state, and therefore most β-decay reactions also emit γ-ray photons arising from the relaxation of these excited states (the higher energy levels in Fig. 7.1b). It is these γ-rays (whose photon energies are characteristic of the *daughter* nuclide (^{60}Ni), not the activated analyte isotope itself (^{60}Co)) that provide the analytical signal used in activation analysis. Any element with an isotope producing a radionuclide whose decay emits γ-rays is amenable in principle to activation analysis, and this is true of over 60 elements. The elemental sensitivity, however, depends on various factors and varies considerably from one element to another.

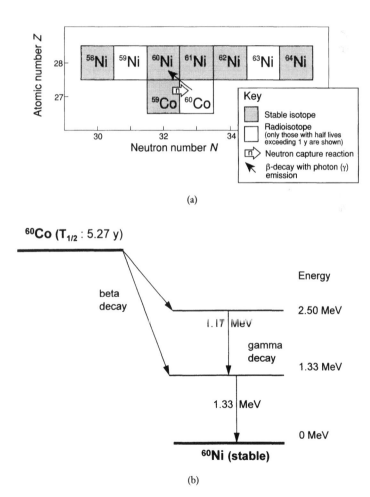

(a)

(b)

Figure 7.1   The decay scheme for ^{60}Co (based on data from Browne and Firestone, 1986).

Box 7.1 The energy-dispersive photon detector

An energy-dispersive (ED) detector is one that distinguishes between the various components of a spectrum by their *photon energy* rather than by wavelength. The last two decades have seen great advances in the design of semiconductor detectors of γ-rays and X-rays that produce output pulses closely related in size to the quantum energy of the incoming photons; this allows the spectrum to be registered and analysed by purely electronic means, without the expense and instrumental complexity associated with wavelength analysis. How such detectors operate is shown in Fig. 7.1.1

The energy levels available to electrons in crystals are grouped in *bands*. Semiconducting properties arise when a narrow energy 'gap' exists between a lower ('valence') band of filled energy levels, and an upper ('conduction') band of empty energy levels. In its normal state the semiconductor crystal behaves as an insulator, because all electrons are accommodated in the valence band where they are unable to migrate.

A high-energy photon entering the crystal, however, will deposit sufficient energy to promote a number of electrons across the energy gap into the conduction band, where they are free to migrate; this leaves an identical number of electron vacancies in the valence band, which are also able to migrate. The arrival of the photon is therefore marked by a short period in which the crystal behaves as a conductor, lasting until the promoted electrons eventually fall back into the lower band. With a potential difference of several hundred volts applied across the crystal, this brief interval of electrical conduction generates a pulse of current through the crystal, which can be amplified by a low-noise charge-sensitive (**FET**) amplifier and stored in a computer.

The magnitude of each current pulse is proportional to the number of electrons promoted (typically 200–2000 in the case of an X-ray photon, and millions for a γ-ray photon), which depends on the quantum energy of the incoming photon. The pulses accumulated during a standard counting interval are sorted by the computer, according to their size, into several thousand pulse-height channels. Displaying the number of counts in each channel as a function of channel number produces an image of the original spectrum, as illustrated for an X-ray spectrum in Fig. 7.1.2. This shows photons of three different energies arriving at the detector. The low-energy oxygen photons ('soft' X-rays) are absorbed before they reach the sensitive region of the detector and are not detected. The Mg and Si quanta register at pulses approximately proportional in magnitude to photon energy. Because proportionality is imperfect (owing to statistical fluctuation in the number of electrons promoted by identical photons), the resulting peaks have appreciable width. The signal recorded for each element is the integral of a small range of channels accumulated during a preset counting period (shown in black). These peak areas are used to calculate element concentrations.

At room temperature thermal fluctuations promote a proportion of electrons up to the conduction band regardless of incoming photons. This thermal excitation causes a background current that obscures low photon intensities. To minimise such problems, both the semiconductor detector crystal and the **FET** preamplifier providing the first stage of amplification are cooled by liquid nitrogen via a cold finger (see Fig. 7.3 and 14.5).

The semiconductor material used depends on the application. Silicon (Si, $Z = 14$) is used for X-ray detectors, whereas germanium (Ge, $Z = 32$) is preferred for γ-rays because its higher atomic number offers greater stopping power and makes it a more efficient absorber of the penetrating γ-radiation (in order to be detected, a photon must be absorbed within the sensitive volume of the detector, and not pass through without depositing energy).

Traces of impurities like gallium in an Si crystal degrade its performance by introducing spurious energy levels into the gap between valence and conduction

Figure 7.1.1 Detector energy levels.

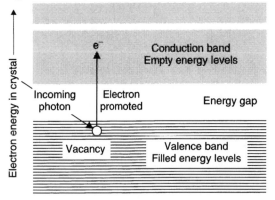

Neutron activation

The activation of the target nuclide will depend on the number of interactions between neutrons and nuclei. Therefore the induced radioactivity will be proportional to the total number of neutrons per unit area per second striking the sample (the neutron flux, ϕ_n) and the

bands. Impurity effects may be counteracted by diffusing lithium (Li) into the active volume of the crystal, producing what is called a 'lithium-drifted' detector. For X-ray measurement an Li-drifted silicon (Si(Li)) device is used. The detector surface is often protected by a thin beryllium window, though 'windowless' detectors are increasingly common. Recent advances in semiconductor manufacture have produced Ge crystals of sufficient purity (total impurities <1 part in 10^{12}) to make Li drifting unnecessary.

Li-doped detectors may deteriorate if allowed to stand at room temperature, as the distribution of Li may be altered by continuing diffusion. The rate of diffusion is insignificant at liquid nitrogen temperatures, providing a second reason for continuous liquid nitrogen cooling.

RG and SJP

Figure 7.1.2 Energy-dispersive analysis of an X-ray fluorescence spectrum of the mineral forsterite.

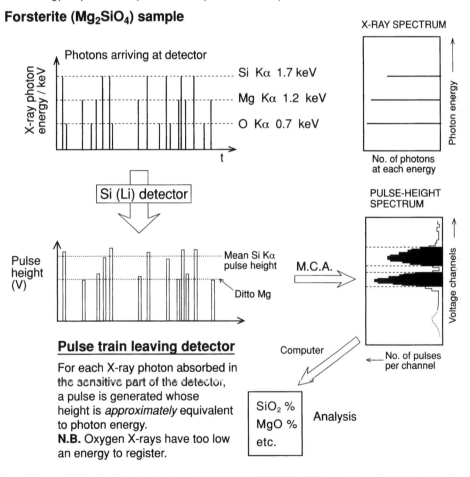

number of target nuclei present in the sample (N). This proportionality is expressed by a physical constant called the neutron–capture cross section (σ_n), which is the probability of an interaction occurring for a particular target nuclide. The activation rate is expressed by:

$$\text{Activation rate} = \sigma_n \phi_n N \qquad\qquad [7.2]$$

The growth of activity during irradiation depends not only on the production rate, but also on the rate of decay of the activation product (i.e. the decay constant λ) during the activation period. The activity of the activated radionuclide (A_t) at the end of irradiation time t_i is given by

$$A_t = \sigma_n \phi_n \, N \, (1 - e^{-\lambda t_i}) \qquad [7.3]$$

The number of target nuclei, N, is a function of the mass of the element of interest m in the sample, Avogadro's number (L), the isotopic abundance (θ_i) and the relative atomic mass of the element concerned (M):

$$N = L\theta_i \, m/M \qquad [7.4]$$

So substituting for N in Equation 7.3

$$A_t = \sigma_n \phi_n L\theta_i \, m \, (1 - e^{-\lambda t_i})/M \qquad [7.5]$$

The activity is measured on a γ-ray detector as counts s^{-1} and corrections are made for the counting geometry, for the number of γ-rays emitted per disintegration of the radioactive isotope and for decay of the active nuclide before counting begins:

$$\text{Activity} = C_\gamma/P_b \epsilon e^{-\lambda t_d} \qquad [7.6]$$

where C_γ is the count rate (s^{-1}), ϵ is the efficiency of the detector for the γ-ray energy in question with the relevant source geometry, P_b is the branching ratio (a constant for the decay scheme) and t_d is the 'cooling time' between activation and counting.

Substituting in Equation 7.5:

$$A_t = \sigma_n \phi_n L\theta_i \, m \, (1 - e^{-\lambda t_i})/M = C_\gamma/P_b \epsilon e^{-\lambda t_d} \qquad [7.7]$$

and therefore

$$C_\gamma = \sigma_n \phi_n L\theta_i \, m \, P_b \epsilon (1 - e^{-\lambda t_i}) \, e^{-\lambda t_d})/M \qquad [7.8]$$

Quantitative analysis

In theory the activation equation (Equation 7.8) may be used to calculate the mass of the element (m) from the measured activity, but this approach requires accurate knowledge of the neutron flux (ϕ_n), the detector efficiency (ϵ) and other physical constants, some of which are imperfectly known.

In routine analysis, therefore, the activation obtained under specified conditions is calibrated by the use of standards of known analyte concentration. The simplest method, used by the majority of analysts in geochemical applications, is to include the standards with the samples during irradiation and counting to simplify the calculations. If the samples and standards are irradiated in a uniform neutron flux, for the same length of time and counted with the same detector geometry, then for a long-lived radionuclide of a particular element the factors in Equation 7.8 are all constant and the activity ratio can be written as

$$\frac{(C_\gamma)_{\text{sample}}}{(C_\gamma)_{\text{standard}}} = \frac{m_{\text{sample}}}{m_{\text{standard}}} \qquad [7.9]$$

In practice the neutron flux varies to a small extent between sample and standard, and 'flux monitors' (capsules containing identical amounts of analyte) are interspersed with samples and standards during irradiation to allow the flux gradient to be measured and appropriate correction applied. Samples and standards are usually counted serially, and therefore a small decay correction must also be made to allow for differences (between standard and sample) in the elapsed time between irradiation and measurement. These corrections are small and can be made very accurately.

Instrumentation

Irradiation facilities

There are a number of neutron sources which may be used for activation, including accelerators and isotopic sources, but for laboratory-based multi-element analysis a nuclear reactor provides the most convenient and versatile source of neutrons. A typical reactor (Fig. 7.2) consists of a core of enriched uranium surrounded by a moderator (water or graphite) and coolant (usually water). The power output of a research reactor may range from kilowatts to megawatts, but the neutron flux is generally in the range of 10^{15} to 10^{18} neutrons $m^{-2} s^{-1}$ in the irradiation positions. Neutron fluxes of 10^{15} to 10^{16} neutrons $m^{-2} s^{-1}$ are sufficient for most activation analysis applications; higher fluxes bring problems associated with high temperature and radiation damage to the samples.

Irradiation facilities inside reactors are of various kinds. In manually loaded irradiation tubes, the samples are lowered on a stringer device with one irradiation container sited vertically above the next. Alternatively pneumatic devices may be used to transfer the sample to the irradiation position; they are designed primarily for short irradiations, but provide a valuable method of loading and transferring irradiated samples by fully automated means. Figure 7.2 shows the design of the Imperial College reactor system and its irradiation facilities. More elaborate irradiation devices may be available in certain types of commercial reactor, such as the rotating flux-equalisation system in the US-designed TRIGA reactor.

γ ray detector

The gamma rays emitted by the sample are analysed to identify the element, to quantify its activity and thereby to determine the amount present. The principle of the germanium semiconductor detector used is outlined in Box 7.1. In current commercial analyser systems, the high-voltage supply for the detector, the pulse amplifier and the ADC are combined in a single unit connected to a personal computer for all calibration and data processing.

A typical detector configuration is shown in Fig. 7.3. The sample container is either placed manually in a sample holder with predetermined shelf positions to ensure reproducible sample–detector distance, or loaded into a motor-driven sample changer (with a fixed sample–detector distance) under the control of the computer for automatic counting. Sample changers take the form of a moving belt or carousel or a pneumatic system; in each case

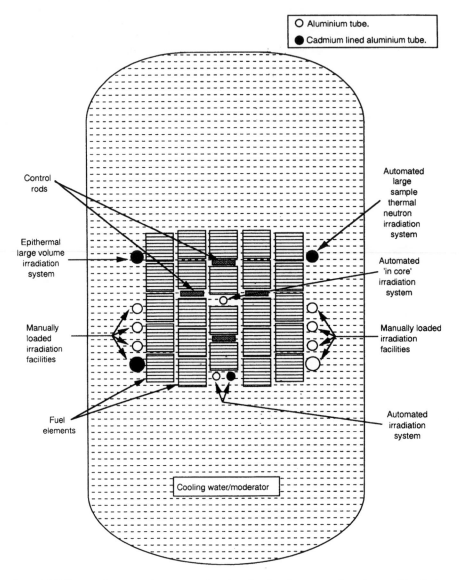

Figure 7.2   Imperial College nuclear research reactor, a plan view showing the pneumatic and manual irradiation facilities (original drawing provided by R Benzing).

successive samples are positioned over the detector for a predetermined counting interval or 'live time' (see below).

Other detectors   The activation analysis of uranium is usually carried out by counting the delayed neutrons emitted from the irradiated sample. When a sample containing uranium is irradiated, the ^{235}U undergoes fission and the resulting fission product decays with the emission of neutrons. These delayed neutrons can be counted and compared with a standard to provide a quantitative measure of the uranium in the sample. As the standard Ge detector does not register neutrons, a special detector constructed of boron trifluoride is used. Figure 7.4 shows an automated pneumatic irradiation system for carrying out such analyses.

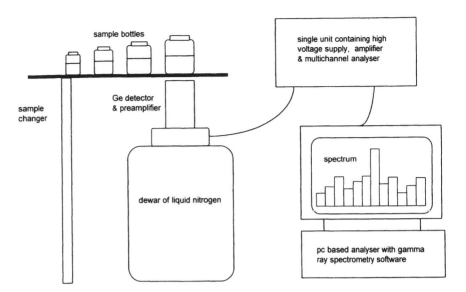

Figure 7.3   γ-ray spectrometry system for quantitative multi-element analysis, showing the detector, cooled with liquid nitrogen, and associated electronics, sample changer and PC-based analyser.

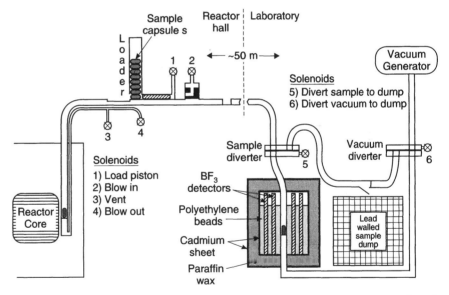

Figure 7.4   An automated pneumatic irradiation system for uranium determination. Delayed neutrons from the fission of uranium are measured with an array of boron trifluoride neutron detectors (reproduced with permission, from Imperial College Reactor Centre Report, 1990).

Data processing

Software packages are now available for processing γ-ray spectra on a personal computer. The signal from the detector preamplifier consists of a train of pulses (or of voltage steps) representing γ-ray photons detected, the voltage of each pulse being nearly proportional to the energy of the corresponding photon (Box 7.1). The pulses are sorted according to their voltage into an array consisting of several thousand 'channels', representing successive narrow intervals of pulse height. The peaks in the resulting pulse-height spectrum (Fig. 7.5) have a Gaussian shape with exponential tails, superimposed on a background continuum caused by **Compton scattering**.

The complex γ-ray spectra are processed using software that calculates in turn:

(1)   the peak position (in terms of channel number);
(2)   photon energy;
(3)   the identity of the daughter nuclide and hence the parent nuclide;
(4)   the peak integral (area).

The channel number of the peak maximum is located and a predetermined energy–calibration function is used to determine the energy of the γ-ray. The source nuclide is identified using a computer library of radionuclide energies. Peak integation consists simply of adding together the contents of contributing channels after subtraction of an interpolated background value. Where one peak overlaps another, more complex peak-fitting routines are used to separate the two signals.

The count rate of an γ-ray detector system departs from a linear response at high count rates owing to **dead time**, the origin of which is explained in Chapter 6. As with X-ray energy-dispersive systems (Chapter 14), this is dealt with by extending the counting period until the *live time* (elapsed time minus the accumulated dead time) has reached the predetermined nominal counting time. Finally the peak area, the total number of counts contributing to the peak, is divided by the live time to give the count rate for the radionuclide of interest. A typical γ-ray spectrum may contain over a hundred peaks from the radionuclides of over 60 elements.

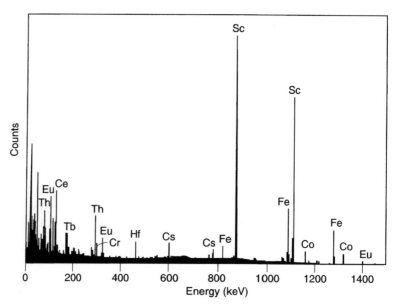

Figure 7.5 γ-ray spectrum of a granite which was irradiated for 22 hours in a thermal neutron flux of 10^{16} neutrons $m^{-2} s^{-1}$, allowed to decay for 30 days and counted for 1.4 hours (spectrum provided by R Benzing).

The resulting data are reported in the form of a list of peak energies and net counts under the peak (or as counts s^{-1}) for each standard and unknown sample. Most programs check whether the peaks from a particular radionuclide all appear with the expected ratios of intensities. The program then calculates the specific activity of the analyte (i.e. counts s^{-1} per unit mass of analyte present) for the standards, after applying appropriate decay and neutron flux corrections. This calibration factor is used to calculate mass concentrations of analyte elements in the unknown samples from the count rates.

Precision When measuring radioactivity the standard deviation (σ) in measuring n counts is $n^{0.5}$ (Box 6.4), so a signal of 10,000 counts will have a precision of 1%. However, the peak area S is evaluated by subtracting the background (B) from the total area under the peak ($S + B$), to give the signal (S). Therefore the error on S is the sum of the errors on $S + B$ and B:

$$\sigma_S = [(S + B) + B]^{0.5} = [S + 2B]^{0.5} \qquad [7.10]$$

For example, a peak of 1000 counts on a background of 2000 counts will have an error of 7%.

Methodology

Sample preparation

The major advantage of activation analysis is that little, if any, sample treatment is required. Solid samples are preferred for analysis and these may be powders, grains, or even chunks of material, provided that the geometry can be replicated in the standards. These solids are encapsulated in containers made of polyethylene, aluminium or quartz glass, and placed together with standards and flux monitors inside an outer irradiation can of polyethylene or aluminium for activation. Rocks, soils and sediments are typically analysed as powders. River or sea water may be measured in the liquid form (irradiation by normal neutron fluxes does not result in a significant rise in temperature), or freeze dried to preconcentrate larger volumes. Vegetation can be analysed directly, but large volumes are dried, ground and ashed before analysis. Air-filter papers for particulate analysis are simply folded and encapsulated.

Standards

The usual procedure is to prepare multi-elemental standard solutions by mixing single-element standard solutions in the appropriate proportions. Often it is a solid standard that is required in activation analysis, and so the solution is dried on filter paper or another suitable support, or is added as **spike** to the sample itself. Alternatively, secondary **reference materials** calibrated against **certified reference materials** may be used as standards. This method is particularly suitable when the secondary reference material has a similar matrix to the samples. Standards can also be used as flux monitors, since any neutron flux variation with irradiation site will be reflected by differences in activation of the standards.

Certified reference materials may be used occasionally to validate the method and procedure. Those materials with a matrix similar to the sample are most appropriate. Using **CRMs** or **geostandards** as primary standards in the analysis of geological materials is not good practice as CRMs are too valuable to be used for this purpose.

Irradiation methods

The samples for irradiation are packed in polyethylene or aluminium containers. Polyethylene is used because the elements carbon and hydrogen do not give any significant radioactivity on irradiation but it is prone to radiation damage on extended exposure to neutrons and so aluminium is used for long irradiations. Aluminium is chosen because its activation product, ^{28}Al, has a half-life of only 2.3 minutes and has largely decayed before counting begins.

The factors which are most important in irradiation of the sample are the neutron flux and the irradiation time (Equation 7.7). Sequential irradiations, even of less than a minute, can be timed very precisely in a pneumatic system, and every sample and standard will be located in exactly the same irradiation position. Samples and standards loaded together in a manually loaded system will have the same irradiation time, which need not be known precisely, though any flux gradient across the length of the irradiation container must be corrected for.

The daughter activity induced in the sample is proportional to neutron flux, the neutron-capture cross-section of the analyte nuclide, and the number of nuclei of that nuclide in the sample (Equation 7.7). The capture cross-section varies widely between nuclides. Elements with high cross-sections (and therefore high analytical **sensitivity**) include gold, rhodium and dysprosium. Elements with very low cross sections include potassium, iron and magnesium. (It is a fortunate accident that most major elements in rocks have low capture cross-sections and/or short-lived activation products, so they constitute a surprisingly favourable matrix for INAA trace element analysis in spite of their chemical complexity.)

If necessary, the sensitivity of an element may be increased by using a higher neutron flux but this increases the background activity as well as the signal, and the peak to background ratio may even be reduced. There is little advantage in irradiating for more than about two half-lives of the radionuclide of interest since longer irradiation will produce little more activity, and may also increase significantly the activity of a longer-lived interferant.

Most activation for analytical purposes is carried out by **thermal neutrons**, but in cases where this creates a serious inter-element interference it may be advantageous to screen the sample from thermal neutrons and activate using **epithermal neutrons** alone (Alfassi, 1985). For example, the neutron activation of a gold-bearing rock produces a spectrum dominated for the first few days by ^{24}Na, from the ^{23}Na(n,γ) reaction, with a half-life of 15 h. Even when left to decay for a week, it is difficult to detect ^{198}Au (the activation product of the ^{197}Au(n,γ) reaction), which has a half-life of only 2.3 days. If the thermal neutron component is filtered out by irradiating the sample behind a 1 mm thick cadmium sheet, the activation of ^{23}Na is reduced by a factor of $30\times$ or more (the 'cadmium ratio'). ^{197}Au readily captures epithermal neutrons and so it is barely affected by the cadmium, and the signal from the decay of ^{198}Au is only reduced by a factor of 2 or 3. Consequently the signal from the gold is enhanced over the sodium-induced background by about 10 in this case. Epithermal activation benefits the analysis of a number of elements (Au, Ag, As and Sb) that suffer severe background interference from ^{24}Na or ^{46}Sc, the usual sources of interferences in samples of soil, rock and sediment.

Counting techniques

The differences in half-lives between radionuclides can be used to optimise the element of interest. If ^{56}Mn (half-life 2.6 hours) is to be determined in the presence of ^{24}Na (half-life 15 hours), it is advisable to count the sample as soon after irradiation as possible. On the other hand the measurement of ^{207}Hg (half-life 30 days) is best measured after all the shorter-lived radionuclides have decayed. For example, several weeks are allowed to pass before measuring Hg in air-filter samples.

In the case of a radionuclide with a half-life of less than 1 minute, there would be little advantage in counting for more than 2 minutes, but the number of counts accumulated in such a short time may not be sufficient to give good counting statistics. It is possible to increase the accumulated counts in the peak by reirradiating the sample and counting again, adding the spectrum to the previous one. This method of cyclic activation is useful for short-lived radionuclides of elements such as Se, Hf, Rh and Ag (Spyrou, 1981). The only disadvantage of the technique is that the longer-lived activity builds up in the background, reducing the signal to background ratio until it is no longer practical to count. An alternative method which overcomes that problem is cumulative activation where a series of replicates of the same sample are irradiated and the spectra accumulated to form one final spectrum, avoiding the build-up of background activity since each replicate is only irradiated once (Guinn, 1980).

The detection of γ-rays may be optimised by careful choice of detector. A thin planar germanium detector (Box 7.1) has the best energy resolution at the low-energy end of the spectrum (30–200 keV), and records a lower background because most high-energy γ-rays pass through without interacting with the sensitive volume of the detector; this kind of detector is used to measure most of the rare earth elements including Sm, Eu and Gd, providing a valuable instrumental technique for preparing chondrite-normalised REE plots. However, such detectors have lower efficiency for higher-energy photons (above 200 keV) and cylindrical germanium detectors offering much larger sensitive volumes may be used to improve sensitivity for elements such as I.

'Prompt' γ-rays (those emitted by the (n,γ) activation reaction itself rather than by the subsequent β-decay ('delayed' γ-rays)) can be used as a more specific way of detecting some radionuclides (Peisach, 1981). In fact this is the only way that boron can be measured by neutron activation, since it produces no delayed γ-rays. Prompt γ-rays have high energies and in some cases are easily detected. The sample must be counted while it is being irradiated, so in this case a neutron beam must be used to activate the sample outside a reactor.

Another very specific method is delayed neutron counting (Amiel, 1981b). Any fissile material gives fission products on irradiation with neutrons. Some of the fission products, with half-lives of a few seconds, decay with the emission of neutrons, known as delayed neutrons, which are measured with an array of neutron detectors (Fig. 7.1). It is a very useful technique for the determination of uranium, and has become an established method for exploratory INAA studies since the procedure requires only 60 s irradiation, 20 s decay and 60 s count. It can be totally automated for the routine analysis of soils, sediments, vegetation and water.

Health physics

Care must be taken when handling samples which have been irradiated in a neutron source. There are two main hazards. One is exposure to an external beam of radiation such as beta particles, X-rays and γ-rays, which are penetrating and can harm the body from an external dose. The other is ingestion of radioactive material via the mouth or through a cut, when even shallow- penetrating alpha particles can damage the tissue. In NAA the need to handle samples is minimised; in most cases the samples remain sealed in their containers throughout the analysis and exposure to radioactivity is limited by the use of lead shielding and automation, where appropriate.

Separations

If adequate detection cannot be obtained by purely instrumental methods, then it is possible to separate the elements of interest chemically before analysis. Removal of one or two interferences may be sufficient to improve detection by several orders of magnitude. For example, if a group separation of the rare earth elements, such as the ion exchange method used for inductively coupled plasma–atomic emission spectrometry (Chapter 4), is applied to a rock sample it will improve the determination of all the elements to below 1 mg kg^{-1} (1 ppm).

Radiochemical NAA, in which chemical separation is carried out after irradiation, has several advantages over other techniques involving chemical separation because chemical reagents cannot contaminate the sample once it has been irradiated, and the recovery efficiency of the separation can be determined by using an inactive carrier.

Applications

There are over 60 elements which are detectable in favourable conditions using a germanium detector and normal counting procedures as described above. Figure 7.6 indicates the approximate detection limits of the technique for these elements in an ideal matrix. Elements can be measured over a wide concentration range from the detection limit up to percentage levels. Table 7.1 gives the detection limits measured in a variety of sample types to show the effect of different types of matrix on the determination of the elements.

NAA has found wide application in mainstream geochemistry for many years, chiefly for the determination of representative rare earth elements (La, Ce, Nd, Eu, Sm, Gd, Tb, Tm, Yb, Lu) and other trace elements of petrogenetic interest such as Hf, Ta, U and Th that are not easy to analyse by other methods. In the examples that follow, however, the usefulness of the method is illustrated in a more applied context.

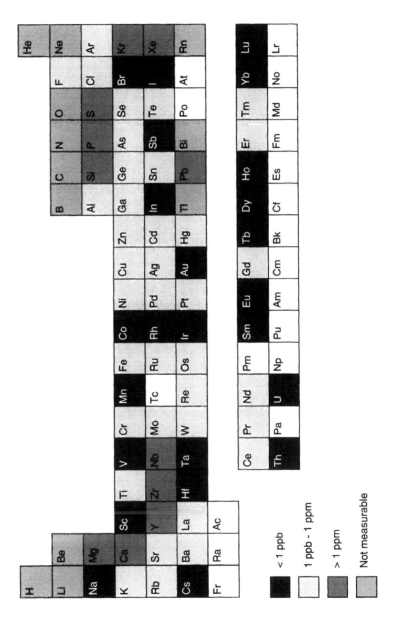

Figure 7.6   3s detection limits for neutron activation analysis in an ideal matrix.

Table 7.1   3*s* detection limits for routine multi-element neutron activation analysis

Element	Air* (ng m^{-3})	Water[†] (μg l^{-1})	Vegetation[‡] (mg kg^{-1})	Sediment[§] (mg kg^{-1})	Basalt[‖] (mg kg^{-1})
Na	40	50	0.16	363	10
Mg	600	800	19	812	–
Al	8	20	–	130	100
Si	–	–	–	4.6%	–
S	5000	–	–	–	–
Cl	100	500	1.4	276	–
K	7.5	200	230	907	–
Ca	200	2000	17	2970	200
Sc	0.004	0.002	0.00051	0.019	0.05
Ti	40	–	–	431	100
V	2	0.2	0.012	4.8	1
Cr	0.25	0.2	0.180	1.2	0.5
Mn	0.6	0.5	0.042	16	100
Fe	20	0.6	3.0	105	50
Co	0.025	0.02	0.0085	0.14	0.1
Cu	20	–	1.3	–	–
Zn	1	2	0.20	220	10
Ga	1	–	–	–	20
As	4	0.4	0.10	9.7	1
Se	0.1	0.05	0.19	4.3	0.5
Br	2.5	1	0.23	4.9	–
Rb	–	0.5	0.32	1.1	0.1
Sr	–	10	–	100	100
Zr	–	–	–	90	100
Mo	–	0.8	–	–	2
Ag	1	0.05	0.064	–	2
Cd	–	0.5	0.50	–	–
In	0.04	–	–	0.60	0.2
Sb	1	0.01	0.020	0.18	0.1
I	20	0.7	–	3.2	–
Cs	–	0.01	0.26	0.24	0.2
Ba	–	10	–	91	20
La	0.2	0.05	–	0.85	0.1
Ce	0.25	0.05	–	0.26	3
Nd	–	–	–	2.4	5
Sm	0.005	0.01	–	0.15	0.1
Eu	0.01	0.003	0.003	0.059	0.5
Gd	–	–	–	2.4	3.9
Tb	–	–	–	0.11	0.1
Yb	–	–	–	0.69	0.2
Lu	–	–	–	0.025	0.05
Hf	–	–	–	0.33	0.2
Ta	–	0.02	–	0.077	0.03
W	0.5	–	–	–	1
Au	0.1	0.003	–	–	0.005
Hg	0.1	0.05	–	–	–
Th	0.04	0.003	–	1.6	0.2
U	–	0.2	–	0.39	0.1

*Dams R., Robbins, J. A., Rahn, K. A. and Winchester, J. W. 1970. Nondestructive neutron activation analysis of air pollution particulates, *Anal. Chem.* **42**(8), 861–868.

†Capannesi, G., Cecchi, A. and Mando, P. A. 1984. Trace elements in suspended particulate and liquid fraction of the Arno river waters. *Proc Fifth Int. Conf. on Nuclear Methods in Environmental and Energy Research, University of Missouri, CONF-840408*, pp. 311–321.

‡Cunningham, W. C. and Stroube, W. B., Jr 1987. Application of an instrumental neutron activation analysis procedure to analysis of food. *Sci Total Environ.* **63**, 29–43.

§Fong, B. B. and Chatt, A. 1987. Characterization of deep sea sediments by INAA for radioactive waste management purposes. *Radioanal. Nucl. Chem.* **110**(1), 135–145.

‖Parry, S. J. 1991. *Activation Spectrometry in Chemical Analysis.* New York: Wiley.

Gold exploration studies

Gold has the lowest detection limit of all the elements analysed by neutron activation and it can be detected down to $1\ \mu g\ kg^{-1}$ (1 ppb) in soils, sediments and vegetation. Sampling for gold can be difficult owing to its occurrence in many rocks and soils as discrete native gold particles, which give Au a very heterogeneous distribution. This 'nugget' effect also leads to difficulties with other analytical techniques which require dissolution of the sample. Large samples are therefore required to give representative analyses. Samples of about 30 g are readily analysed using NAA, although sample sizes of several hundreds of grams have been used. The only preparation required, drying and grinding the sample, is for the benefit of accurate sampling. The powdered sample is packed into a polyethylene container and irradiated for 30 min in a neutron flux of 5×10^{16} neutrons $m^{-2}\ s^{-1}$. Table 7.2 lists the detection limits in vegetation for elements that can be determined at the same time as gold.

A recent example of the use of NAA in gold exploration is illustrated by McConnell and Davenport (1989). They sampled organic lake sediment from 65 sites in four areas in southwest Newfoundland, two areas of which had economically significant Au prospects. Samples of organic-rich lake-bottom sediment were taken from a helicopter on the lake surface, each sampling taking less than a minute. The samples were air dried at $60°C$, then disaggregated and sieved to remove organic debris. The NAA analysis was carried out on 5–10 g of each sample, with a nominal Au detection limit of $1\ \mu g\ kg^{-1}$; 27 other trace elements were also determined. Of these As, Sb and Zn were shown by their association with Au in some deposits to act as pathfinders, demonstrating the value of simultaneous determination of a wide range of elements using NAA. Most samples measured had levels of Au between 1 and $10\ \mu g\ kg^{-1}$. The alternative method for determining Au in sediments (aqua regia leach and solvent extraction followed by graphite furnace AAS), though capable of a lower detection limit for Au, does not offer this multi-element advantage.

Biogeochemical prospecting has become an established method of exploration since plants can take up gold through the entire volume covered by the root systems and this integration overcomes the effect of the concentration of gold as metal particles in the soil. A

Table 7.2   Detection limits ($3s$) for gold and other elements in 30 g samples of vegetation (Source: Hoffman, 1992)

Element	Detection limit (ppb)	Element	Detection limit (ppb)
Au	0.1	Rb	1000
Ag	300	Sb	5
As	10	Sc	10
Ba	5000	Se	100
Br	10	Sr	10,000
Ca	100,000	Ta	50
Co	100	Th	100
Cr	300	U	10
Cs	50	W	50
Fe	50,000	Zn	2000
Hf	50	La	10
Hg	50	Ce	100
Ir	0.1	Nd	300
K	10,000	Sm	1
Mo	50	Eu	50
Na	500	Tb	100
Ni	5000	Yb	5
		Lu	1

study by Busche (1989) in California made use of the creosote bush, burr bush, sagebrush, juniper, digger pine and scrub oak for exploration purposes. It is important to sample a reasonable amount of the vegetation but samples of 150 g were sufficient to achieve an Au detection limit of 0.2 µg kg^{-1}, lower than for sediment. The vegetation was dried (but not ashed to avoid loss of volatile constituents) and pelletised prior to irradiation. Anomalous values for Au were above 0.9 µg kg^{-1} for juniper leaves and 2.0 µg kg^{-1} in scrub oak leaves.

Air pollutants

Air particulates present a challenging analytical problem for those techniques requiring the sample in liquid form. There are a number of problems particularly related to losses of volatiles such as Se, Sb and Hg on dissolution of the sample. Particulates are usually collected over a time period on media such as cellulose paper, glass fibre, organic membranes or polystyrene. Because the concentrations of some of the elements to be measured are very low, the trace element composition of the filter can contribute to background effects.

The method of analysis for air-filter samples by INAA is very straightforward: the filters are irradiated with no prior sample preparation, avoiding all potential losses and contamination (provided blank determinations are made on filter media). Synthetic standards can be prepared by doping blank filters with standard solution and drying, but there are no suitable reference materials for multi-elemental analysis of filters and so method validation is not easy. This is a problem, however, that is common to all methods of analysis.

The method of Dams *et al.* (1970) remains the procedure used routinely for air pollution studies. The samples are irradiated with standards for 5 minutes in a thermal neutron flux of 2×10^{16} neutrons m^{-2} s^{-1} (to activate elements with half-lives of less than 1 hour) and counted after 3 minutes to determine Al, V, Ti, S and Ca; they are recounted after 15 min-

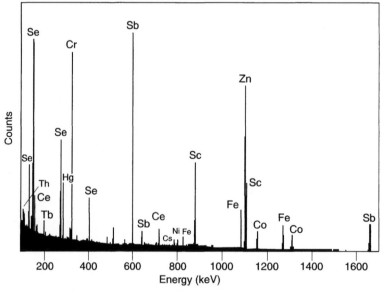

Figure 7.7   γ-ray spectrum of an air filter which was irradiated for 15 hours in a thermal neutron flux 10^{16} neutrons m^{-2} s^{-1}, allowed to decay for 30 days and counted for 3 hours (spectrum provided by R Benzing).

utes for Mg, Br, Cl, I, Mn, Na and In. The samples are reirradiated for 2–5 hours in a flux of 1.5×10^{17} neutrons m^{-2} s^{-1} to activate elements with longer half-lives and counted after 20 – 30 hours for K, Zn, Br, As, Ga, Sb, La, Sm, Eu, W and Au. They are counted again after a further decay of 20–30 days to determine Sc, Cr, Fe, Co, Ni, Zn, Se, Ag, Sb, Ce, Hg and Th without interference from short-lived radionuclides. Figure 7.7 is the spectrum for a typical air-filter sample, showing the γ-ray energy lines of Se, Hg, Cr, Sb, Sc, Fe, Zn and Co.

Another way of monitoring airborne pollutants is by collecting snow from rooftops and roads using plastic tools (Zikovsky and Badillo, 1987). When it has thawed the sample is filtered to collect the solid fraction, and the liquid portion dried to measure dissolved solids. The filter paper is irradiated for 1 min and 100 min and as many as 23 elements can be detected in the samples. Another interesting collection medium is pine needles. Wyttenbach *et al.* (1987) dissolved the wax off spruce and pine needles using toluene/tetrahydrofurane and collected the solid residue, weighing about 10 mg, for analysis by neutron activation, including epithermal neutron activation. A total of 33 elements can be measured by this method.

Mosses are also a valuable means of monitoring the atmospheric environment and have been used for many years to measure deposition of heavy elements. *Hylocomium splendens* was used in the national survey of trace element atmospheric deposition in Norway in 1990, when data for 26 elements were obtained for 514 locations scattered evenly over Norway. The samples were collected in polyethylene bags, dried and sub-sampled, and always

Table 7.3 Some typical results obtained by analysis of moss samples using INAA. (Source: Steinnes, 1980)

Element	Radionuclide used*	Concentration range observed (ppm)	Approximate detection limit (ppm)
Na	^{24}Na	80–550	1
K	^{42}K	1800–4000	100
Rb	^{86}Rb	2–23	0.5
Cs	^{134}Cs	0.04–1.1	0.01
La	^{140}La	0.1–1.5	0.02
Sm	^{153}Sm	0.01–0.27	0.01
Th	^{233}Pa	0.05–0.18	0.02
V	^{52}V	0.8–10.1	0.5
Cr	^{51}Cr	0.8–5.2	0.1
Mo	^{99m}Tc	0.1 0.4	0.1
U	^{239}Np	0.03–0.12	0.03
Mn	^{56}Mn	30–700	1
Fe	^{59}Fe	130–1700	20
Co	^{60}Co	0.1–1.2	0.01
Ag	^{110m}Ag	0.02–0.6	0.02
Hg	^{197}Hg	0.05–0.20	0.03
Al	^{28}Al	30–600	5
As	^{76}As	0.13–2.3	0.07
Sb	^{122}Sb	0.05–0.94	0.03
Se	^{75}Se	0.05–1.1	0.05
Cl	^{38}Cl	60–400	5
Br	^{82}Br	1.6–19	0.1

* 'm' denotes γ-emission by decay of a metastable excited state (isomer).

handled with disposable polyethylene gloves. NAA was used to determine the majority of the elements (Steinnes, 1980). Subsamples weighing 0.2–0.3 g were irradiated in a thermal neutron flux of 1.5×10^{17} neutrons $m^{-2} s^{-1}$ for a period of 30 s for Al, V, Mn, Cl and I and then reirradiated for 20 h for a wide range of other elements (see Table 7.3). Multi-variate analysis of the results, supplemented with values for Ni, Cu, Cd and Pb by flame atomic absorption spectrometry, identified several sources of pollution (Schaug et al., 1990). Zn, Cd, Sb, Pb, As, Se, Ag, Mo and V in southern Norway were due to local industrial activity and traffic, predominantly a zinc-refining factory in southwestern Norway; V and Cr were associated with local smelters in western Norway; and Ni and Co, particularly in northwesten Norway, were due to smelters in the former Soviet Union.

Future developments

There are no current developments in irradiation devices which will improve the capabilities of NAA but the manufacture of even larger crystals of pure germanium will provide higher-efficiency detectors, which will enhance the sensitivity for elements measured in a clean matrix. Most advances in recent years have been due to automation of procedures and the use of powerful software to improve the analysis of γ-ray spectra. There is a wide area of potential applications to environmental radioactivity, and expert systems software is now being applied to the interpretation of γ-spectrum analysis (Aarnio et al., 1992). In future it is to be expected that expert systems will also be developed specifically for γ-ray spectra resulting from neutron-activated samples.

Chapter 8

Thermal ionisation mass spectrometry (TIMS)

M. F. Thirlwall

What is mass spectrometry?

The objective of mass spectrometry is to separate a **sample** into its constituent parts, such as atoms or molecules, on the basis of their mass. This allows us to investigate the relative proportions of different molecular, elemental or isotopic species in a sample, whether graphically as a mass spectrum (Fig. 8.1), or numerically as ratios of different masses. The parts of a highly simplified mass spectrometer are illustrated in Box 8.1. All mass spectrometers possess three essential components:

A. ion source;
B. mass analyser;
C. ion collector.

However, the design of each component differs according to sample type and the analytical objectives: Chapters 8 to 12, 15 and 16 describe the various types of mass spectrometer used in the geosciences, which are distinguished primarily by the type of source they employ. Recent advances in instrument design have to some extent broken down these traditional source-based divisions; some of these developments are summarised below.

Objectives of thermal ionisation mass spectrometry (TIMS) in the geosciences

With the reservation that not all elements may usefully be ionised by this method, geological uses of TIMS analysis include:

- Measurement of the isotopic compositions of elements where natural variations occur (primarily as a result of radioactive decay). These are used to provide absolute age determinations of rocks and minerals, and to determine the origins of geological materials by reference to reservoirs of known isotopic composition.
- **Precise** measurement of element **concentrations** by **isotope dilution** (Box 8.3).

Both uses require measurement of **isotopic ratios**: that is, the relative abundances, in numbers of atoms, of two **isotopes** in a sample.

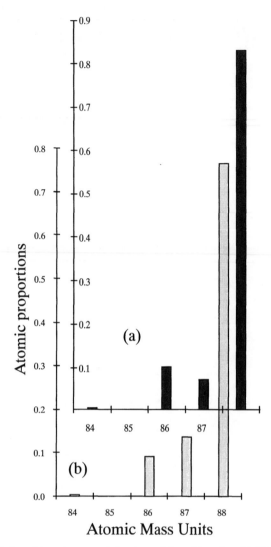

Figure 8.1   Mass spectra of Sr from (a) modern oceanic basalts, and from (b) a mica with Rb/Sr concentration ratio of 50, separated from a 400 million year old granite.

Instrumentation

The Thermal Ionisation Source

In a thermal ionisation source, the **analyte** is emitted from a heated metal filament of high **work function** W and low volatility (e.g. Re, Ta, W). This is welded onto steel posts set in a ceramic or glass insulator (Fig. 8.2), and, prior to use, is typically purged of volatile contaminants by heating at high temperatures (greater than analytical) under vacuum. A chemically purified salt of the analyte, or group of analytes, chemically separated from the sample, is evaporated onto the filament. The source is pumped to high

Box 8.1 Principle of a mass spectrometer

3. Collector slit

Located diametrically opposite to the source slit, on a straight line through the centre of curvature of the analyser tube, where the magnetic sector analyser focuses ions of the selected mass.

4. Collector assembly

This may consist of a single ion detector into which an ion beam may be focussed, or two switchable detectors, or an array of several detectors that may simultaneously collect several ion beams within a narrow range of m/z ratio ('multicollection'). Types of collector are decribed in Box 8.2.

Together with the mass analyser, the collector assembly must be able to resolve the ion beams of interest, and should have equal sensitivity to different masses (or be easily mass-calibrated).

6. Computer

On most modern mass spectrometers, the computer controls a range of instrument functions in addition to processing data.

5. Signal processing

The DC signal from the collector(s), after amplification, is integrated over an interval of up to 10s and recorded as the number of ions arriving at each collector during this period, or the mean ion current into each collector. This measurement is passed to the computer for isotope ratio calculations or other forms of analytical output.

Signal out

Electronics

2. Magnetic sector mass analyser

The analyser tube is curved into an arc of a circle, radius r, by being passed between the poles of a large electromagnet. The magnetic field deflects the ion beam into a curved path of radius r according to the equation:

$$\frac{m}{z} = \frac{B^2 r^2}{2U} \qquad [8.1.2]$$

where m/z is the ionic mass:charge ratio, B is the magnetic flux density, U the accelerating voltage through which ions pass. Ion beams of successively higher m/z may be deflected into the detection system by increasing B or decreasing U (stepwise or continuously) to generate a mass spectrum.

Not all mass spectrometers employ a magnetic sector. Other mass analysers in current use are:
- quadrupole mass filter (Box 10.1)
- time of flight analyser.

Paths of ions for which Equation 8.1.2 is satisfied

Magnet
(pole pieces above & below analyser tube)

Paths of heavier (less deflected) and lighter (more deflected) ions.

Analyser tube (high vacuum)

Source slit and accelerator slits

All parts of the mass spectrometer are maintained at high vacuum during analysis.

1. Ion source

Here ions in gas form are generated from the sample, and accelerated and collimated to a narrow ion beam. Can be:

1) Thermal ionization (TIMS) (Ch. 8)
2) Electron bombardment of gas (Ch. 9)
3) Plasma source e.g. ICP-MS (Ch. 10)
4) Spark source (SSMS) (Ch. 11)
5) Secondary ionization (SIMS) (Ch. 15)
6) Glow discharge (GDMS) (Ch 16)

Acceleration is by means of a high voltage ($U =$ several kV) applied between the source assembly and an accelerator slit. This imparts a kinetic energy to an ion of mass m and charge z, giving a velocity v:

$$K.E. = zU = \tfrac{1}{2}mv^2 \qquad [8.1.1]$$

Sample in
(gas, liquid or solid)

vacuum ($\approx 10^{-5}$ Pa, 10^{-7} mbar), achieved over 1–2 h using an ion pump or turbomolecular pump. The filament is heated to a temperature T that causes evaporation of n neutral atoms from the filament, which are accompanied by n^{+} positive ions according to the equation

$$n^{+}/n = e^{(W-E)/kT} \qquad\qquad [8.1]$$

where E is the first **ionisation energy** of the material evaporated, and k is Boltzmann's constant.

The ratio of ions produced to atoms loaded is known as the **ionisation efficiency** and this is a principal limitation on the sensitivity of TIMS. Volatile elements with ionisation energies less than the filament metal work function (e.g. alkalis) produce many ions at low evaporation temperatures ($\approx 700°$C), whereas elements with higher ionisation energies (e.g. U, rare earth elements) have low ionisation efficiencies which can only be improved by increasing the filament temperature. This causes rapid exhaustion of sample from the filament, which can only be reduced by using a double- or triple-filament assembly (Fig. 8.2). Here, the element is evaporated at around 1200°C from the side filament(s), and ionised by a separate ionisation filament at about 2000°C.

The resultant ions are accelerated towards the mass analyser by a high voltage applied between the source assembly and an accelerator slit (Box 8.1), and **collimated** into a tight ion beam by applying smaller potentials to a series of slits. For some elements that do not readily form positive ions, it is possible to mass-analyse negative compound ions (e.g. OsO_4^{-}, Martin, 1991) by reversing the acceleration voltage polarity and magnet current.

The relative abundances of isotopic masses of a given element in this ion beam are not in general identical to the abundances in the original sample owing to **mass fractionation**. Methods of correcting for this are discussed below.

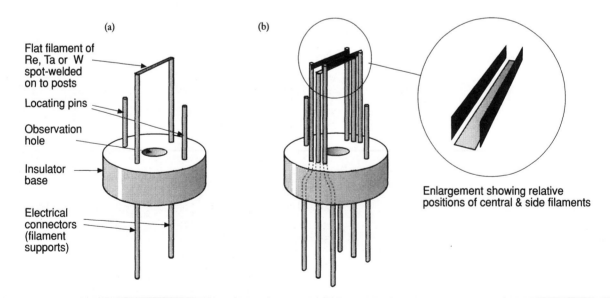

(a)   (b)

Flat filament of
Re, Ta or W
spot-welded
on to posts

Locating pins

Observation
hole

Insulator
base

Electrical
connectors
(filament
supports)

Enlargement showing relative
positions of central & side filaments

Figure 8.2   Single- and triple-filament assemblies in glass 'beads'. The integral bead design shown here, though now superseded in many instruments, shows the filament geometry most clearly.

The magnetic sector analyser in TIMS

Mass spectrometers used in the geosciences are based on the designs of Nier (1940). The TIMS source is coupled to an evacuated analyser tube that is curved into an arc of a circle between the poles of an electromagnet (see Box 8.1). In most instruments this produces a final ion path at 90° to that entering the electromagnet. The applied magnetic field will deflect the ion beam along this analyser tube if appropriate magnitudes of accelerating potential and magnetic flux density are chosen for the ionic mass of interest (see the equation in Box 8.1). The mass spectrometer must resolve ion beams of closely similar mass/charge ratio and focus these beams separately into the detector system. The ability of a mass spectrometer to resolve adjacent ion beams depends on the width of the ion beam, which is controlled by the width of the source slit, and the physical dispersion between adjacent masses, which is determined by the radius of curvature of the analyser tube, commonly 27 cm. High **mass resolution** can only be attained at the expense of sensitivity. Thermal ionisation mass spectrometers are usually set up to resolve clearly ion beams 1 **amu** apart at a mass double that of the heaviest naturally occurring isotope, because the thermal ionisation source usually produces singly charged monatomic ions with minimal mass overlap. The analyser tube must be pumped to high vacuum (10^{-6}–10^{-7} Pa, 10^{-8}–10^{-9} mbar, achieved by ion or turbomolecular pumps – Appendix A).

Owing to variability in the ionisation process, ions leaving the source show some variation in both energy and direction of travel. Efficient collection of the ions requires that as many as possible ions of a given m/z are focused by the electromagnet onto the collection slit; this is achieved by placing the source and collector slits symmetrically opposite at specific distances dependent on the radius of the analyser tube. For some applications, an electrostatic filter is required to perform energy focusing (e.g. Belshaw and O'Nions, 1990; see Box 11.1).

Ion beam detection

The principles and characteristics of the main ion beam detectors are summarised in Box 8.2. For isotope ratio analysis by TIMS, a major requirement of the collector–analyser configuration is that it should generate 'flat-topped peaks'. This means that, as a mass spectrum is scanned across the collector slit, the entire ion beam of a specific isotope should pass into the collector over a significant range in magnetic flux density. The ion current will then remain constant despite small fluctuations in accelerating voltage and magnetic field. Precise isotope ratio analysis requires use of a **Faraday cup**, because of its linear response; **electron multipliers** are primarily of use in TIMS work for isotope dilution analysis and relatively low-precision analysis of very small ion beams. Most TIMS machines have both types of detector, and many use the Daly multiplier system.

Automation

Most current TIMS instruments use a computer to control the major instrument functions: sample selection from a barrel with 6–20 samples, heating up the filament, peak searching, source focusing, setting the magnetic field, data collection and reduction. This allows analysis of all the samples in the barrel over several days without operator intervention.

Box 8.2 Ion current detectors

Ion detection systems must be designed so that the ionic masses of interest are resolved, and their ion currents amplified to an extent where useful precision is attained. The detector should have an equal sensitivity to different masses, and should respond rapidly to changes in beam current. Three principle types of detector are used in the geosciences, which detect different ionic properties.

Faraday Cup

A Faraday Cup

The Faraday cup detects ions as positive electric charges, and hence has essentially equal response to different ionic masses. It consists of a carbon-coated metal box with an inclined metal electrode that collects incoming ions. Ions enter the box through a slit: a narrow slit gives better mass resolution, but a wider slit gives greater sensitivity. Positive ions striking electrode cause an accumulation of positive charge, which is neutralised by an electric current flowing from earth through a large resistor (ca. $10^{11}\,\Omega$). According to Ohm's law, the potential difference across the resistor, measured by a high-impedance voltmeter (electrometer), is proportional to the ion current (the sensitivity is typically 10^{-11} A per volt).

This detector has (in principle) an equal response to all ions, is robust, and has low electrical noise. However, the ion signal takes a significant time to build up or decay after switching an ion beam in or out of the cup. Electronic noise in the resistor restricts the Faraday cup to measurement of ion beam currents greater than 10^{-15} A.

Electron Multiplier

A schematic
electron multiplier

The electron multiplier utilises the ability of ions and electrons to generate secondary electrons. Incoming ions are accelerated at 2-5 kV towards the first electrode of the multiplier, the 'conversion **dynode**', from which they liberate electrons. The resulting electron beam then bombards a sequence of further electrodes at successively higher potentials (up to 20), resulting in magnification of the electron beam by a factor of up to 10^7. The final electron beam is collected by an anode and measured using an electrometer.

Electron multipliers have very low electronic noise, and may be used to measure ion beams as low as 10^{-19} A, or 1 ion/second. Secondary electron emission is however dependent on the mass, energy, chemical form and charge of the ions, so the output from the detector is mass-dependent. Mass calibration and correction are therefore needed for accurate isotope ratio measurement.

The Daly multiplier system uses an aluminized cathode at negative potential (ca. 30 kV) instead of a conversion dynode to generate secondary electrons at a rate of ca. 6 per incident ion. These are deflected towards a scintillator (Box 6.3) which emits photons that are amplified by a **photomultiplier**. Relative to conventional electron multipliers, the Daly system provides better peak flat, greater detector efficiency and stable gain. Many TIMS machines use a Daly in tandem with a Faraday, using the cathode potential as the switch to deflect ions away from the Faraday.

It is common in modern mass spectrometers to combine several Faraday cups, Faraday/multiplier combinations or multiple **channeltrons** in *multicollector* systems capable of collecting several ion beams *simultaneously*. Multicollection offers three major advantages:

- the analysis time is reduced because more of the available ions are being collected.
- short-term instabilities in ionisation in the source may be almost entirely overcome.
- very large isotope ratios can be measured (up to 10^8) by Faraday and electron multiplier combinations.

Photographic plates

A photoplate exploits the ability of ions to blacken photographic emulsion, with the degree of blackening indicating ion beam intensity. This technique is particularly suited to SSMS, and is described in detail in Chapter 11.

MFT

Preparation for TIMS analysis

Suitable elements

TIMS may be used to measure the isotopic composition of any element that may be introduced into the source in a solid form and can be thermally ionised. There is obviously little point in attempting to ionise monoisotopic elements, except for isotope dilution analysis where a long-lived radioisotope is available as **spike** (Box 8.3). TIMS is in general unsuited to the determination of small, natural, mass-dependent isotopic variations in the light elements, unless they can be ionised as a high-mass polyatomic ion (e.g. $Cs_2BO_2^+$), as mass fractionation during ionisation swamps natural fractionation effects. TIMS has been most commonly employed in the geosciences for the measurement of isotopic variations of heavy elements produced by radioactive decay, compared with which fractionation effects are negligible, or may be easily corrected.

It is primarily the elements with **radiogenic** isotopes that have natural isotopic variations large enough for an analysis to be geologically useful (Sr, Nd, Pb, to a lesser extent Hf, Ce, Ba, Ca, and with recent modification to hardware, Os and Th). A far wider range of elements can be analysed for elemental abundance using isotope dilution.

Sample preparation

For almost all TIMS applications it is essential to separate the element of interest from a geological sample prior to isotope analysis. This is both to minimise or eliminate potential **isobaric interference** from other elements, and to reduce the proportion of other constituents of the sample to a level where they do not inhibit ionisation of the element of interest. Sample preparation usually involves chemical pretreatment, acid digestion, addition of a spike solution if isotope dilution is to be used (Box 8.3), and chemical separation of the element of interest.

Chemical pretreatment (or **leaching**) is commonly used where the analyst suspects that isotopic heterogeneity in a bulk sample may arise from processes other than those being investigated. For example, the Sr isotopic composition of submarine lavas may commonly be changed from that in the original molten rock by reaction of fine-grained parts of the solidified rock with Sr-rich sea water. Leaching in hot 6M HCl can preferentially dissolve the fine-grained material that has reacted with seawater (including the seawater-derived component of Sr), allowing analysis of Sr in unreacted parts of the rock.

Acid digestion and element separation methods for isotope analysis are similar to those outlined in Chapter 3. The following criteria are used to select the most appropriate techniques:

- Minimisation of contamination (**blank**). All sample preparation procedures unavoidably result in mixing between the analyte in the sample and the same element in reagents and the laboratory environment. This problem is particularly severe in a TIMS laboratory because contaminants will be of different isotopic composition to the analyte, and isotopic ratios can be analysed very precisely in very small amounts of analyte (1 ng– 1 µg). Techniques should thus be chosen that minimise reagent quantities. A clean laboratory environment (Box 3.2) is vital in minimising contamination. The level of con-

Box 8.3 Isotope dilution

Isotope dilution is potentially the most precise and accurate elemental analysis technique available, with 1% **RSD** on concentrations relatively easily attained. Unlike any other analytical technique, precision on trace element ratios can be substantially better than on concentrations, with a reproducibility of 0.1% commonly attained on Sm/Nd ratios.

Principles

Isotope dilution is an extension of **internal standard** methods of concentration determination, where the internal standard is the same element as that to be analysed, but artificially enriched in one (or more) of its isotopes. It is essential that the artificially enriched isotope (the **spike**) is fully homogenised with the analyte, but afterwards any form of chemical processing and purification may be undertaken provided that it does not induce mass fractionation.

A known mass m_s of the spike, which may be in solid form (for ID–SSMS, see Chapter 11), liquid form (for TIMS and ID–ICP–MS, see Chapter 10) or gaseous (for gas source mass spectrometry, see Chapter 9), is mixed with the sample to be analysed. For ID–SSMS and gas source work, the mixture is then introduced directly into the mass spectrometer, while for solution work a chemical process (see text) may be used to concentrate the elements of interest. The isotopic composition of the element(s) of interest in the mixture is then determined, and expressed by the ratio of two isotopes A and B: $R_{mix} = A/B$. This is simply related to the mass of element E present in the original sample m_E by the equation:

$$m_E = m_s A_E C_{B,S}(R_{mix} - R_{spike})/[X_{A,E}(1 - R_{mix}/R_{sample})]$$

where : A_E = relative atomic mass of element E in unspiked sample (g mol^{-1})

$C_{B,S}$ = concentration of isotope B in spike (mol g^{-1})

$X_{A,E}$ = atomic fraction of isotope A in element E in unspiked sample

R_{sample} = isotopic ratio of A/B in unspiked sample

R_{spike} = isotopic ratio of A/B in pure spike

This may be related to the **concentration** of element E in the sample by dividing by the mass of sample to which spike was added.

*The **precision** of ID measurements* is not merely a function of the precision of R_{mix}, but also of the differences between the mixture, sample and spike isotopic compositions. If one of these differences is small, uncertainties in isotope ratio measurement may be magnified by the equation to near-infinite levels. It is thus essential to know accurately both natural sample and spike isotopic compositions.

Spike isotopes are chosen from the catalogues of commercial suppliers so that:

- the difference between mixture, natural and spike isotope ratios is maximised – the spike should therefore be rich in an isotope having low natural abundance and be available in as pure a form as possible;

- the spike is relatively inexpensive;

tamination may be determined by performing the entire chemical procedure in the absence of sample, and measuring by ID (Box 8.3) the mass of element that is obtained. This blank contribution is typically 10 pg–4 ng Sr, 2–300 pg Nd or 10 pg–2 ng Pb, depending on procedure cleanliness. It is best to reduce this so it has insignificant effect on the measured isotope ratio, but it may sometimes be necessary to make a blank correction, using equations similar to that in Box 8.4.

- Purity of the separation of the analyte from other elements is essential to minimise isobaric interferences and maximise ionisation efficiency of the analyte. For isotope ratio analysis, it is usually unnecessary to recover all of the analyte from the chemical procedure: frequently a narrow collection interval will be used from an **ion exchange** column to obtain the purest separate.

- ID analysis requires that the whole sample is taken into solution. Small amounts of minerals resistant to dissolution (e.g. zircon) can host a large part of the sample budget for the analyte. If these are not fully dissolved, the spike solution (Box 8.3) cannot equilibrate with them and an inaccurate low concentration will be determined. Dissolution of

- the spike has no isobaric interferences in the mass spectrometric procedure being used, or ones that can be easily corrected for;

- the spike does not form part of an important natural radiogenic isotope ratio such as $^{87}Sr/^{86}Sr$, $^{143}Nd/^{144}Nd$, etc., so as to avoid the possibility of spike contamination of natural isotopic ratios.

Choice of isotope is often a compromise between these requirements. For example, for ID analysis of Ce, the lowest-abundance mass is 136 (0.19 atom %). Use of this would fulfil the first objective, but it is only commercially available at present at 50% enrichment, and is about 40 × more expensive than ^{142}Ce (11.1 atom %). This, however, has a large potential interference from ^{142}Nd, which is not easily chemically separated from Ce, unlike the Ba interference on mass 136. In ID–ICP–MS, Ba and Ce are ionised with roughly equal efficiency, and hence ^{136}Ce is probably the better choice of spike. In ID–TIMS, Ba is ionised much more efficiently than Ce, and often very irregularly, while it is possible to make a highly accurate interference correction for ^{142}Nd (Thirlwall, 1982), so that ^{142}Ce is a better choice of spike. Further, ^{138}Ce and ^{136}Ce form the radiogenic ratio $^{138}Ce/^{136}Ce$, which is used in La–Ce isotope studies, and hence are not good spike choices.

For monoisotopic[1] elements, isotope dilution is often still possible using artificial radioisotope spikes, providing that appropriate radiochemical precautions are taken. These may also be a better choice for spikes than naturally occurring isotopes (e.g. for Pb), where all natural isotopes are used to configure radiogenic isotope ratios, but both radioactive ^{202}Pb and ^{205}Pb may be utilised.

Spike calibration is the main source of systematic error in ID procedures, involving the determination of the $C_{B,S}$ parameter in the ID equation. Spikes may be supplied as metals, their oxides or salts or as gas ampoules, and must be diluted to permit accurate weighing of the quantities required for trace analysis. For both ID–TIMS and ID–ICP–MS the spike is added as a few drops of solution (weighed), while for ID–SSMS the spike is usually added as a dilute mixture with graphite. $C_{B,S}$ is then the concentration of the spike isotope in the solution or in the graphite. This can be estimated by careful weighing of the supplied spike compound used in preparing the solution, but it is preferable to analyse a standard sample to determine $C_{B,S}$, prepared with great care from stoichiometric compounds of the analyte. Regular recalibration of spike solutions is necessary to monitor evaporative concentration change.

Concentration ratio determination is greatly enhanced by using spike and standard solutions containing both elements of interest. Evaporative concentration change may occur in either solution, but the ratio between the two elements cannot change. This is especially advantageous in **geochronology**, where error on the parent/daughter concentration ratio is important. An Sm/Nd standard solution has been circulated by Wasserburg *et al.* (1981), which allows Sm/Nd ratios measured worldwide to be consistent to about 0.1%. This can be extended to any number of elements in principle.

---

[1] Here meaning elements with only one *naturally occurring* isotope.

MFT

Box 8.4 Equations for correction of blank

- Correction of trace element concentration (in ppm) determined by isotope dilution:

  True ppm × sample wt(g) = measured ppm × sample wt(g) − blank contribution (µg)

- Correction of isotopic ratios requires measurement of the isotopic composition of the blank, and the correction equation is complex, as sample and blank Sr, Pb etc. have different atomic weights. If R represents the $^{87}Sr/^{86}Sr$ ratio and Sr the number of nanograms of Sr contributed to the measured (meas) ratio by sample (s) and blank, then

$$R_{true} = \frac{826.881 Sr_{blank}(R_{meas} - R_{blank}) + (826.88 + 86.909 \times R_{blank})R_{meas}Sr_s}{(826.881 + 86.909 \times R_{blank})Sr_s - 86.909 Sr_{blank}(R_{meas} - R_{blank})}$$

MFT

such minerals can be achieved in 'bombs' (Fig. 3.1b), bayonet-fit pressure capsules encased in steel jackets that can reach internal temperatures up to 190°C.

Sample loading involves evaporation of a solution of the purified analyte onto one or more metal filaments (Fig. 8.2). The chemical form of the analyte, any additional reagents added, the nature of the filament metal, the configuration of the filaments and the temperature used for evaporation all rather unpredictably influence the nature of the ionic species formed, the ionisation efficiency and the ion beam stability (short-term fluctuations in ion beam **intensity** at constant temperature). For example, 200 ng of $PbCl_2$, loaded with silica gel and H_3PO_4 onto a single Re filament, will give a high-intensity low-fractionation Pb^+ ion beam (e.g. Cameron *et al.*, 1969), while the same technique used with Nd and Ce nitrates yields respectively high- and low-intensity NdO^+ and CeO^+ ion beams (Thirlwall, 1991a).

Isotope ratio measurement

The best methodology for isotope ratio analysis depends on the end use of the ratio. Lower precision (0.1–1.0% **RSD**) is usually acceptable for ID measurements, but for applications such as **Sr isotope stratigraphy** and Sm–Nd **geochronology**, the highest precision and accuracy are desired for the isotope ratio, and possible sources of error should be thoroughly investigated.

Three principal measurement modes are used: single collector, static multi-collector and multi-dynamic. Mass spectrometers fitted with multi-collectors (Box 8.2, Fig. 9.3) will use the latter two modes almost exclusively, but consideration of single-collector methodology is essential, since a substantial amount of data in the literature is still generated by this technique.

Production of an ion beam

Following insertion of the sample into the source, the filament current is raised to a level where ionisation is expected based on previous experience. Either the magnetic field is set to the value expected for the most abundant isotope present and the collimator focus voltages adjusted, or the focus voltages are set to expected values and the magnetic field is scanned across the expected field position. This initial process of peak searching may be carried out using an electron multiplier if there is relatively little analyte available. Once a peak has been located it must be **centred**, that is the exact magnetic flux density determined for the ion beam to pass symmetrically into the collector.

The intensity of the ion beam is then maximised by adjusting voltages on the collimator plates, which are reoptimized at intervals during the analysis. Because the **random error** on an isotope ratio measurement is a function of the number of ions collected (cf. Box 6.4), the ion beam intensity should be increased (by raising filament current) to a level where acceptable precision is attained, but at which the sample loaded on the filament is not exhausted too rapidly. It is not easy to predict what ion beam intensity can be achieved for a particular sample, and maintained long enough for an acceptable analysis. Even when the filament current is held constant, ion beam intensity will change with time, either as a slow systematic increase (stable 'growth', e.g. Fig. 8.3) or decrease, or as erratic variability over short time intervals (**instability**, e.g. Fig. 8.3).

Acquisition of mass spectra, single-collector mode

Isotope ratio measurements involve switching the magnetic field so that the different isotopic masses fall in turn into the single collector (Fig. 8.3). Where possible, at least three mass peaks of the analyte are cyclically monitored, a unique mass of any element that has **isobaric interference** on the analyte, and a between-peak position to estimate the intensity of the **baseline**. The ion beam intensity for each mass in the cycle is measured (integrated) during the **integration time** (typically 5 s), and before each measurement a **delay time** (typically 2 s) is waited to allow the magnet current to stabilise and the previous ion signal to decay in the resistor. Since precision of isotope ratios improves as the square root of the number of ratios measured, this cycle is repeated a large number of times. If higher integration times are used, fewer cycles are needed for a given precision. The final isotope ratio result is the **mean** of data collected at close to optimum ion beam intensity. Typical analytical cycles for Sr, Nd and Pb are given in Table 8.1, together with typical voltmeter readings (in V) for a pair of scans across the Sr spectrum.

Corrections to measured peak intensities

Calculation of isotope ratios from data collected in mass cycles such as those of Table 8.1 requires correction of the data for **baseline**, **isobaric interference**, change in ion beam intensity with time and **mass fractionation**.

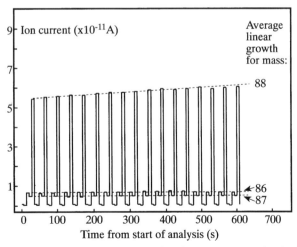

Figure 8.3 Chart record of several peak jump scans across part of the Sr mass spectrum, showing ion beam growth and minor instability (e.g. 88 peaks at 520 and 560 s are substantially less intense than expected for linear growth of the ion beam – the dashed line). Interpolation to constant time can be carried out using the gradient of average growth.

Table 8.1   (a) Typical mass cycles for isotope ratio measurement; (b) typical peak intensities (in V) for an Sr isotope determination

	Element	1	2	3	4	5
(a)	Pb	203.5*	204	206	207	208
	Nd	143	144	146	147†(Sm)	147.5*
	Sr	85†(Rb)	86	87	88	91.5*
(b)	Sr voltage scan 1	0.00005	0.50870	0.36066	4.24667	0.00003
	Sr voltage scan 2	0.00003	0.51051	0.36183	4.26193	0.00002

*Baseline measurements.
†Monitor of interference: interfering species in parentheses.

Baseline consists of two components:

(1)   scattered ions of the analyte, the intensity of which increases with the total ion beam intensity;
(2)   electronic noise in the detector which would be present with no ion beam.

The former is caused by inelastic collisions between ions and residual gas molecules in the analyser or flight tube, and results in a tail to each peak (Fig. 8.4). The effect is worst next to the largest ion beams (Box 8.5); for example, ^{88}Sr commonly has a measurable tail at mass 87.5, and can contribute to mass 87 if the analyser vacuum is sufficiently poor. The electronic noise is an inherent function of the detector system, and amounts to about 4×10^{-14} A on a Faraday cup, or about 1 ion per second on an electron multiplier. It determines the minimum ion beam intensity that can be measured on a specific collector. The magnitude of Faraday noise may vary substantially with resistor temperature.

Isobaric interference during isotope ratio analysis is largely determined by the purity of the chemical separation procedure, although in cases where the isobar has a much higher ionisation efficiency than the element of interest (e.g. ^{87}Rb on ^{87}Sr, ^{138}Ba on ^{138}Ce), very small amounts of the interfering element in the loading solutions or in the mass spectrometer source can cause problems. When an element with relatively low ionisation energy is an interferent, this may be wholly burnt off prior to analysis. The correction for interference

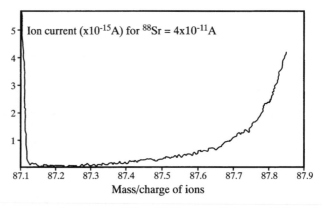

Figure 8.4   Scan of Sr mass spectrum between mass 87.1 and 87.9 using an electron multiplier to show tailing from a 4×10^{-11} A ^{88}Sr peak.

Box 8.5 Abundance sensitivity

How much the 'tail' of a large mass peak interferes with its neighbours is quantified by 'abundance sensitivity', usually reported as the proportion of mass ^{238}U detectable 1 amu away. This is about 2–5 ppm of ^{238}U on most modern machines, but can be improved to 20 ppb by fitting an **electrostatic analyser (ESA)** or a retarding lens between magnet and collector, and values as low as 7 ppb have been reported on the VG Isolab120 (Belshaw and O'Nions, 1990), which incorporates two ESAs.

The abundance sensitivity of a mass spectrometer limits how large an isotope ratio it can measure. Sr (^{88}Sr/^{87}Sr ratio = 10) is usually unaffected on a conventional machine, Th (^{232}Th/^{230}Th ratio = 200,000) is possible with an ESA or retarding lens, and Be (^{9}Be/^{10}Be ratio = 10^{10}–10^{11}) is only possible on the Isolab120 (or by AMS – Chapter 12). An approximate correction for tail can be made by measuring the baseline on both sides of an analysed mass and assuming that the tail is linear. This is easy in static multi-collector mode, but impractical in single-collector analysis as far too much time must be spent integrating baseline.

Maintenance of a good analyser vacuum makes the tail negligible for conventional Sr–Nd–Pb analysis.

MFT

---

is usually made by monitoring an interference-free isotope of the interfering element, for example ^{85}Rb for ^{87}Sr, and ^{147}Sm for ^{144}Nd. The correction equation is of the form

$$^{87}\text{Sr}_{\text{corr}} = {}^{87}\text{Sr}_{\text{meas}} - ({}^{87}\text{Rb}/{}^{85}\text{Rb}) \times {}^{85}\text{Rb}_{\text{meas}} \qquad [8.2]$$

where (^{87}Rb/^{85}Rb) is the known isotopic ratio of Rb in the sample. In some cases, for example when REEs are analysed by ID in one triple-filament loading, it is impossible to find an interference-free monitor, and either a complex multiple correction must be made, or the data scrutinised for internal consistency (Thirlwall, 1982).

Time interpolation is essential in a single-collector analysis, as each isotopic mass is measured at a different time, and simple ratios of successive peak measurements would be inaccurate because of changing beam intensity (Fig. 8.3). Linear change in the ion beam with time is usually assumed. So, for example, the intensity of ^{87}Sr increases from scan 1 to scan 2 in Table 8.1. If each scan takes 5 s, the ^{87}Sr intensity whilst ^{86}Sr was measured in scan 2 is predicted to be $(4/5)(87_2 - 87_1) + 87_1$, and the ^{87}Sr/^{86}Sr ratio is calculated to be 0.708288. The major problem with time interpolation is the common short-term **instability** of ion beams, which can result in severe imprecision in the mean measured isotope ratio. This is the limiting factor on the precision of single-collector measurements, although careful sample loading and heating can reduce instability to manageable proportions. Multi-collector analysis, where the masses are measured simultaneously, almost completely overcomes instability problems.

Mass fractionation is frequently observed in systematic changes in time-interpolated isotope ratios during the 1–2 hour duration of the analysis. This is because of preferential thermal ionisation of lighter masses from the filament. The residual material is depleted in the lighter isotopes, and consequently the measured ratio usually becomes progressively heavier through time. Sometimes reverse fractionation can take place, owing to ion emission from multiple points on a filament, and sometimes ratios stay constant, at values substantially different from the true value. The magnitude of fractionation is greater for more volatile elements (e.g. the alkalis) and for **lighter elements**, where the relative mass difference is greater. It also depends on the loading method, with triple-filament loads showing generally less fractionation.

Thus it is impossible to measure the absolute value of an isotope ratio by TIMS. In practice, this does not matter for most elements which have three or more isotopes. If variation in ratio is controlled by mass fractionation, a graph of any two isotope ratios for a single analyte (e.g. Fig. 8.6) will show near-linear correlations, with the gradient controlled by the mass differences between denominator and numerator isotopes. Adoption of a standard value for one ratio necessarily fixes the other ratios in that sample. The appropriate value to adopt may be estimated by integrating the entire ion beam from a sample (integrating until the sample on the filament is exhausted), or by mixing known masses of pure isotopes and comparing the calculated ratio with that measured under highly standardised conditions. It is more usually taken to be a value near the middle of the range observed during normal TIMS measurements. The radiogenic isotope ratio that is of interest in the analyte may then be simply corrected for mass fractionation by taking its value at the intersection of the correlation with the adopted value (e.g. Fig. 8.6), or by calculation (Box 8.6). This correction process is called internal **normalisation**. For Sr, the normalising value $^{86}Sr/^{88}Sr = 0.1194$ has been accepted for a long time (Steiger and Jäger, 1979), but for Nd, some workers assume $^{146}Nd/^{144}Nd = 0.7219$, while others assume $^{150}Nd/^{142}Nd = 0.2096$; $^{143}Nd/^{144}Nd$ ratios determined on the latter assumption are 0.15% lower. This discrepancy covers much of the observed range in natural $^{143}Nd/^{144}Nd$ ratios; and thus these data should always be interpreted in the context of the normalising value used.

This type of normalisation cannot be used when only two isotopes are present (e.g. Rb[2] ID), or where there is no non-radiogenic isotope pair (e.g. Pb[3]). Instead, a loading method can be used that minimises mass fractionation, and an isotopic **standard** of known or accepted composition (e.g. the SRM981, 982, 983 series of Pb standards, Fig. 8.6) can be run with the unknowns to measure the average mass fractionation. This is assumed to be equal for samples and standards. Alternatively, an artificial radioisotope spike can be used.

Random errors

The detected ion intensity is subject to random uncertainty in the same way as X-rays or γ-rays (Box 6.4), and the **standard deviation** s expected from multiple determinations of the same isotope ratio can be easily calculated from the square roots of ion beam intensities. s can be reduced by a factor $1/F$ by increasing the intensity or integration times by a factor F^2. However, reported isotope ratio measurements are usually means of a large number n of individual ratio determinations, for which s stays constant, but we become more confident of the mean the more ratios we measure. This extra confidence is expressed by the **standard error on the mean** se, equal to $s/\sqrt{n}$ (Box 2.1); $2 \times se$ is usually reported as the analytical error (or 'internal precision') on TIMS isotope ratio measurements in the literature. Where a lot of sample is loaded on the filament, high ion beam intensities allow n to be small; where little sample is available, a large number of ratio determinations at low intensity is required to attain comparable precision on the ratio. The $2se$ error can be simply predicted from knowledge of the ion beam intensity, the Faraday noise and the number of ratios. If the observed $2se$ on n ratios is substantially greater than this predicted error, problems with interference correction or ion beam stability can be suspected.

---

[2] Rb has only two naturally occurring isotopes, mass numbers 85 and 87.
[3] Pb has four naturally occurring isotopes, mass numbers 204, 206, 207 and 208. Only 204 is non-radiogeneic.

Box 8.6 Correction for mass fractionation

If the intensities of three or more isotopic masses are measured, one of which is radiogenic, the radiogenic isotope ratio can be algebraically **normalised** to an assumed value for the non-radiogenic ratio (e.g. $^{86}Sr/^{88}Sr = 0.1194$) if we know how the extent of mass fractionation varies with mass difference. Wasserburg *et al.* (1981) gave equations for linear, power and exponential law relationships, and showed that, for the extent of mass fractionation normally observed in TIMS, the difference between linear and power law was negligible, while slight exponential behaviour could be observed in Nd at the extremes of mass fractionation. Thirlwall (1991b) showed that, with the higher precision attainable in multi-collector analyses, an exponential law correction described Sr ionisation much better than the power law. However, if the measured normalising ratio is close to the assumed value, no distinction between the laws is possible. The relevant equations are

$$(A/B)_{l.l.} = (A/B)_{meas} \times \{1 + [(M_A - M_B)/(M_C - M_D)] \times [(C/D)_{l.l.}/(C/D)_{meas} - 1]\}$$
$$\text{(linear law)}$$

$$\left[\frac{(A/B)_{meas}}{(A/B)_{e.l.}}\right]^{\ln(M_C/M_D)} = \left[\frac{(C/D)_{meas}}{(C/D)_{e.l.}}\right]^{\ln(M_A/M_B)} \qquad \text{(exponential law)}$$

where l.l. and e.l. signify the linear-law- and exponential-law-normalised ratios, meas the measured ratio of A/B and C/D, and M_i is the isotopic mass of isotope i.

   This technique of normalisation for mass fractionation can also be used for isotope dilution measurements, provided that three or more isotopes are available, one of which is used for the spike (Box 8.3) and the other two are used for normalisation. If the spike contains the other isotopes as impurities (almost inevitable), the assumed normalising ratio must be corrected for these. This is done by solving two simultaneous equations, one being the normalisation equation and the other being a mixing equation between spike and natural isotopic ratios, which is linear if a common denominator isotope is used.

MFT

Multi-collector ratio measurement

A multiple collector instrument allows simultaneous measurement of the intensities of several ion beams over an array of two to nine collectors. These are usually Faraday cups placed each side of the central (axial) Faraday cup/electron multiplier combination used in a single-collector instrument. Since ion beam dispersion is a function of relative mass, the spatial separation of ion beams 1 amu apart decreases strongly with isotopic mass from 5.4 cm for ^{10}B to 2.3 mm for ^{238}U with a 54 cm radius analyser. This means that either the Faraday cups must be preset to fixed positions suitable for a particular element (fixed multi-collector), or they must be adjustable by hand or by motor external to the vacuum (variable multi-collector). In the latter case the collector positions are adjusted so that the ion beams in each collector are **centred** at a given magnetic flux density. Thus, to analyse all Sr isotopes on a five-collector machine, the collectors would be set so that, when the magnetic flux density gave the centre of the ^{86}Sr beam in the axial collector, ^{84}Sr would be centred on the lowest-mass collector, ^{85}Rb in the next lowest, and ^{87}Sr and ^{88}Sr centred on the high-mass collectors.

Static multi-collection

This is the simplest form of multi-collector analysis and involves measuring baselines at half-mass offsets either side of the aligned magnetic field position, and then measuring the peaks themselves. The magnetic field stays constant (static) during peak measurement. Commonly, the same time is spent integrating baseline as peaks, as there is often a low-intensity ion beam being measured for which high-precision baseline measurement is necessary (e.g. ^{85}Rb during Sr analysis). The baseline correction is simply performed by subtracting the mean of the two baselines on opposite sides of the peak. Compared with single-collector measurements:

- one-fifth of the time is spent on peak analysis for Sr, and no time delay is needed for the magnet current to settle;
- baseline drift is minimised because peak analysis takes so little time;
- a linear correction for tailing is made using the half-mass baselines;
- the monitor isotope for interference corrections is measured simultaneously with the interfered isotope so any pulsed ionisation (Fig. 8.5) is corrected for;
- time interpolation is unnecessary, as isotope ratios are calculated directly from ratios of the baseline-corrected integrated counts;
- small ion beams can be measured on an electron multiplier at the same time as large ion beams are measured on a Faraday (e.g. ^{232}Th/^{230}Th = 200,000 can be precisely measured).

Provided that interference is minor, the observed precision of static multi-collector isotope ratios is thus close to the theoretical precision estimated from beam currents, in about a quarter of the time required for single-collector analysis.

Despite these major advantages, ensuring accuracy and reproducibility of static multi-collector measurements is a lot harder. This is because each ion beam intensity is deter-

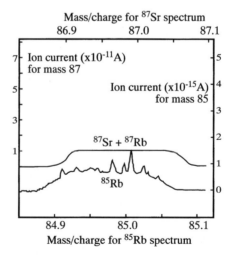

Figure 8.5   Scan of Sr mass spectrum between mass 84.85 and 85.12 using an electron multiplier to show pulsed ionisation of ^{85}Rb, measured simultaneously with mass 87 on a Faraday. Note that no pulsed ionisation is visible on mass 87: this is because ^{87}Rb forms an almost negligible proportion of the mass 87 ion beam. This type of pulsed ionisation can give rise to substantial error in single collector measurements, but is corrected accurately during static multicollector analysis.

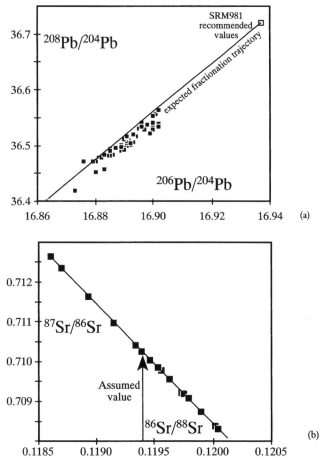

Figure 8.6   (a) Mean measured isotopic ratios for Pb isotope standard SRM981 over a year. Each point represents a complete Pb analysis. The data show fairly constant average mass fractionation of +0.14%/amu, a correction applied to analyses of unknowns. The data lie slightly below the expected fractionation trajectory, possibly because the recommended values are incorrect. (b) Mean data, uncorrected for mass fractionation, for blocks of 15–25 ratios measured on three separate runs of Sr isotope standard SRM987. The tight correlation is due to mass fractionation; correction of these data by power law gives a mean $^{87}Sr/^{86}Sr$ of 0.710239 ± 34 (2sd), and 0.710241 ± 24 (2sd) by exponential law (Box 8.5).

mined on a different Faraday cup, with associated differences in amplifier gain, and each ion beam follows a different **ion optical** trajectory, so that there are **transmission** differences between each collector. The amplifier gain problem is usually overcome by regular measurement of an electronic constant current in each of the amplifiers. Transmission differences can be minimised by careful design of the magnetic analyser so that all ion beams are focused onto the collector plane, and can be checked by comparing the intensity of a stable ion beam in each of the Faradays. Experience with the multi-collector at Royal Holloway suggests that transmission differences may exist from sample to sample, and that systematic changes in ion beam transmission occur with use of the Faraday buckets over a 6 month period (Thirlwall, 1991b). Similar changes have been reported on other multi-collectors (Makishima and Nakamura, 1991). This may result in a gradual +0.04% change in static $^{87}Sr/^{86}Sr$ data. Even if this systematic error is corrected, inter-sample differences in transmission mean that the reproducibility of static Sr and Nd isotope ratios may be no better than using a single collector (about $\pm 0.008\%$ $(2s)$). Nevertheless, static multi-collector analysis offers substantial time advantages over single collection. It is used in most laboratories for isotope dilution and Pb isotope ratio analyses, and some use it for Sr and Nd, either because their scientific application does not need the best reproducibility, or because their instruments are less affected by transmission problems.

Multi-dynamic analysis

This yields much improved reproducibility. It involves calibrating the relative gains and transmissions of the collectors by successively switching the same ion beam from the sample being analysed into all of the collectors. The term 'dynamic' means that the magnetic field is switched during the analysis, in contrast to static analysis. As with single-collector measurement, baselines can be measured before or during the main mass cycle. Interference corrections can be made using simultaneous measurement of interference monitors.

Examples of multi-dynamic routines for Sr, Nd and NdO analysis on a five-collector machine have been given by Thirlwall (1991a, b). Using Sr as an example, where the collectors are set at 1 amu separation, mass 87 can be used to calibrate two adjacent collectors (A and B) that measure 86 and 87 ion beams at one field setting, and 87 and 88 at a second field setting. From these four measurements:

$$^{87}Sr/^{86}Sr = \frac{87_B \times 87_A}{86_A \times 88_B \times 0.1194} \qquad [8.3]$$

Transmission and gain differences between collectors, and temporal changes in ion beam intensity, are cancelled out by the division, and mass fractionation is also corrected. The peak jumping sequence used by Thirlwall (1991b) takes as long as a single-collector measurement, but twice as many isotope ratios are collected. Multi-dynamic measurements are also more precise than single-collector ratio determinations (for the same reasons as in static multi-collection), and more accurate than static multi-collection (because of the near elimination of transmission problems). Multi-dynamic analysis is capable of long-term reproducibility of approximately $\pm 0.003\%$ $(2s)$ for Sr and $\pm 0.002\%$ $(2s)$ for Nd (Thirlwall, 1991b), although prolonged use of collectors can result in deterioration of this reproducibility, sometimes over time periods of as little as 6 months. The only solution to this is to replace the Faraday cups, and cups with much greater expected lifetimes have recently been developed (Palacz et al., 1995).

Recent equipment developments

The most important recent developments involve the breakdown of the traditional source-based divisions of mass spectrometry. This has been made possible by the development of multi-collection, which allows isotope ratio measurements on strongly fluctuating ion beams. For example, the ISOLAB 120 (e.g. Belshaw and O'Nions, 1990) employs a source that is capable of thermal ionisation, secondary ionisation involving **sputtering** of the sample using an Ar^+ or O_2^+ primary ion beam (Chapter 15), or photoionisation of neutral molecules by laser. The **SIMS** capability allows ionisation and measurement of $^{10}Be/^9Be$ (Belshaw *et al.*, 1995) and $^{18}O/^{16}O$ (Saxton *et al.*, 1995) amongst others.

Another development is that an ICP ion source coupled to a magnetic sector analyser and multi-collector is now commercially available. Compared with TIMS, this provides much better sensitivity for elements with high thermal ionisation energies (e.g. Lee and Halliday, 1995); the ion beams do not undergo progressive mass fractionation, so that a simple correction may be made for **mass bias** (Chapter 10) using for example Tl to correct Pb mass bias (Walder *et al.*, 1993). *In situ*, spatially resolved analysis is available using a laser to ablate sample into the **plasma** (e.g. Christensen *et al.*, 1995, Thirlwall and Walder, 1995).

Applications

TIMS has a very wide range of applications in all branches of the geosciences. This is because of its ability to provide precise numeric ages for geological events, and because isotope ratios can be used as tracers for interactions between reservoirs of material with different isotopic composition. These reservoirs may range in scale from planetary bodies down to small shell fragments. The use of TIMS for age determinations is illustrated in Fig. 8.1 which shows a mass spectrum for a 400 Ma old mica. This has elevated intensities at mass 87 because mica contains a high concentration of ^{87}Rb, and relatively low Sr, so that over time radioactive decay of ^{87}Rb leads to unusually high abundance of ^{87}Sr relative to the other Sr isotopes. The age may be calculated from the measured $^{87}Rb/^{86}Sr$ and $^{87}Sr/^{86}Sr$ ratios of the mica if we can estimate the $^{87}Sr/^{86}Sr$ ratio that the mica had when it crystallised. This is best done by analysing a low-Rb mineral from the same rock as the mica, in which change in $^{87}Sr/^{86}Sr$ over 400 Ma may be negligible. The biggest source of error on the age is usually in the $^{87}Rb/^{86}Sr$ measurement. Rb and Sr concentrations are measured using a mixed spike (Box 8.3) by ID, and converted to atomic proportions of ^{87}Rb and ^{86}Sr, but because Rb has only two isotopes, it is impossible to correct for mass fractionation of this very volatile element, and at best an **RSD** of 0.3% is given.

To illustrate interactions between reservoirs, consider the case of a permeable rock unit near the coast. Water contained in the pore spaces may be a mixture of sea water, which has a constant $^{87}Sr/^{86}Sr$ of 0.70918 and high Sr content (about 7 ppm), and potable groundwater, which has low Sr content (<0.5 ppm) and may have much higher $^{87}Sr/^{86}Sr$ through dissolution of Sr from high-$^{87}Sr/^{86}Sr$ sediments. Even very minor contamination of groundwater by sea water will result in easily measurable changes in $^{87}Sr/^{86}Sr$, because of the much greater Sr concentration of sea water. Thus isotopic methods can be used in parallel with chemical methods for assessing groundwater provenance. More detailed discussion of the wide range of applications may be found in many textbooks, such as Faure (1987) and Dickin (1995).

Chapter 9

Gas source mass spectrometry: isotopic composition of lighter elements

David P. Mattey

Introduction

The elements hydrogen, carbon, nitrogen, oxygen and sulphur possess two or more naturally occurring isotopes that are not involved in any form of radioactive decay process (Table 9.1). These elements are key constituents of rock-forming minerals as well as the hydrosphere, biosphere and atmosphere and their stable isotope ratios are powerful tracers of geological processes, especially those involving fluids. For light elements, the relative mass differences among their isotopes are large enough so that the physical and chemical properties of molecules containing one or other isotope are slightly different. For example, molecules of water containing the heavier isotopes of hydrogen, ^2H (deuterium), or oxygen, ^{18}O, have a slightly lower vapour pressure than water composed of the common light isotopes ^1H and ^{16}O. During a physical process such as evaporation, water molecules substituted with the lighter isotope pass into the vapour phase more rapidly than water molecules substituted by heavier isotopes, such that rainwater eventually becomes slightly

Table 9.1 Natural abundances of stable isotopes and reference standards used in delta notation. Reference standards: V-SMOW – standard mean ocean water prepared by the International Atomic Energy Agency Vienna ($^2H/^1H = 0.00015595$; $^{18}O/^{16}O = 0.0020052$); PDB – belemnite from Pee Dee formation ($^{13}C/^{12}C = 0.00112372$); Air – atmospheric nitrogen ($^{15}N/^{14}N = 0.0036765$); CDT – Canyon Diablo Troilite ($^{34}S/^{32}S = 0.0450451$). Oxygen isotopes can be reported relative to PDB (carbonates) or V-SMOW (water and silicates)

Element	Analysed as	Isotope	Natural abundance (%)	Reference standard	Terrestrial range (%)
H	H_2	^1H	99.984	V-SMOW	$\delta D = -450$ to $+50$
		^2H	0.01557		
C	CO_2	^{12}C	98.888	PDB	$\delta^{13}C = -120$ to $+15$
		^{13}C	1.1112		
N	N_2	^{14}N	99.634	Air	$\delta^{15}N = -20$ to $+30$
		^{15}N	0.366		
O	CO_2 (O_2)	^{16}O	99.759	V-SMOW or PDB	$\delta^{18}O = -50$ to $+40$
		^{17}O	0.037		
		^{18}O	0.204		
S	SO_2 (SF_6)	^{32}S	95.0	CDT	$\delta^{34}S = -65$ to $+90$
		^{33}S	0.76		
		^{34}S	4.22		
		^{36}S	0.014		

more enriched in ^1H and ^{16}O than sea water. Systematic isotopic variations in $^2H/^1H$ and $^{18}O/^{16}O$ ratios develop within the water cycle and different forms of natural waters possess characteristic stable isotopic signatures. Similarly distinctive isotopic fractionations are associated with the geochemical cycles of C, N and S and stable isotopes provide unique information on the source, temperature and mass balance of chemical fluxes between the surface and interior of the Earth.

Gas source mass spectrometry

The methodology of stable isotope analysis involves two distinct stages: chemical processing of the sample to convert the element of interest into a gas such as H_2, CO_2, N_2 or SO_2, followed by isotope ratio measurement by gas source mass spectrometry. Gas source mass spectrometers were first designed and operated in the late 1940s and although refinements to the electronics, the inlet and ion source have led to improvements in analytical precision and sensitivity, the basic design of gas source mass spectrometers has changed little.

The gas source mass spectrometer

The gas source mass spectrometer comprises an electron impact ion source, a flight tube with a magnetic sector mass analyser, and an ion collector assembly (Fig. 9.1). The body of the mass spectrometer is maintained under a high vacuum to avoid scattering of the ion beam by collisions with residual gas molecules. An operating vacuum of at least 10^{-5} Pa is normally required which is attained by pumping with turbomolecular or oil diffusion pumps, backed by rotary-vane pumps (Appendix A). Cleanliness of the vacuum system is of paramount importance to avoid unwanted interferences with the ion species being detected.

The electron–impact ion source consists of a small chamber into which the gas is admitted from the inlet system and is shown diagrammatically in Fig. 9.2. A beam of electrons with an energy of $\sim$50–100 eV is extracted from a hot filament and accelerated across the chamber at right angles to the ion exit slit to an anode (the 'trap'). Small magnets cause the electrons to follow a helical trajectory to increase the ionisation path length within the ion source. An electrode (the 'ion repeller') normally at a small +ve potential relative to the source chamber deflects ions towards the exit slit whilst electrons not involved in ion–molecule collisions are collected by the trap plate. Ions passing through the source exit slit are accelerated and focused by a series of plates at variable high potentials (typically 2–5 kV) to emerge as a collimated ion beam into the flight tube.

The flight tube passes between the poles of a magnet which is used to separate ion beams according to mass and charge (see Box 8.1). The magnet may have specially shaped pole pieces to focus the image of the source slit onto the collector mass resolving slits. Both permanent and electromagnets are in common use but electromagnets allow the field intensity to be adjusted to maximise the ion accelerating potential in the source, thereby increasing extraction efficiency and sensitivity. The ion optics are designed to give broad ion beams with flat tops when the beam is scanned across the collectors (Fig. 9.3). The collectors consist of Faraday cups placed behind a mass defining slit and the charge accumulated on

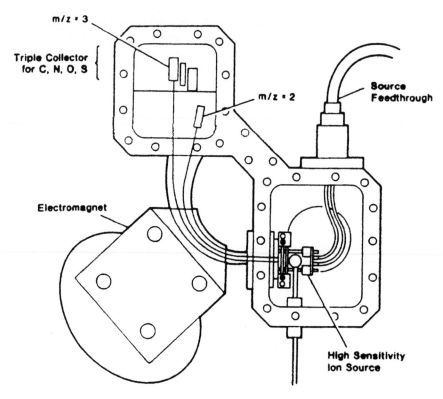

Figure 9.1   Schematic layout of a gas source mass spectrometer. Diagram of OPTIMA mass spectrometer courtesy of Micromass UK Ltd.

the Faraday cup is leaked to earth through a precision resistor ($\sim$100 MΩ or more). The voltage drop across the resistor, proportional to the intensity of the ion beam, is amplified and monitored via voltage to frequency converters by a computer. Gas source spectrometers use double- or more usually triple-collector assemblies to detect multiple ion beams simultaneously so that isotope ratios are independent of fluctuations in ionisation efficiency (Fig. 9.3).

The source is tuned for maximum efficiency and stability by adjusting the voltages applied to the ion repeller, trap and beam steering plates to give the strongest possible ion beam. The quality of the ion beam is monitored by examining its peak shape obtained by scanning the beam across the collectors by changing the accelerating potential or magnetic field intensity (Fig. 9.3). The objective is to obtain well-defined symmetrical flat-topped peaks combined with the good sensitivity for the gas species of interest. Source tuning can be quite an art as the various electric fields within the source interact such that changing the potential of one affects another, and there may be several different sets of conditions that combine to produce an acceptable peak shape.

The gas inlet system

Isotopic compositions are measured by comparing the ratio of ion beam intensities of the sample gas with those of a reference gas of known isotopic composition. The process of

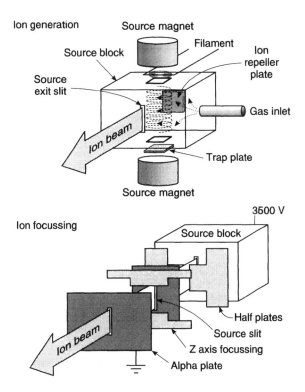

Figure 9.2 Schematic diagram of an electron impact gas ion source.

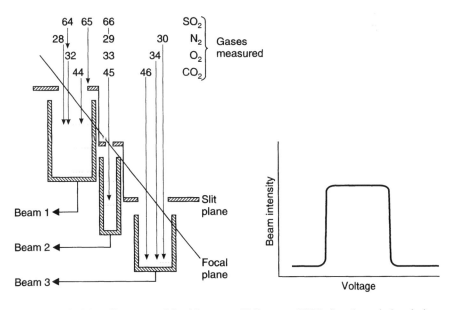

Figure 9.3 The universal triple-collector assemblies (Courtesy of Micromass UK Ltd) and a typical peak shape obtained by scanning an ion beam across the narrow slit beam 2 collector.

reference–sample comparison is controlled by the gas inlet system of which three configurations have evolved according to the type of analysis required.

Dual-inlet 'dynamic' mass spectrometry   The dual-inlet system is designed to provide the highest possible analytical precision by allowing multiple comparisons of sample and reference gas. The gases are leaked into the ion source under their own pressures and a relatively large reservoir of sample gas is required to maintain a sufficient gas flow into the source for the duration of the analysis. The inlet consists of identical reservoirs to store the reference and sample gas (Fig. 9.4) and a change-over valve is operated such that one or other of the gases is admitted to the mass spectrometer, while the other is bled to a vacuum pump at precisely the same rate. The dual inlet is connected to the ion source via matched capillaries such that equal gas pressures in the inlet give equal ion beams in the source. An important function of the capillaries is to provide a viscous flow of gas into the ionisation chamber which minimises mass fractionation of isotope species. The flow rates of the reference and sample gases are equalised by adjusting the pressure of gas in each side of the inlet using a metal bellows or 'variable volume'. After equalising the pressures the variable volume can be closed off from the rest of the inlet immediately prior to an analysis to ensure that each inlet contains an equal volume of gas at an equal pressure. By equalising the operating conditions while the reference and sample gas are admitted to the ion source any isotopic fractionation that might take place during the analysis is cancelled out.

Two features of the dual inlet are used for the analysis of smaller samples. The bellows can be used as a simple pump whereby large ion beams can be obtained from small samples by expanding the gas into fully open bellows, closing off the inlet and decreasing the bellows' volume to increase the sample gas pressure. For even smaller samples, a liquid-nitrogen-cooled cold finger is used to concentrate condensable gases such as CO_2, SO_2 or SF_6 (i.e. gases with boiling points higher than the boiling point of liquid nitrogen at $-196°C$) into the inlet. 'Non-condensable' gases such as O_2, H_2 and N_2 cannot be easily concentrated using a cold finger using liquid nitrogen (see Box 9.2); hence the attraction of continuous

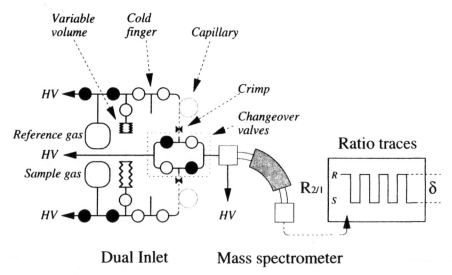

Figure 9.4   Schematic diagram of a dual-inlet system.

helium flow techniques for the isotope analysis of very small samples of non-condensable (and condensable) gas species (see below).

Modern dual-inlet systems are constructed to have as small an internal volume as possible to provide the best possible sensitivity and precision. The object is to generate the highest possible gas pressure in the inlet for a given sample size to give large noise-free ion beams that are well above any background contributions. As the sample size is decreased, ion beams become more noisy and eventually the flow regime through the capillaries changes from viscous to molecular flow causing unpredictable mass fractionation effects. The smallest samples that can be analysed using a dual inlet are effectively governed by the minimum pressure required to maintain a viscous flow of gas to the ion source. In practice, the lower limit is in the order of 0.1 μmol which nevertheless does not pose any restrictions on the great majority of stable isotope analytical tasks.

Continuous helium flow mass spectrometry

Continuous flow techniques for isotope ratio analysis use a stream of He carrier gas to transport sample gas directly into the ion source of the mass spectrometer, by-passing the conventional dual inlet. A schematic continuous flow inlet is shown in Fig. 9.6. A stream of He carrier gas is continuously fed into the ion source through a capillary and the sample and reference gases are injected at appropriate intervals into the flowing stream. The He flow into the mass spectrometer is in the order of 0.2 ml per minute (about three orders of magnitude greater than the flow rate of sample gas from a dual inlet) and the background pressure in the instrument is much higher, $\approx 10^{-3}$ Pa. The only requirement of the mass spectrometer is that it has sufficient pumping capacity to deal with the large throughput of helium carrier gas.

Samples can be introduced into the flowing He stream in a number of ways, for example by injection, but by far the most attractive aspect of this technique is the ability to link the mass spectrometer with the output of a gas chromatograph, which is used to isolate the chemical species of interest. Comparison of isotope ratios of sample and reference gases is carried out sequentially and the analysis is usually timed relative to the output of sample gas from the gas chromatograph. While the sample gas is passing through the GC column an aliquot of reference gas is injected into the He stream and its isotope ratios measured over a predetermined time interval. The sample gas arrives as a broad peak and the ion beams are integrated across the entire peak as the isotopic composition of the effluent changes with time (Fig. 9.6). Analytical precision is inherently lower than that obtained using a dual inlet, largely because only one sample–reference comparison is made per sample, but this technique offers significant advantages over the capabilities of dual-inlet systems. Samples in the order of 1 nanomole give a usable instrumental reproducibility of <0.3‰ with further scope for analysis of samples as small as 20 picomoles. As the entire sample is flushed by the He carrier into the ion source for isotope analysis the net sample requirement can be at least three orders of magnitude smaller than dual-inlet systems. Another key advantage is that both condensable (CO_2, SO_2 or SF_6) and non-condensable gas species (N_2 and O_2) can be introduced into the ion source for isotope analysis without resorting to cryogenic trapping using cold fingers. Techniques for the analysis of CO_2, N_2, O_2 and SO_2 have been developed and continuous flow technology is set to become a major technique in the future.

Static vacuum mass spectrometry

The stable isotope analysis of the very smallest samples has been performed by static vacuum mass spectrometry based on techniques developed for the isotope analysis of noble gases. The principle of the technique is that after the sample is introduced into the ion source, the instrument is isolated from the pumping system so that the instrument itself acts as the sample container during isotope analysis. The sample is only pumped away after the analysis is complete. The static vacuum instrument is more efficient than continuously pumped 'dynamic' instruments where the sample is continuously pumped away even when the reference gas is being analysed.

Unlike noble gases, active gases decompose over the hot filament and the amount of time available to measure isotope ratios is limited, in the case of CO_2 to only a few minutes. Gases such as N_2 or CH_4 have longer half-lives and perform better in a static mass spectrometer. Only one opportunity is allowed for the comparison of the sample and reference gas ratios and inlets have a means of accurately matching the amount of sample admitted into the instrument. The reproducibility of isotope analyses by static vacuum mass spectrometry is less than 1‰ at the nanomole level, increasing to $\approx$2‰ for samples in the picomole range. The instrument has mostly been used in conjunction with stepped heating and laser-desorption gas extraction systems (see below).

Data corrections and calculation of the δ value

Gas source mass spectrometers compare the isotopic ratio of the sample gas with that of a standard of known composition. Dual-inlet systems allow multiple comparisons of large samples at the best precision; continuous flow and static vacuum instruments provide only a single reference–sample comparison but of much smaller samples. In all cases the output of the instrument is the difference in the ratio of beam intensities of the sample and working standard, normally expressed as permil (‰) and referred to as the 'raw delta'. Before the raw δ value can be expressed relative to the international standard in the form of the standard δ notation (Box 9.1), the 'raw delta' is corrected for any instrumental effects or **isobaric** interferences.

An example of an instrument-specific correction is associated with hydrogen isotope determinations. The D/H ratio is determined from the intensity ratio of the H_2^+ and HD^+ peaks at masses 2 and 3 respectively. Ion–molecule reactions also produce unwanted H_3^+ as an interfering species. H_3^+ production is related to the hydrogen pressure in the source and the $m/z = 3/2$ ratio varies as a linear function of the intensity of the major ion beam at $m/z = 2$. An H_3^+ correction factor can be obtained from a plot of mass 3/2 ratio versus $m/z = 2$ peak height where the $m/z = 3/2$ ratio extrapolated to zero pressure gives the true D/H ratio. This correction, specific to a given instrument and source operating parameters, is normally recalibrated prior to a series of sample measurements.

Isobaric corrections are necessary when more than one isotopic species contributes to the ion beam and isobaric corrections are necessary for the analysis of CO_2 and SO_2 owing to the presence of ^{17}O and ^{18}O. The isotopic composition of CO_2 may consist of any combination of the isotopes ^{12}C, ^{13}C, ^{16}O, ^{17}O and ^{18}O. Natural CO_2 will consist overwhelmingly of molecules substituted by the common isotopes $^{12}C^{16}O^{16}O$, with much lower abundances of molecules substituted with one or other of the minor isotopes (e.g. $^{13}C^{16}O^{16}O$, $^{12}C^{16}O^{18}O$ etc.) (Table 9.1 and Fig. 9.3). The probability of molecules substituted by more than one rare isotope is negligible. Singly charged ions of CO_2 will therefore give a large ion beam corresponding to $m/z = 44$ ($^{12}C^{16}O^{16}O$) and two minor ion beams with $m/z = 45$ ($^{13}C^{16}O^{16}O$ or $^{12}C^{16}O^{17}O$) and $m/z = 46$ ($^{12}C^{16}O^{18}O$ or $^{13}C^{17}O^{16}O$). For natural terrestrial CO_2 approximately 6% of the mass 45 peak is due to ^{17}O and 0.2% of the peak at mass 46 is a contribution by species containing both ^{13}C and ^{17}O. Thus measured 45/44 and 46/44 ratios have to be corrected for these isobaric interferences in order to obtain the

---

Box 9.1 The δ notation and fractionation factors

The absolute abundances of light stable isotopes are difficult to determine with good precision and stable isotope ratios are reported as a difference function, δ, relative to an international standard measured at the same time. The standard δ notation is a magnifying glass for small isotopic variations and gives the deviation in parts per thousand (‰) of the ratio from that of the reference material. For example, $\delta^{18}O$ would be defined as follows:

$$\delta^{18}O = \left[\frac{^{18}O/^{16}O_{sample}}{^{18}O/^{16}O_{standard}} - 1 \right] \times 1000(‰)$$

The magnitude of isotopic fractionation between two phases is expressed by the fractionation factor, α_{A-B}. For example, for the oxygen isotope exchange reaction between calcite and water, α is defined as

$$\alpha = \frac{(^{18}O/^{16}O)_{calcite}}{(^{18}O/^{16}O)_{water}} = \frac{1000 + \delta^{18}O_{calcite}}{1000 + \delta^{18}O_{water}}$$

Fractionation factors have values that are close to unity and the value of α for example might be 1.008 at 200°C. A useful mathematical approximation, $1.00X \approx 10^3 \ln X$ (i.e. $1.00X \approx X$ permil) is widely used in the literature to express fractionation factors as a permil value, which for small values (< 30‰) approximates the measured permil difference between the two species. Thus

$$10^3 \ln \alpha_{calcite-water} \approx \delta^{18}O_{calcite} - \delta^{18}O_{water} \approx \Delta^{18}O_{calcite-water}$$

It is more convenient to refer to fractionations as permil values and geochemists would read $\alpha_{calcite-water} = 1.008$ as meaning an 8‰ difference between the oxygen isotope composition of calcite and water.

DPM

---

true $^{13}C/^{12}C$ and $^{18}O/^{16}O$ ratios of the gas. Similar corrections necessary for the determination of $\delta^{34}S$ using SO_2 gas have stimulated interest in using SF_6 (F being monoisotopic) as a gas species for sulphur isotope analysis.

Standards and normalisation of δ values

Stable isotope data are reported relative to an international standard using the standard δ notation (Box 9.1). Although it should only be necessary to calibrate the zero point of the δ scale using an internationally approved standard there is a tendency for different mass spectrometers to expand or contract the δ scale such that δ values reported for the same material by different laboratories may disagree by a few tenths of a permil unless the δ scale is pinned by another standard of different composition. In hydrological studies $\delta^{18}O$ and δD scales are fixed at two points by measuring the compositions of two water standards V–SMOW ($\delta^{18}O = 0‰$ and $\delta D = 0‰$) and V–SLAP (Standard Light Antarctic Precipitation), defined as having $\delta^{18}O = -55.50‰$ and $\delta D = -428‰$. A comprehensive list of recommended δ values for international standards has been published. For oxygen there are two standards in common use: V–PDB and V–SMOW (Table 9.1). The V–PDB standard is used for palaeotemperature studies based on carbonate analyses; the V–SMOW scale is used to report isotopic compositions of all other forms of oxygen-bearing species. The V–PDB and V–SMOW scales are related as follows:

$$\delta^{18}O_{V-PDB} = 1.03039\delta^{18}O_{V-SMOW} + 30.39 \tag{9.1}$$

Stable isotope fractionation and its application to geological problems

Stable isotopic fractionations (i.e. enrichment or depletion of an isotope in one material relative to another) develop as a result of differences in the strength of chemical bonding between each isotope in a molecule or crystal lattice. Bond strength varies as a function of atomic mass, ionic charge and oxidation state, and in general terms, bonds to ions with low atomic mass *or* high ionic charge *or* high oxidation state favour the heavy isotope. **Equilibrium** isotope fractionations develop between materials that are in chemical equilibrium but have different molecular structures. The constituent minerals in a metamorphic or igneous rock will each possess slightly different oxygen isotope ratios related to the way that oxygen is co-ordinated in different crystal structures. Diffusion at high temperatures, recrystallisation during diagenesis and metamorphism and solution/precipitation from fluids are all processes that facilitate isotopic equilibrium, and the magnitude of the isotope fractionation of stable isotope ratios between different minerals at isotopic equilibrium is inversely proportional to temperature. **Kinetic** isotope effects occur during fast, unidirectional processes such as diffusion or for chemical reactions that result in *partial* transfer of a light element from one form to another. Kinetic isotope fractionations are largely independent of temperature and are associated with many biologically mediated reactions such as photosynthesis or bacterial oxidation.

The difference in isotopic composition between two minerals or a mineral and a fluid at isotopic equilibrium is defined as the fractionation factor, α (Box 9.1). The inverse relationship between isotopic fractionation and temperature provides a powerful and versatile geothermometer. Stable isotope geothermometers have been calibrated from experiments or by thermodynamic calculations for a wide variety of materials and one of the first, and geologically most useful, systems to be studied is that of the distribution of oxygen isotopes between calcite and water. The equation for the exchange of oxygen isotopes between calcite and water can be written as follows:

$$CaC^{16}O_3 + H_2{}^{18}O = CaC^{18}O^{16}O_2 + H_2{}^{16}O \qquad [9.2]$$

Oxygen isotopes are exchanged when calcite is precipitated or during recrystallisation in the presence of a fluid and the calcite–water oxygen isotope geothermometer provides, for example, a means of determining ocean temperatures from the oxygen isotope composition of foraminifera skeletons, or the temperature of mineralisation from the oxygen isotope composition of calcite in mineral veins. Alternatively, if the temperature can be constrained independently, for example by using fluid inclusion data, the oxygen isotope composition (hence the source) of a fluid can be identified from the oxygen isotope composition of calcite cements or mineral veins. This key feature of stable isotope behaviour provides a powerful and versatile means of determining the temperature of mineral assemblages or the isotopic signature of fluids associated with geological processes.

Sample preparation systems

Geological samples normally require chemical or physical processing to obtain a suitable gas for mass spectrometric analysis. The diverse occurrence of H, C, N, S and O in geological samples has meant that a wide range of preparative techniques have been developed for stable isotope analysis. For the great majority of stable isotope studies, knowledge of the bulk isotopic composition of an element in a material may be sufficient to solve a particular problem but other problems may require stable isotope data for individual components in a mixture (e.g. for each of the different forms of carbon in a sample), or for a given spatial region of a sample (e.g. for different generations of calcite cements in a limestone).

Most traditional sample preparation techniques are designed for the bulk analysis of milligram quantities of sample, and individual components must be separated prior to analysis. In practice, as few materials are conducive to the physical separation of different components, there has been considerable research into developing techniques for stable isotope analysis of multi-component mixtures, or for spatially resolvable or *in situ* analyses of mineral surfaces. These specialised techniques demand high performance from the mass spectrometer as the amounts of material available of a given component or *in situ* analysis are often much smaller. These requirements have been the prime driving force for developing a new generation of techniques such as continuous He flow processes and laser sampling.

Conventional sample processing is carried out using a vacuum preparation system constructed of tubes and valves, normally in conjunction with dual-inlet mass spectrometers for routine stable isotope analysis. The line is evacuated to eliminate contamination by air and enable gases extracted from the sample to be purified and collected. The preparation system is designed to serve three functions: (a) to release the element of interest from the sample and convert it into a suitable gas for isotope analysis; (b) to purify the gas; and (c) to determine the amount of gas recovered in order to calculate the extraction yield.

Stable isotope elements in mineral and rock samples are converted into the appropriate gas suitable for mass spectrometry (Table 9.1) using a variety of physical and chemical means. The most commonly used techniques include:

- isotopic exchange, for example equilibration of water with CO_2 of known isotopic composition to determine the oxygen isotopic composition of water;
- acid attack using orthophosphoric acid to generate CO_2 from carbonates;
- vacuum pyrolysis to release structurally bound components such as H_2O or CO_2 from minerals and fluid inclusions, or SO_2 from sulphates;
- oxidation in the presence of O_2 or other oxidising agents such as CuO or V_2O_5, to prepare CO_2 from organic carbon, SO_2 from sulphides, etc;
- reduction, for example using zinc or uranium to prepare H_2 from water;
- reaction with fluorinating reagents (F_2, ClF_3, BrF_5) to liberate O_2 from silicates or SF_6 from sulphides.

To illustrate some of the techniques that are commonly used in sample preparation systems a typical line for the extraction of water from silicate minerals is shown in Fig. 9.5. The sample to be analysed is placed in the quartz glass finger which is then evacuated. After the sample vessel has pumped down to high vacuum, the sample is heated using a radiofrequency furnace induction heater. As the mineral is heated H_2O and other gas species (potentially H_2 as well as condensable species CO_2 and SO_2 and non-condensable species such as N_2 and Ar) are liberated as the mineral decomposes. Hot CuO is present as an oxidant to ensure that H_2, CO and hydrocarbons are converted to water and CO_2. Condensable gas species are collected in the cold finger held at $-196°C$ and at the end of the heating step the unwanted non-condensable gases are pumped away, leaving a mixture of H_2O, CO_2 and SO_2 frozen in the cold finger.

The next stage of the extraction procedure in this example is to remove unwanted CO_2 and SO_2 by vacuum distillation leaving pure water for reduction to H_2. The mixture of

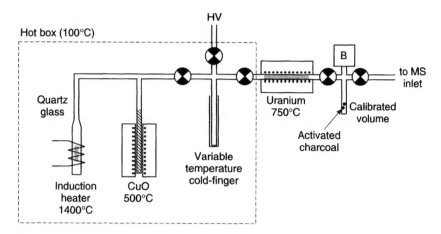

Figure 9.5   Schematic layout of a preparation line for the extraction of water from silicate minerals for D/H ratio
analysis. Water is a problematic phase to work with in a vacuum system as it tends to freeze inside the
vacuum line even at room temperature. To avoid loss of water the line is permanently heated to around
100°C to maintain water in a vapour state when being processed in the line. 'B' is a capacitance manometer for measuring pressure.

gases frozen under liquid nitrogen at $-196°C$ is warmed by electrical heating to $-100°C$
to drive off CO_2 and SO_2. The finger containing water ice is now warmed to 100°C and
the water vapour is then reduced to H_2 gas in the hot uranium furnace. The final step is
to collect the hydrogen gas in the calibrated volume using a cold finger containing activated charcoal as an adsorbent and the amount of hydrogen, and hence the water content
of the sample, is obtained from the pressure measured in the manometer from the gas law:

$$n = \frac{PV}{RT}$$
[9.3]

where n is the number of moles of gas, P the pressure, V the volume, R the gas constant and
T the temperature in K.

Quality of analyses   The greatest errors in stable isotope analysis are those induced during the chemical processing of the sample and for this reason the sample preparation schemes are kept as simple as
possible. The quality of a stable isotope analysis is governed by the system blank, presence
of memory effects, gas purity and quantitative yield. A small background component is
derived from the preparation system or introduced during sample handling and will be analysed with the sample. The amount of sample analysed is adjusted so that the blank contribution is nominally less than 1% of the amount of sample prepared for analysis. For the
analysis of very small samples system blanks pose a significant problem and ultimately
limit the amount of sample that can be analysed to give a reliable result. System memory is
essentially a system blank whose composition is inherited from samples that have previously
been processed through the system. Memory effects tend to be noticed when samples having widely different isotopic compositions are processed but can be eliminated by keeping
the preparation system clean and with regular baking to outgas the parts of the system that
are subject to heating or cooling during the analytical cycle.

Gas purity and quantitative yields are the most important requirements for accurate
stable isotope analysis. Gas used for mass spectrometry must be 100% pure, as the presence

of other gas species (e.g. water) can cause undesirable interferences in the mass spectrometer. The yield of gas, when the stoichiometry of the sample is known, is the ultimate check on the quality of the analysis. Incomplete reactions resulting in low yields invariably cause isotopic fractionation and yields that are higher than expected point to other sources of error such as contamination.

Continuous helium flow processing

Continuous flow systems are pressured with He and operate at pressures greater than atmospheric pressure. Continuous flow systems use narrow bore or capillary tubing to transfer gas through the system and into the mass spectrometer and are designed to avoid 'dead volumes' that are not efficiently purged by the carrier gas. Continuous flow preparation systems use gas chromatography to separate and purify gas species of interest and use multiport valves to control the flow of gas to different parts of the system (Fig. 9.6). Unwanted gas species are removed from the flowing He stream using chemical scrubbers (e.g. magnesium perchlorate to remove water) or by diverting the peak by 'heart cutting'. Gas chromatography is ideally suited to the separation and purification of small amounts of sample gas and is configured with a combustion system to generate gas for mass spectrometry.

Combustion–gas chromatography–isotope ratio mass spectrometry (C–GC–IRMS) This system is used for the rapid isotopic analysis of C, N and S in geological materials and consists of an elemental analyser and isotope ratio mass spectrometer for the continuous flow isotope analysis. Samples are placed in pure tin foil capsules and are sequentially dropped by an autosampler into the combustion furnace which is flushed with oxygen-enriched He. The tin foil capsule 'flash combusts' generating temperatures in excess of $1800°C$ to ensure complete and rapid combustion of the sample to carbon dioxide, oxides of nitrogen, oxides of sulphur and water. The combustion products are passed through a reduction column to reduce oxides of nitrogen to N_2 gas and water is chemically removed prior to the separation of N_2 and CO_2 in a packed GC column. The output from the column is passed through a thermal conductivity detector (Box 16.1) to determine the yield of N_2 and CO_2 and gases are directed to the mass spectrometer for isotope ratio measurement. This technique provides a rapid and precise technique for the analysis of organic and inorganic liquid and solid samples, with a unique ability to produce a choice of three gases for isotopic analysis. The system requires only a small amount of sample ($<100ng$) and also identifies C and N compositions from the same sample in one analysis.

Stepped heating–mass spectrometry

Many geological materials contain several forms of carbon (e.g. organic carbon, carbonates, CO_2 fluid inclusions, etc.) and are characterised by different thermal release or combustion temperatures. Isotope analysis of CO_2 released by incremental gas extraction over a series of temperature intervals provides a powerful means of resolving the abundances and isotopic compositions of individual molecular components. Stepped pyrolysis/combustion techniques can be used to resolve carbon species such as organic carbon (combustion at 300–600°C), graphite or diamond (combustion at 800–1000°C), carbonates (thermal breakdown

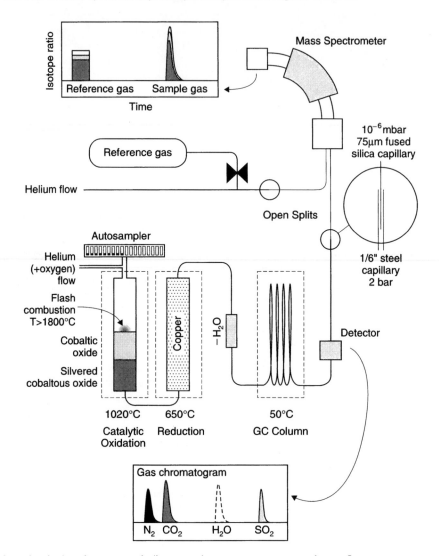

Figure 9.6   A combustion/gas chromatography/isotope ratio mass spectrometry continuous flow system.

at 600–700°C), fluid inclusions (thermal decrepitation at 200–800°C depending on mineral host, inclusion size and fluid density) or dissolved carbon in silicates (>1000°C to fusion temperature).

Stepped heating techniques were developed and refined to study complex meteoritic material contaminated by terrestrial carbon in order to obtain the isotopic composition of carbon components that are otherwise difficult to separate physically and has been used extensively for the analysis of carbon at low levels of abundance where sample contamination by non-indigenous carbon species poses serious problems in the analysis of carbon-poor materials such as dissolved volatiles in basalt glass. The contaminant component can be resolved and removed by stepped combustion at 400°C to 600°C.

A fruitful application of stepped heating is to measure $^{39}Ar/^{40}Ar$ ratios in igneous rocks for dating purposes (Box 9.2).

Box 9.2 $^{40}Ar/^{39}Ar$ geochronology

$^{40}Ar/^{39}Ar$ dating is a refinement of the the K–Ar dating system that commonly involves stepped-heating gas source mass spectrometry. Many geological materials ranging in age from 10 ka to 4.6 Ga can be dated with the $^{40}Ar/^{39}Ar$ technique which is based on the branched radioactive decay of ^{40}K to ^{40}Ar.

Traditional K–Ar dating involves measurement of the K content (flame photometry or XRF spectrometry) and ^{40}Ar content (isotope dilution gas source mass spectrometry) on two separate aliquots of a rock or mineral sample. In $^{40}Ar/^{39}Ar$ dating the sample is first irradiated with neutrons in a nuclear reactor, which transforms ^{39}K to ^{39}Ar by neutron capture (Chapter 7) and thus allows the parent:daughter ratio $(^{40}K:^{40}Ar)$, and therefore age, to be determined from mass spectrometric isotope ratio measurements on a single sample aliquot.

Irradiated samples are typically heated in a furnace at various temperature steps from about 500 to more than 1500°C and the released gas transferred through a clean-up line into a static vacuum mass spectrometer. For the gas increment released at each step, five isotopic measurements at masses 40, 39, 38, 37 and 36 are made by the mass spectrometer and, after initial correction for postirradiation radioactive decay of ^{39}Ar and ^{37}Ar, the various Ar measurements are corrected for:

- irradiation-produced, interfering isotopes of Ar not produced from ^{39}K, by measuring production ratios on pure K_2SO_4 and CaF_2 salts $(^{40}K \rightarrow {}^{40}Ar; {}^{42}Ca \rightarrow {}^{39}Ar; {}^{40}Ca \rightarrow {}^{36}Ar)$;

- additional interfering isotopes of Ar produced from irradiation of Cl;

- the presence of atmospheric Ar.

The ratio of radiogenic ^{40}Ar to K-derived ^{39}Ar (R) is then used to calculate the age of the gas using the equation

$$Apparent\ age = \frac{1}{\lambda}\ln(JR + 1)$$

where λ is the decay constant for ^{40}K and J is a neutron flux parameter

$$J = \frac{(e^{\lambda t_m} - 1)}{(^{40}Ar_R/^{39}Ar_K)_m}$$

where t_m is the age of a flux monitor mineral (i.e. a standard) that is irradiated with the samples, and $(^{40}Ar_R/^{39}Ar_K)_m$ is the measured ratio of the standard corrected as above ('R' and 'K' standing for radiogenic and potassium-derived respectively).

The advantage of the $^{40}Ar/^{39}Ar$ step-heating technique over conventional K–Ar dating is that samples can be step heated in a furnace, or using a laser, and each aliquot of gas released can be analysed in a mass spectrometer and an apparent age calculated for each temperature step. The ages calculated for each step are then plotted versus the percentage of ^{39}Ar released during the heating experiment – a plot referred to as an age spectrum. A smooth age profile in igneous rocks or minerals typically indicates that a sample has undisturbed Ar systematics and may permit calculation of a plateau age (Fig. 9.2.1a). However, many samples have age spectra that indicate the following: (a) the presence of inherited or excess Ar, perhaps introduced by alteration of a sample, resulting in old apparent step ages in low- and/or high-T steps (Fig. 9.2.1b) – a sample containing inherited or excess Ar would yield an erroneously old K–Ar date; (b) post-crystallization loss of radiogenic Ar resulting in young apparent step ages in low-T steps (Fig. 9.2.1c) – a sample which has undergone Ar loss would yield an erroneously young K–Ar date. More complex age spectra are also often observed for minerals from metamorphic and igneous plutonic rocks and these can glean useful P–T–t information in the study of structural geology, plate tectonics and magmatic processes, for example the timing and rates of lithospheric exhumation or cooling deduced from the apparent ages of minerals, with different argon closure temperatures, taken from the same sample.

JAB

Figure 9.2.1 $^{40}Ar/^{39}Ar$ age spectra of flood basalts from Yemen (Baker *et al.*, 1996) illustrating (**a**) sample giving a well-defined plateau age, (**b**) age spectrum showing excess argon and (**c**) the effect of argon loss.

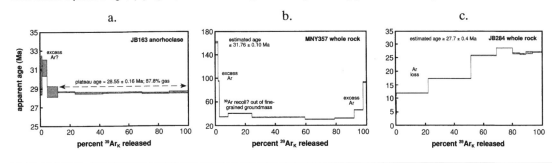

Laser-assisted extraction techniques

Laser heat sources can be considered as variations on conventional methods of stable isotope analysis but a laser heat source can be directed onto the sample through a suitable window in the evacuated sample chamber and, as only the sample itself is heated, blanks are inherently lower as the sample chamber remains at room temperature. The obvious attraction of lasers is that they provide a means of sampling small selected areas of a specimen for *in situ* stable isotope analyses. The physical principles of lasers are summarised in Box 9.3.

The layout of a laser heating system is shown in Fig. 9.7. The laser is directed onto the sample contained in a small chamber fitted with a window and a port for attachment to the vacuum system. A microscope/video camera system is used to view the sample and combined with the laser optics with a beam splitter or mirror. Motorised translation stages are used to position the target under the laser, by moving the sample chamber, the laser, or a beam steering mirror.

Laser heating techniques have been developed for isotope analysis of elements where they occur as major components of minerals – sulphur in sulphides, carbon and oxygen in carbonates and oxygen in silicates and oxides – whereby sufficient gas can be released for high-precision analysis by dual-inlet mass spectrometry. The main applications of laser heating for stable isotope analysis have been for carbon and oxygen isotope analysis of carbonates by thermal decomposition to produce CO_2, for oxygen isotope analysis of silicates by laser heating silicate or oxide minerals in the presence of a fluorinating agent, and sulphur isotope analysis of sulphides by laser heating samples in the presence of O_2 to generate SO_2 or F_2 to generate SF_6.

The most commonly used lasers in stable isotope analysis are CO_2 gas lasers operating in the far-infrared region at 10.6 μm and Nd–YAG lasers which generate light in the near-infrared region at 1.066 μm. CO_2 gas lasers are probably the most versatile as silicate and carbonate minerals have a strong absorption peak near 10 μm. The wavelength of Nd–YAG lasers corresponds to absorptions caused by transition metals and is suitable for the

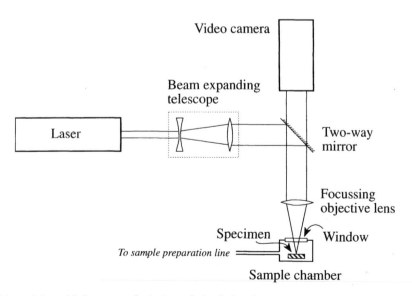

Figure 9.7.   Schematic laser ablation system for *in situ* analysis of minerals.

Box 9.3 How a laser works

Lasers (the name is an acronym for Light Amplification by Stimulated Emission of Radiation) are sources of intense light which can be focused onto a small area and used as an intense heat source. Quantum physics describes electrons in an atom as belonging to precise energy levels (Box 4.1). Light (i.e. a photon) is absorbed when an electron in an atom moves from a lower energy level (the 'ground state' of the atom) to a higher energy level (an 'excited state'), and a photon is released when the electron returns to its original energy state. The frequency and wavelength of a photon associated with a transition are well defined and the light absorbed or emitted is monochromatic. For an atom at a high energy state the emission process can be either spontaneous (random) or *stimulated* by an incoming photon of identical energy (Fig. 9.3.1a). The photon emitted by stimulated emission has the same energy and phase as the stimulating photon and the two photons add constructively, increasing the amplitude of the light wave: light is *amplified*.

For there to be a reasonable probability of stimulated emission occurring the population of atoms at the higher energy state must be greater than those at the lower energy state (a population inversion, Fig. 9.3.1b). To create a population inversion the laser medium must be 'pumped' with energy from an external source, for example by using a flash lamp to irradiate a solid laser medium (e.g. Nd–YAG glass, see below) or by electrical discharge through a gas laser medium (e.g. CO_2 – see text). The level of light amplification attained in a pumped laser medium is nevertheless quite small and to obtain useful amplification mirrors are placed at each end of the laser medium, forming an optical cavity or resonator (Fig. 9.3.1c). The light from stimulated emission bounces back and forth, gaining amplification each time the light passes through the laser medium. One of the mirrors is designed to have slightly lower reflectance to allow a small component of the beam to emerge from the optical resonator to form the external laser beam.

Compared with other sources of **monochromatic** radiation, lasers possess three unique properties that hold the key to their many novel uses:

(1) *High intensity* – A laser beam can deliver a much greater concentration of energy at its focus than other light sources, making possible applications such as laser ablation (increasingly widely used as a tool for microanalysis) and even nuclear fusion.

(2) *Coherent, parallel beam* – All rays in a beam of laser light are **in phase** (Box 4.3) with each other and form a narrow beam of negligible divergence. These attributes are essential for applications such as holography and laser ranging.

(3) *Narrow spectral bandwidth* – The width of a spectral line produced by a laser is far narrower than the corresponding linewidth produced by other spectral sources.

DPM, RG

Figure 9.3.1. (**a**) Light amplification when an excited atom is stimulated into emission by a photon of appropriate energy. (**b**) Creating a population inversion (a preponderance of atoms in excited states) by optical pumping, using an intense source of light to excite laser atoms. The initial (equilibrium) energy distribution is shown by the dashed line. (**c**) Construction of an Nd–YAG (neodymium-doped – yttrium aluminium garnet) laser showing the resonant cavity.

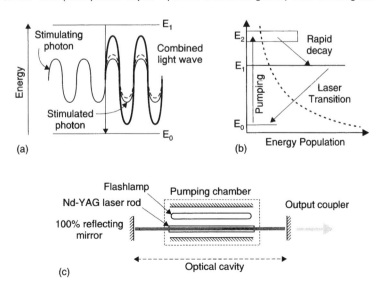

heating of ferromagnesian minerals and sulphides. The effective minimum beam diameters obtained by focusing CO_2 and Nd–YAG lasers are $\approx$50 μm and $\approx$10 μm respectively, although owing to the nature of the photothermal heating process the area affected by laser radiation is often much larger. UV laser radiation has interesting properties that are advantageous to *in situ* laser fluorination. UV lasers are capable of being focused to spot diameters substantially less than 1m and ablation pits as small as 2 μm diameter are readily achieved with a simple focusing objective. The key difference between IR and UV is that the low temperature of the photochemical ablation process means that reaction haloes are virtually absent during *in situ* laser heating and the analytical precision of UV laser fluorination will be far less dependent on the size of the sampled domains allowing greater spatial resolution.

Spatially resolvable stable isotope analyses   The ability to perform *in situ* stable isotope analysis is one of the greatest challenges in stable isotope geochemistry and is an area that is undergoing rapid growth and development. To perform stable isotope analyses that are spatially resolvable, in the same way that electron microprobes provide spot analyses of minerals, opens new horizons of research in isotope geochemistry, providing a means to study mineral isotope systematics on a microscopic inter- and intra-granular scale. These data would provide new information on diagenesis, mineral reaction kinetics, thermometry and fluid processes in a wide range of geological problems.

Spatially resolvable analyses have been obtained by laser heating in one of two ways: by analysing portions of a mineral excised from thin sections or wafers, or more elegantly, by direct *in situ* laser heating of a mineral surface. *In situ* heating is the most attractive, but not necessarily the most precise, means of obtaining spatially resolvable $\delta^{18}O$ analyses. The laser is used as a spot heat source to provide thermal energy for the decomposition reaction to proceed, leading to the formation of a reaction pit surrounded by a reaction halo. In order to obtain acceptable levels of precision the ablation pits have to be relatively large, enforcing an effective limit on linear spatial resolution of 500 to 1000 μm. Best results have been obtained by laser heating thin sections such that the reaction pit forms a hole through the slide rather than creating reaction pits with a tightly focused beam on the surface of a specimen. Recognising the inevitable analytical vagaries of direct *in-situ* laser heating, some workers analyse mineral fragments drilled or excised from thin sections. The obvious advantage of this technique is that as weighed amounts of material are analysed, precision is not compromised by unknown oxygen yields. Spatial resolution is subject to the physical limitations on preparing and handling mineral fragments in addition to those described above and resolutions lower than sub–millimetre levels are not easily attainable.

Chapter 10

Inductively coupled plasma–mass spectrometry (ICP–MS)

Kym E. Jarvis

Introduction

In this recently developed multi-element technique, positive ions generated in an inductively coupled plasma (Box 4.2) are extracted, via a differentially pumped air–vacuum interface, into a low-resolution mass analyser designed to scan a wide mass spectrum very rapidly. This provides near-simultaneous determination of most elements in the periodic table at levels down to 10 pg ml^{-1}. ICP–MS, although expensive in terms of capital outlay (£160,000), provides a rapid and very versatile means of trace element analysis of solutions, slurries and even solid samples over a wide concentration range (Houk and Thompson, 1988).

Instrumentation

A schematic of a typical ICP–MS system is shown in Fig 10.1. The sample is usually introduced in the form of a solution which is converted by a nebuliser into an aerosol dispersed in a stream of argon gas which sweeps it through the spray chamber, where larger droplets settle out, into the plasma torch. The torch is constructed of three concentric glass tubes between which an inert gas, normally argon, is passed. The design, construction and operation of the ICP torch are explained in Box 4.2. In the present configuration, the purpose of the ICP is to serve as an ion source for a mass spectrometer, and the sample injector is of relatively narrow bore (1.5 mm diameter). The sample aerosol on entering the high-temperature region of the ICP is rapidly volatilised, dissociated and ionised. The sample emerges from the mouth of the torch as a mixture of atoms, ions, undissociated molecular fragments and unvolatilised particles. Ions are extracted due to a decrease in pressure into the mass spectrometer from the axial zone of the torch.

Unlike the optical spectrometer of an ICP–AES instrument (Chapter 4), a mass spectrometer must operate under high vacuum. The critical design feature upon which ICP–MS depends is the differentially pumped interface region which transmits ions from the atmospheric pressure plasma into the mass spectrometer (Fig 10.2). This consists of two conical nickel apertures, the *sampling cone* and the *skimmer*, which allow ions to pass into the mass spectrometer but deflect away a majority of uncharged molecules and atoms. The sampling cone and skimmer are arranged co-axially one behind the other, about 6–7 mm apart. The low-vacuum region between them is evacuated by a rotary pump to a pressure of about 0.5 kPa (0.005 atmospheres), whereas the focusing region behind the skimmer, and the mass spectrometer itself, are maintained at moderate and high vacuum respectively by oil diffusion, cryogenic or turbomolecular pumps.

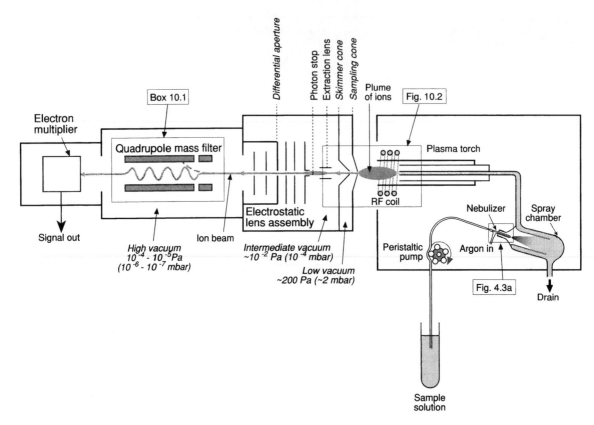

Figure 10.1   Schematic of an inductively coupled plasma–mass spectrometry instrument.

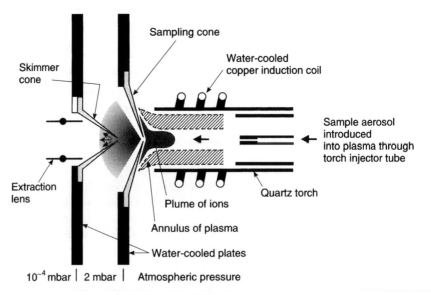

Figure 10.2   Schematic of the interface region between the ICP ion source (contained within the quartz torch) and the ion lenses.

Focusing of the ion beam beyond the skimmer is achieved using a set of electrostatic lenses. A physical obstruction called the 'photon stop' is placed directly in the ion beam path to prevent light reaching the ion detector and hence contributing to the background signal; the ion beam is deflected around the stop by a lens and continues its passage into the mass analyser.

The commonest form of mass analyser used in ICP–MS is a quadrupole mass filter and its operation is described in Box 10.1. The quadrupole allows ions of only one mass/charge ratio (m/z) through to the detector for each combination of potentials applied to opposing pairs of rods. By sweeping these potentials the mass spectrometer can be scanned very rapidly. For example, a single scan from $m/z = 4$ to 240 can be collected in as little as 0.06 s. During an analysis, many scans (typically 100–1000) are added together and an aggregate peak integral accumulated for each mass number. The relatively low mass resolution of the quadrupole analyser is sufficient to separate adjacent elemental mass number peaks completely but it cannot resolve the small differences in mass which separate elemental peaks from polyatomic ions that may interfere at the same nominal mass number; how such interferences are dealt with is discussed in a later section.

Ion detection is usually accomplished using electron multiplier detectors (Box 8.2). The ability to count individual ions, coupled with the very low background signals, results in high sensitivity for most elements. At high count rates **dead time** correction needs to be applied (p. 102). Variable dead times may occur depending on the age of the detector. The amplified signal is transmitted in digital form either directly, or via some form of multi-channel analyser, to a personal computer (PC) where corrections and calculations are carried out. In most modern instruments, the PC is also used to monitor, and often control, many of the instrument functions such as RF power level, gas flow rates, cooling water flow and instrument start-up and shut-down.

Performance

Figure 10.3(a) shows a typical ICP–MS spectrum for a simple solution containing $10\ \mathrm{ng\ ml^{-1}}$ of several elements across the mass range. Concentrations are usually calculated by taking the peak area for a single isotope of the element of interest. In the blank spectrum (Fig. 10.3b), several peaks in the low-mass range ($m/z < 80$) result from polyatomic ion species such as oxides, or traces of the atmospheric or carrier gases present in the plasma, that is H, N, O and Ar. Above $m/z = 80$, the spectrum is clean, with peak integrals of only a few counts.

ICP–MS offers the analytical geochemist several advantages over more traditional methods:

- The spectra collected are simple, and easily identified even in complex matrices, and with a little experience qualitative information can be gained quickly (Fig. 10.3c).
- The sensitivity for most elements across the mass range is uniform.
- Detection limits are exceptionally low, for most elements typically $<0.01\ \mathrm{ng\ ml^{-1}}$, resulting in quantitation limits in the solid of less than $50\ \mathrm{ng\ g^{-1}}$.
- The technique has a wide linear dynamic range over five orders of magnitude allowing simultaneous multi-element determination at concentrations from $0.01\ \mathrm{ng\ ml^{-1}}$ to $1.0\ \mathrm{\mu g\ ml^{-1}}$.
- Precision is between 2 and 5% RSD while accuracy is better than 5% absolute.
- Analysis times are rapid (typically 60 s per sample).
- Using specialised methods of sample introduction, even very small samples ($>10\ \mathrm{\mu l}$) can be analysed.

Box 10.1 Quadrupole mass analyser

Quadrupole mass analysers are useful where relatively low-mass resolution is required and cost is an important consideration. In a quadrupole analyser (mass filter), mass separation is achieved solely by electric fields.

The analyser consists of a set of four parallel rods (Fig. 10.1.1 – typically 125–230 mm in length), usually of circular section, manufactured from stainless steel or molybdenum, and mounted to very close tolerances in precision-ground ceramics. Opposing rods are connected together and to each pair a potential, $U + U_1 \cos(2\pi\nu t)$, with opposite polarities, is applied. U is a fixed dc potential (of up to about 70 volts) and U_1 is the peak amplitude for a radiofrequency (RF) potential (maximum of about 400 volts) of frequency ν (typically 2 MHz). The RF voltages on each pair have the same amplitude but are opposite in sign, that is they are 180° out of phase. Ions are injected into the rod assembly parallel to the axes of the four rods and undergo transverse motion. This motion is described by differential equations of the form known to mathematicians as Mathieu equations. The system operates as a mass band pass filter, such that only ions within a limited range about a mean mass/charge ratio (m/z) are transmitted:

$$\frac{m}{z} = \frac{kU_1}{\nu^2 r_0{}^2}$$

where k is a constant and r_0 is the radius of the cylinder defined by the inner surface of the rods (typically 6 mm). There are two types of motion described by the Mathieu equations. The first is where the ion oscillates about the quadrupole axis and ultimately emerges from the opposite end of the mass filter. The second is an unstable motion in which the ion moves away from the axis and ultimately strikes one of the four rods and is neutralised. The particular motion is determined by the parameters U, U_1, ν, r_0 and m/z.

For simplicity, the ion paths can be considered separately in each of the two rod planes, as depicted in Figure 10.1.2. In the positive rod plane (Fig. 10.1.2a) the lighter ions tend to be deflected too much and strike the rods, while the ions of interest and the heavier ions have stable paths. In this plane the quadrupole acts like a high-pass mass filter. In the negative rod plane (Fig. 10.1.2b), the heavier ions tend to be lost preferentially, and the ions of interest and the lighter ions have stable paths. Thus in this plane the quadrupole acts like a low-pass mass filter. In actuality, the positive and negative planes are superimposed physically, so these two filtering actions occur on the same ion beam at the same time. The result of high-pass and low-pass mass filtering is a structure that trans-

mits ions only at the m/z of interest to the detection system.

The mass resolution power of the analyser is electronically controlled by the U_1/U ratio. The mass transmitted is selected by varying U_1 and U together in such a way that U_1/U remains constant; this ensures the relative peak width is kept constant across the mass range. A ratio of 6:1 results in a mass resolution of about 1 m/z unit.

JGW

Figure 10.1.1 Schematic of a quadrupole mass filter.

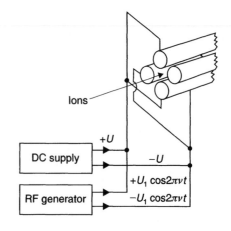

Figure 10.1.2 Side view of the ion separation processes in the two rod planes of a quadrupole.

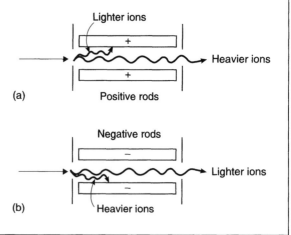

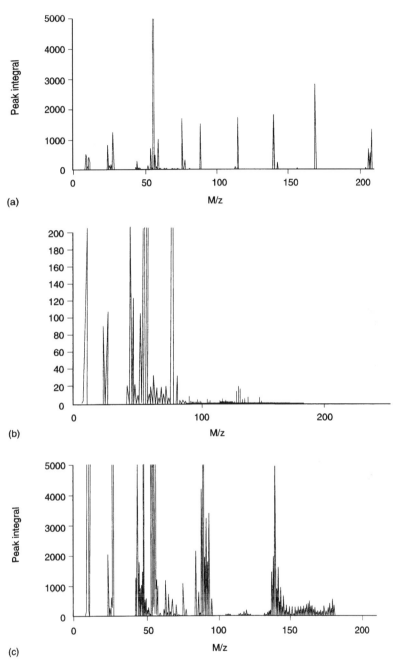

Figure 10.3   Typical ICP–MS mass spectra. The horizontal axis represents m/z but is graduated in terms of relative atomic mass for singly charged ions. The integral shown on the vertical axis is a measure of peak
size in each case. The peaks shown in the lower part of the mass spectrum between $m/z = 23$ and 40
mostly result from polyatomic ions. The peak at $m/z = 80$ in each spectrum is that of the argon
dimer $^{40}Ar^{40}Ar$.
(a) Solution containing 10 ng ml^{-1} of each of ^9Be, ^{24}Mg, ^{25}Mg, ^{26}Mg, ^{27}Al, ^{59}Co, ^{89}Y, ^{115}In, ^{140}Ce,
^{169}Tm and ^{206}Pb, ^{207}Pb, ^{208}Pb;
(b) 2% v/v HNO_3 acid blank;
(c) standard reference material granite AC-E (International Working Group) after digestion of 0.1 g in
100 ml using a mixed HF/HClO$_4$ digest followed by evaporation and dissolution in HNO_3. The
REEs ($m/z = 139$–175) are clearly visible .

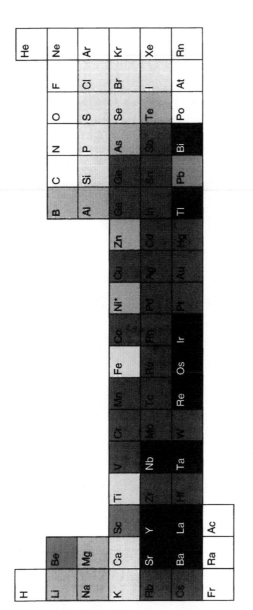

Figure 10.4   ICP–MS elemental limits of detection $(ng\,ml^{-1})$ calculated as the concentration equivalent to three times the standard deviation on the background signal, measured using the most abundant interference-free isotope for each element.

- The instrument is suitable for both elemental and isotope ratio measurement. The best precision attainable for isotope ratio determinations, however, is no better than 0.1% RSD (1 sigma) on two equal abundance isotopes. This is adequate for elemental analysis by isotope dilution (Box 8.3) but not sufficiently precise for most geological isotope ratio studies.

Fortunately it is those trace elements which are of particular interest in many geological studies which display some of the lowest limits of detection (Fig. 10.4) and the fewest interferences including the rare earth elements (REEs), high-field-strength incompatible elements (Zr, Hf, Nb, Ta; Hall *et al.*, 1990) and the platinum group elements (PGEs). Few other analytical techniques can give comparable performance for so many elements in a broad range of sample types and thus ICP–MS has a unique place in geochemical analysis.

Sample introduction

The preparation of geological samples may be a complex procedure with the aim of producing a homogeneous solution containing the analyte elements (Jarvis, 1992). The instrument components which introduce a representative portion of the solution into the plasma, where ions are generated and extracted, play an essential role in determining the reproducibility and sensitivity of analyte signals.

Nebuliser

The sample solution is either sucked or pumped from its container into a nebuliser whose purpose is to generate a fine aerosol carrying the sample. A number of different designs are available which vary in their construction, material and mode of operation. Two of the most widely used are the one-piece concentric glass 'Meinhard' design and the 'Babington' type. The former may be operated free running (i.e. solution is drawn up by a pressure drop generated as the nebuliser gas passes through a small hole) or the sample solution may be pumped through using a peristaltic pump (essential for the latter design). For dilute solutions of low viscosity, containing no particulate material (which can cause blocking), concentric glass nebulisers are ideal. They cannot, however, be used with solutions containing hydrofluoric acid since rapid degradation of the capillary tubes will occur. Meinhard nebulisers are delicate and must be handled with great care.

The Babington-type design is by contrast very robust. Sample solution flows over a small aperture in a curved or grooved surface, rather than passing through a fine capillary. A gas stream punctures the thin film producing a fine aerosol. This design of nebuliser is more tolerant of solutions containing higher levels of dissolved solids than the glass concentric designs. It is usually constructed of acid-resistant plastics such as PTFE or Ryton$^{\circledR}$. They cannot be operated free running.

Spray chamber

The nebuliser in practice rarely produces an aerosol of uniformly small droplet size. In order to prevent larger droplets (>10 µm) reaching the plasma, a spray chamber is placed between the nebuliser and the torch. As the gas flow carrying the aerosol enters the spray chamber, it undergoes sharp changes in direction (Fig. 10.1). The larger droplets strike the walls of the chamber and subsequently run to waste. The spray chamber ensures that only droplets small enough to remain in suspension in the gas flow are carried into the plasma; with most pneumatic nebulisers this means a loss of about 99% of the sample solution.

During the early development of ICP–MS, water was identified as a major source of ions for the formation of polyatomic species. Various advantages of introducing samples in the absence of water, with techniques such as laser ablation, electrothermal vaporisation or arc nebulisation, have been reported. Cooling the spray chamber (usually to about 5°C) reduces the amount of water or solvent introduced into the plasma, resulting in a significant reduction in the levels of some polyatomic ions (Williams, 1992). Stabilising the temperature of the spray chamber also leads to improved analytical precision. With organic solvents such as n-heptane, cyclohexane, methanol and diethyl ether, for example, the spray chamber is operated at between -5 and $-15°C$ significantly reducing the amount of vapour entering the plasma. Without this precaution many organic solvents cause the plasma to be extinguished.

Interferences

Interference effects in ICP–MS fall broadly into two categories: spectral interferences in which overlapping mass peaks add to the analyte signal and non-spectral interferences (matrix effects) in which analyte sensitivity is influenced by reagents or other sample constituents.

Spectral interferences

Isobaric overlap   Interference from an isobaric isotope of another element, although a common occurrence, is in practice rarely a serious analytical problem. With the exception of In, all elements which can be determined using ICP–MS have at least one (e.g. Co), two (e.g. Sm), or even three (e.g. Sn) isotopes that are free from isobaric overlap and one of these isotopes is used for analysis. In has only two isotopes, both of which have an overlap with Sn. Determination of In in samples containing high concentrations of Sn, such as cassiterite for example, is therefore not practical.

Polyatomic ions   Polyatomic ions form in the plasma from reactions between the most abundant ions present, for example Ar, O, N and H. Many different molecules form, but the most abundant are those involving argon (the plasma support gas), for example ArO^+, ArH^+. In addition, the acids used for sample dissolution or to stabilise dilute solutions may contain high levels of elements which form polyatomic species (Cl from HCl and $HClO_4$ and S from H_2SO_4). Combinations such as $ArCl^+$, ClO^+ and ArS^+ cause significant interferences. Some of the most serious polyatomic ion species produced by 1% v/v of each of the common high-purity mineral acids are shown in Table 10.1. Fortunately, the most abundant atoms in the plasma are usually of the light elements, with m/z below 40, and therefore no significant polyatomic species are produced above the argon dimer at $m/z = 80$. The elements most severely affected in routine analysis are V and As in the presence of chloride. These elements have

Table 10.1   Typical levels of polyatomic ion interferences

Mass number (m/z)	Analyte	Abundance of analysis isotope (%)	Equivalent Co concentration (ng ml^{-1})			Dominant polyatomic species
			1%HNO$_3$	1%HCl	1%H$_2$SO$_4$	
31	P	100	6.03	6.22	7.29	$^{15}N^{16}O$, $^{14}N^{16}O^1H$
44	Ca	2.1	8.50	10.3	**23.4**	$^{32}S^{12}C$, $^{14}N_2^{16}O$
48	Ti	73.7	0.62	1.24	**906**	$^{32}S^{16}O$
49	Ti	5.5	0.13	**2.93**	12.4	$^{32}S^{16}O^1H$, $^{35}Cl^{14}N$, $^{37}Cl^{12}C$
51	V	99.7	0.66	**230**	1.57	$^{35}Cl^{16}O$
52	Cr	83.8	0.61	6.58	0.88	$^{35}Cl^{17}O$, $^{37}Cl^{15}N$, $^{40}Ar^{12}C$
53	Cr	9.5	0.17	**75.0**	0.58	$^{37}Cl^{16}O$, $^{35}Cl^{18}O$, $^{35}Cl^{17}O^1H$
56	Fe	91.7	81.9	84.2	52.7	$^{40}Ar^{16}O$
57	Fe	2.14	1.68	2.08	1.24	$^{40}Ar^{16}O^1H$
65	Cu	48.9	0.22	0.84	9.85	$^{32}S^{33}S$
66	Zn	27.8	0.33	**3.59**	6.49	$^{35}Cl^{15}N^{16}O$, $^{32}S^{34}S$
67	Zn	4.1	0.09	0.94	0.63	$^{35}Cl^{16}O_2$
75	As	100	0.20	**43.1**	0.35	$^{40}Ar^{35}Cl$
77	Se	7.5	0.24	**14.8**	0.28	$^{40}Ar^{37}Cl$
80	Se	50.0	678	716	425	$^{40}Ar^{40}Ar$

Concentrations are calculated using Co as a single internal standard reference point. Stippled bars highlight analyte isotopes which coincide with polyatomic ions derived from carrier or atmospheric gases. These interferences apply regardless of the acid matrix used. Figures in bold type highlight interferences specific to one or more acids. HNO$_3$ is clearly the preferred acid matrix.

no alternative isotopes free from interference which can be used for analysis. It is therefore normally necessary to avoid the use of Cl-based acids during sample preparation and in the case of samples which contain high levels of Cl (e.g. sea water) to separate the analytes from the matrix. Sulphur (with four isotopes) is a particular problem, combining with Ar and O to form a large number of polyatomic species, and sulphur-bearing reagents should be avoided at all costs.

Refractory oxide ions   These species occur either as a result of incomplete dissociation of the sample matrix or from recombination in the plasma tail flame. Whatever the origin of these ions, the result is an interference 16, 32 or 48 mass units above the major isotope peak M^+ of the element forming refractory oxides MO^+, MO_2^+ and MO_3^+ respectively. In general, the relative levels of oxides expected can be predicted from the oxide dissociation energy of the molecule concerned. Those elements with the highest oxide bond strengths usually give the greatest yield of MO^+ ions including Al, Ba, Mo, P, REE, Si, Ti and Zr. However, providing that the ICP–MS operating conditions are optimised, modern instruments rarely develop oxide abundances in excess of 1.5% (the oxide signal as a ratio of the M^+ ion signal), insignificant unless the parent ion forms a major component of the sample matrix. For example, Si in limestones may occur at about 1 wt % SiO$_2$ (4675 µg g^{-1} Si). With a dilution factor of 1000 the solution presented for analysis contains only 4.6 µg g^{-1} Si and interferences from refractory silicon oxide species would be insignificant. However, in a pegmatite vein sample containing in excess of 75 wt % SiO$_2$ (350,625 µg g^{-1} Si) the solution would contain 350 µg g^{-1} Si and significant interference could occur at $m/z = 44$ (one of the main Ca isotopes).

In exceptional circumstances, refractory oxide species may cause severe interference problems. The REEs as a group all form oxide species to some degree. For REE distributions which occur in most silicate rocks this does not usually present a practical problem but corrections need to be made when analysing materials with high La/Yb ratios.

Doubly charged ions   Most of the ions which form in the plasma are singly charged (M^+) but some elements such as Ba and the REE produce a small proportion of M^{2+} ions. These may interfere on singly charged analyte isotopes that have the same m/z (i.e. having a mass number equal to half that of the doubly charged species). Under optimum plasma conditions (Jarvis *et al.*, 1989) the proportion of M^{2+} ions is low (0.5–0.8% of the M^+ ion abundance) and the number of elements affected is small. The interference of $^{138}Ba^{2+}$ on $^{69}Ga^+$ is a geologically important example. Methods of overcoming these interferences are discussed by Jarvis *et al.* (1992).

Matrix effects

This second group of interferences can be broadly divided into two categories: (a) physical effects resulting from the dissolved or undissolved solids present in a solution; and (b) analyte suppression or enhancement effects.

Dissolved solids effects   For elemental analysis by ICP–AES, samples are usually diluted to give a final solution containing $\sim$10,000 µg ml^{-1} total dissolved solids. Although the ICP source is common to both instruments, the level of dissolved solid which can be tolerated in ICP–MS is much lower (about 1000 µg ml^{-1}). Ions are extracted from the ICP through a chilled, metal sampling cone with an orifice of only 1 mm. High levels of dissolved solids result in undissociated material being deposited in the sampling cone orifice causing severe signal drift (usually loss) over short time periods and a reduction in precision.

Suppression and enhancement effects   Ionisation effects seem, in general, to be more severe in ICP–MS than in ICP–AES. With increasing Na concentration, a greater suppression of the analyte signal is recorded. The analysis of sea water is therefore particularly difficult, and samples should always be diluted to reduce the level of Na to less than about 1000 µg ml^{-1} prior to analysis.

In addition to analyte suppression from a specific easily ionised element, nearly all matrix elements will cause some degree of signal suppression (or enhancement) if they are present at a high enough concentration. In general, the lower the atomic mass of the analyte, the greater will be the effect of a given matrix element on the ion count rate of the analyte. For a given analyte, the greater the mean atomic mass of the matrix, the greater will be the effect on the analyte count rate (Gregoire, 1987).

In practical analysis, dilution of the sample to reduce the level of any single matrix element to less than about 500 µg ml^{-1} (the exact concentration should be experimentally determined for the element concerned) is usually carried out and ensures that suppression effects are avoided. In cases where dilution would result in the analytes of interest being below the limit of detection, matrix matching of standards and samples or standard additions may be appropriate alternative calibration strategies.

Data acquisition

Qualitative analysis

The capacity to collect signals for all masses during a series of scans can be used for purely qualitative examination of a sample and it is possible to collect information over the complete mass range ($m/z = 4$–240) in less that 60 s. The spectra can then be visually examined for the presence or absence of an analyte, and to identify possible sources of interference. Sample matrices which are unfamiliar should always be subject to qualitative scrutiny prior to full quantification. Software packages provided with commercial ICP–MS instruments usually allow several spectra to be displayed and subsequent manipulation of these spectra to enlarge scales and subtract background signals.

Semiquantitative analysis

On most instruments, a plot of sensitivity against mass yields a relatively smooth curve if degree of ionisation and isotopic abundance are taken into account (Figure 10.5). This response curve can be used to calibrate the instrument, using a single element from the curve, to provide semiquantitative data. In practice, this calibration method is ideal as a survey tool, particularly when the sample matrix is unfamiliar or if only approximate concentrations are required. The accuracy of this method is variable (typically between 5 and 150%) and highly dependent on the element sought and the sample matrix.

Quantitative analysis

External calibration   The most widely used calibration method is that using a set of external calibration standards, usually three plus a blank. For solution analysis these may have a simple aqueous or acid matrix containing the analytes of interest. Several standard solutions are prepared which cover the range of expected concentrations and a calibration line is fitted to the measured data using least-squares regression analysis. During an analytical run the sensitivity

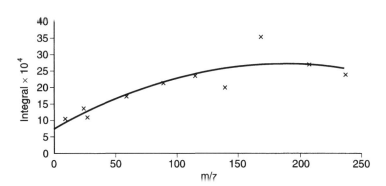

Figure 10.5   A plot of mass vs response yields a relatively smooth response curve providing that isotopic abundance and ionisation potential are taken into account.

may fluctuate, principally owing to the presence of the sample matrix. Short-term fluctuations in signal may occur over a time scale of a few minutes and a gradual loss of sensitivity is not uncommon over several hours. Two approaches, **drift** correction and **internal standardisation**, can be used to correct for changes in sensitivity.

Standard additions   This alternative calibration strategy has been used with some success in ICP–MS analysis. The calibration is performed by taking aliquots (normally three or more) of the sample to be analysed, and adding to each increasing quantities of a reagent containing the element or elements of interest. The calibration set therefore consists of a number of spiked samples plus the unspiked original, all of which have an almost identical matrix. The samples are analysed and a graph constructed of the integral of the isotope of interest versus the concentration of the element added. The intercept of the calibration line on the x axis (a negative number) gives the concentration in the unspiked sample. Although this method can produce highly accurate and precise data, it is time consuming to perform and is best suited to a relatively small number of elements.

Isotope dilution   Stable isotope dilution is a powerful strategy for elemental analysis which can in principle be applied with any mass spectrometric technique. The basis of the method is the measurement of the change in the ratio of signal intensities for two selected isotopes of an element, after the addition of a known quantity of a spike which is enriched in one of these isotopes. The principles of isotope dilution are given in Box 8.3.

Isotope dilution methodology offers a number of advantages over other calibration strategies in ICP–MS. It compensates for partial loss of the analyte during sample preparation, provided that such loss occurs after chemical equilibrium of the spike with the analyte has been achieved. It is also immune to a wide variety of 'matrix effects' which have an identical effect on all isotopes of the same element. The method can be considered the ideal form of internal standardisation since the internal standard for each element is one of its own isotopes. However, isotope dilution is generally not suitable for monoisotopic elements, for example Au (unless radioactive spikes are used), or for elements where only one isotope is free from isobaric overlap, for example Ba. It is expensive and time consuming to undertake (particularly for multi-element analysis) and generally offers little improvement over other forms of calibration methodology for routine analysis. It is most widely used by agencies involved in the production of **certified reference materials** where the highest quality of data is required.

Analytical strategy

The range and combination of elements which can be determined by ICP–MS is very broad. The instrument provides the possibility for the simultaneous determination of most elements in the periodic table at major, trace and ultra-trace levels. However, there are a number of limitations which, in practice, restrict those elements which can be measured simultaneously in a single solution. For example, the precision required for most major determinations is between 0.5 and 1.0% RSD, a figure which is not routinely achievable on current instrumentation. Some volatile trace elements (Hg, As, Cd) may be lost during sample preparation and may therefore not be available for determination. Thus, in practice, analytical methodology is best developed for specific elemental groups, taking into account the matrix and method of sample preparation. Prior to quantitative determination, samples of unknown composition should be qualitatively scanned to identify major and trace components and to identify possible interferences, particularly from dissolution acids. Final sample dilutions should only be made once the decision has been taken to include or exclude one or more internal standards.

With sample preparation concluded, the ICP–MS instrument can be optimised either for general multi-element analysis or for the determination of only a few specific elements. Data acquisition times should be long enough to ensure that counting statistics obtained are adequate for the elemental concentration range expected. Standard calibration solutions may need to be matrix matched for certain applications, although simple aqueous standards are usually sufficient. Frequent wash solutions (2% w/v HNO_3) are included to remove traces of the previous sample. The uptake time is the time required for the new solution to travel from its container into the plasma. Samples should be analysed with up to three replicates. If a signal drift monitor is included, it should be first analysed prior to any of the standards or unknown samples. Repeated analyses of this solution should be made at relatively high frequency, every 5–10 samples, throughout the run, to monitor and correct for changes in analyte signal.

Selected applications

Of the many scientific fields in which ICP–MS has been applied, geological applications have received the most attention. The technique can be applied to multi-element analysis in most sample types, but it is particularly appropriate for the determination of some specific elemental groups which are considered below in detail.

Rare earth elements (REEs)

REEs are an essential part of many geochemical studies. For example, the fractionation of light relative to heavy REEs during the evolution of magmas can provide valuable evidence for crystal fractionation or magma mixing. They are a difficult elemental group to quantify by most instrumental techniques without separation and preconcentration and ICP MS is, in many ways, an ideal technique (Jarvis, 1989). The ICP–MS spectra (Fig. 10.3c) are simple to interpret, each REE has at least one isotope free from isobaric overlap and sensitivity is relatively uniform from [139]La to [175]Lu. The major potential analytical problem is the level of refractory oxide formation. Since the REEs form a continuous group from $m/z =$ 139 to 175, the formation of light REE oxide species can produce significant interferences on the middle to heavy REEs. Corrections may therefore be required. High concentrations of Ba in solution may also cause significant interferences, particularly on Eu.

The detection limits for all 14 REEs are typically between 0.001 and 0.01 ng ml^{-1}, figures which are considerably lower than most other instrumental methods for comparable preparations. However, a comparison with other techniques is best made by comparing quantitation limits in the solid (Fig. 10.6). Separation of the REEs from the rock matrix is not necessary and concentrations down to about 10× chondrite (La = 3 µg g^{-1}, Lu = 0.3 µg g^{-1}) can be

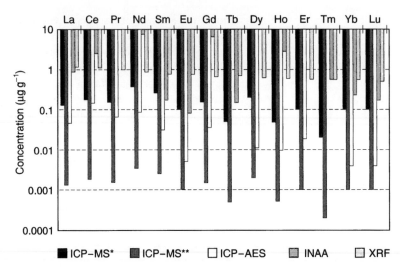

Figure 10.6   A comparison of REE quantitation limits for ICP–MS, ICP–AES, INAA and XRF calculated as the concentration equivalent to 10 times the standard deviation of the background. Concentrations in the solid sample. ICP–AES data assume dilution factor of ×10 (1 g in 10 ml) after separation and pre-concentration by ion exchange, ICP–MS* assumes a 1000 × dilution factor (1 g in 1000 ml), ICP–MS** assumes dilution factor of ×5 (1 g in 5 ml) after separation and preconcentration by ion exchange. For ICP–AES and XRF, REEs are separated prior to analysis.

determined with good accuracy and precision. Below this concentration, separation and preconcentration of the REE is required so that high dilution factors can be avoided. After separation and preconcentration by ion exchange chromatography, concentrations in the region of 1 ng g^{-1} (i.e. >0.001 times chondrite), can be determined. At this level, however, reagent blank concentrations become the limiting factor.

Platinum group elements (PGEs)

PGE (Ru, Rh, Pd, Os, Ir and Pt) levels in silicates and most unmineralised samples are below the limit of detection by ICP–MS once samples have been diluted to reduce the level of dissolved solids. In addition, PGE-bearing minerals are often resistant to acid attack and some form of fusion is normally required to bring them into solution. To compensate for the addition of flux, further dilution is required. It is therefore normally essential to separate the PGEs from the sample rock matrix prior to analysis. Pd, Pt and Rh have traditionally been determined using classical lead fire assay. However, the nickel sulphide fire assay technique is gaining popularity because it provides an effective collection for the whole group. This collection technique has been adapted for ICP–MS determination with some success (Jackson _et al._, 1990, Sun _et al._, 1993, Juvonen _et al._, 1994).

The PGEs lie in two distinct parts of the mass range from ^{96}Ru to ^{110}Pd and ^{186}Os to ^{198}Pt, and each element has at least one isotope free from isobaric overlap. Owing to their position in the mass spectrum, the PGEs are relatively free from interference effects. Caution should be exercised when analysing samples prepared for fusion in zirconium crucible (essential for Na_2O_2 fusions). Significant contamination of the sample solution occurs and the formation of ZrO and ZrOH causes a number of interferences from m/z = 106 to 113.

Although the detection limits are impressively low for all elements $(0.05–0.4\ ng\ ml^{-1})$ most sample types require separation and preconcentration prior to analysis. Apart from fire assay, tellurium co-precipitation and ion exchange chromatography have also been effectively used as separation procedures.

The identification of enhanced concentrations of iridium in Cretaceous/Tertiary boundary sediments has led to the suggestion that extensive vulcanism may have been responsible for the mass extinction event recorded at that time. Iridium spikes have also been reported in rocks at a number of other geological stage boundaries. However, recent work suggests that these enrichments may not be the result of such a dramatic volcanic event, but are controlled by burial or diagenetic processes. Often associated with enhanced iridium levels are spikes in other PGEs. Whatever the cause of these anomalies, the concentrations determined are very low, for example $<1\ ng\ g^{-1}$ Ir or a few $ng\ g^{-1}$ of Pt. ICP–MS has recently been applied to the determination of Pt, Ir and Re in abyssal plain sediments from the North Atlantic using the technique of flow injection for sample introduction (Colodner *et al.*, 1992). This work clearly demonstrated that observed spikes in Pt concentration profiles are related to the position of the redox front (Fig 10.7).

Although data for all of the PGEs are normally required, in certain applications only selected elements are of interest. For example, Re and Os are fractionated during natural processes and therefore a range of Re/Os ratios are to be found in nature. These elements are of particular interest in geological studies and have been used to determine the age and genesis of ore deposits, the origin of meteorites and as indicators of catastrophic events in

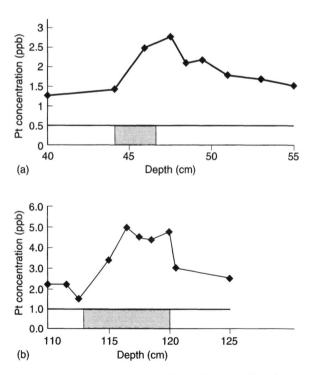

Figure 10.7   The distribution of Pt surrounding an oxic–sub-oxic transition zone (shaded area) in two cores from North Atlantic abyssal plain sediments determined by flow injection ICP–MS after preconcentration by anion exchange chromatography (from Colodner *et al.*, 1992).

the geological record (e.g. Lichte *et al.*, 1986). Detection limits are typically better than 0.02 ng ml^{-1}.

Ultra-trace-level gold determination may be carried out in solutions preconcentrated from sea water. The concentration range of Au in sea water is typically between 0.02 and 0.2 pg g^{-1} and contamination is therefore potentially a serious problem. Separation and pre-concentration are essential in this application to eliminate the NaCl matrix and to bring concentrations to above the detection limit (Bakowska *et al.*, 1989).

Sea water analysis

Sea water, particularly from near-shore sources, contains a wide range of trace elements at concentrations from a few μg ml^{-1} to less than 0.1 ng ml^{-1}. Open-ocean sea water is typically depleted in metal content by comparison. The accurate determination of the elemental constituents of such samples is required for studies of water movement and evolution, while evaluation of near-shore pollution may be essential for environmental monitoring purposes. The major components of sea water, Na and Cl, hinder accurate analyte determination not only by ICP–MS but also by other spectroscopic techniques. However, the excellent sensitivity of the ICP–MS technique allows the direct determination of some elements (e.g. Ca, Sr and Mg) after simple dilution by a factor of between 10 and 100. Concentrations down to about 10 ng ml^{-1} in the seawater sample can be determined in this way with good accuracy and precision, particularly where isotope dilution can be applied. For elements occurring at much lower concentrations, such as As, Cd, Cr, Co, Cu, Fe, Pb, Ni and Zn for example, the analytes must be separated from the NaCl matrix and preconcentrated prior to determination. McLaren *et al.* (1993) have developed an on-line preconcentration method using columns packed with silica-immobilised 8-hydroxyquinoline. Using this method, accurate measurements can be made down to 0.022 ng ml^{-1}.

Analysis of solid samples

A majority of geological samples occur in the solid form and must be brought into solution prior to analysis. However, a number of real advantages can be identified with respect to the direct analysis of solids. Dilution of a sample normally results during a preparation procedure, reducing the actual determination limits available. Some solid sample introduction techniques (e.g. laser ablation), allow the sample to be analysed directly with little or no dilution. The time required for some preparation schemes can be extensive, from a few hours to days. This preparation time may be considerably reduced for solid sample introduction where samples are typically ground to a fine powder and suspended in a fluid medium (slurry nebulisation), mixed with a binder and pelletised (arc nebulisation or laser ablation) or analysed directly (laser ablation). Other advantages include reduced contamination, retention of volatile elements, and finally the ability to analyse highly refractory materials which are extremely difficult to bring into solution. In addition to bulk analysis, laser ablation offers the possibility of carrying out discrete profiling. In this way, detailed elemental or isotopic distributions through a sample can be investigated (e.g. mineral zoning).

A laser is used to remove material from a solid sample, such as a mineral grain, and the resultant vapour or vapour/particulate mixture is transported into the ICP. Owing to limitations with calibration, laser ablation is currently used to study elemental zoning in single minerals in a qualitative or semiquantitative manner. It is feasible to determine readily the sense of REE zoning in monazites or Ni zoning in olivines, for example. The major advantage of LA–ICP–MS over existing microprobe techniques is the very low detection limits available, particularly for elements with $m/z > 100$, and rapid analysis times of <60 s. Fully quantitative element determination, with a few exceptions (Jarvis and Williams, 1993), is currently difficult to achieve. Future developments with laser ablation ICP–MS include a reduction in the ablation spot size needed for a single analysis from ~ 100 µm to about 10 µm and the possibility of automated analysis.

Summary

ICP–MS is an analytical technique which has application in a wide variety of geochemical studies. Although the instrumentation is capable of near simultaneous determination of most elements in the periodic table, it is clear from the above that there are a number of analytical considerations to be made. Appropriate sample preparation methodology is essential and in many ways is the major limiting factor with modern instrumental methods. Both elemental and isotope ratio measurements can be made rapidly and with minimal sample pretreatment, making the technique both cost effective and versatile.

Chapter 11

Elemental analysis by spark source mass spectrometry

Klaus P. Jochum

In spark source mass spectrometry (SSMS), a powdered solid sample (mixed with graphite if not itself conducting, and pressed into electrodes) is vaporised under vacuum by a high-voltage radiofrequency spark. Ions in the resulting plasma are analysed using a high-resolution mass spectrometer in which mass peaks may be recorded photographically or electrically. The advantages of SSMS include wide element coverage, an extensive concentration range and analysis of solid materials without dissolution.

Apparatus

Sample preparation

One of the principal advantages of SSMS over other methods is the simplicity of sample preparation. Typically, 50 mg of powdered rock sample is mixed with ultrapure graphite (containing an **internal standard** element or isotopic **spikes**) and then compressed into rod-shaped electrodes.

Ion source

SSMS is based on the generation of ions within a plasma formed by a radiofrequency electrical discharge between two electrodes containing the sample. The RF high-voltage supply for the spark (typically 25 kV at 1 MHz) is modulated in short pulses (length 200 μs, repetition rate 300 Hz). The total power density is extremely high (about 10^{12} W m^{-2}). Ramendik *et al.* (1988) have described the ion formation process as a complex series of steps (atomisation, ionisation, plasma dispersal) that rapidly progress from one to another. This results mainly in the formation of singly charged and neutral species. The electrodes are adjusted periodically from outside the vacuum to maintain a constant spark gap and position throughout an analysis.

The ions pass through accelerating slits (20–25 kV) into the mass spectrometer.

Analyser

The ions produced in the spark gain a wide spread of kinetic energies (about 2 keV). To avoid consequent loss of **mass resolution**, all commercial spark source mass spectrometers (e.g. AEI MS7 and MS702 and Jeol JMS-01B) are double-focusing instruments with Mattauch–Herzog geometry (Box 11.1).

The spark source produces a fluctuating ion beam, and a beam current monitor intercepting a fixed fraction of the beam between the electrostatic and magnetic analysers provides a measure of the total charge accumulated in the course of a measurement (Fig. 11.1).

Ion detection

The *photographic plate* is the most widely used ion detection system in SSMS (Fig. 11.1). The bombardment by energetic ions results in the production of a latent image on the AgBr grains of the photosensitive emulsion. Photographic development blackens the grains, rendering this latent image visible. About 5000 ions are needed to obtain a perceptible mass line on the photoplate.

The photoplate has many advantages over other detection systems, for example: (a) many mass lines can be registered simultaneously; (b) it provides a permanent visual record of an analysis; (c) a high mass resolution of about 2000–10,000 can be achieved.

Unfortunately photoplates suffer the disadvantage of having a low **dynamic range**, which is around 100. In practice a series of 15 exposures with increasing total ion charges (as recorded on the beam current monitor) are taken in order to accommodate the range of elemental abundance variations present in the sample. A typical set of exposures would range from 0.001 to 300–2000 nC, corresponding to measuring times of 0.5 to 3 hours.

Electrical ion detection is suitable for sensitive, precise and fast analysis of a limited number of elements. Ions of a specific mass are selected by an appropriate choice of magnetic or electric field strength, and registered by a Faraday cup or an electron multiplier (Box 8.1). Recently a system has been developed consisting of 20 separate **channeltrons** for simultaneous ion-counting measurements.

Working techniques

Internal standard elements

For quantitative analysis, an **internal standard** element has to be present in the sample incorporated in the electrodes. Its abundance must be independently determined by some other method, or some suitable element (e.g. Lu; Taylor and Gorton, 1977) must be added. The multi-element **isotope dilution** technique (Jochum et al., 1988) allows the simultaneous determination of many internal standard elements of different geochemical behaviour, distributed in different mass ranges, and over a wide range of concentration (Figs 11.1 and 11.3).

Box 11.1 Double-focusing mass spectrometers

Because the trajectory of an ion in a conventional mass spectrometer is dependent on its kinetic energy (Box 8.1), the width of a mass peak depends on the energy spread of the ion source. If very high **mass resolution** is required (e.g. for organic compounds of high relative molecular mass – Chapter 16), or if a source having a wide energy spread is being used (e.g. SSMS), it is necessary to limit, or to compensate for, the range of ion energies accepted by the mass spectrometer. A double-focusing mass spectrometer is one that uses an *electrostatic analyser* in series with the magnetic field to accomplish this.

The need for high resolution in SSMS is illustrated by the analysis of trace amounts of niobium. The spectrum at **mass number** 93 consists of several peaks: the peak of interest $^{93}Nb^+$ ($M = 92.906$ amu) and certain molecular peaks caused by major elements in the sample, such as $^{29}Si^{16}O_4^+$ ($\underline{M} = 92.956$ amu). The mass difference ΔM between these neighbouring ions is 0.050 amu; an instrument with a mass resolution $M/\Delta M$ significantly greater than $92.9/0.05 = 1860$ is required to resolve the Nb peak from this molecular interference.

The electrostatic analyser is an energy-selecting or energy-focusing device; it either rejects ions whose energies lie outside a predetermined range (Fig. 11.1.1a), or modifies their trajectory to cancel out energy dispersion arising in the magnetic analyser (Fig. 11.1.1b). The commonest electrostatic field geometry is the radial field between two co-axial cylinder segments (Fig. 11.1.1) held at equal but opposite electric potentials. The combination of accelerator and analyser fields is such that r_e, the radius of curvature for a particular ion, depends on its kinetic energy, *not* on its mass.

Two alternative designs of double-focusing mass spectrometer are used, with somewhat different properties. The Nier–Johnson configuration (Fig. 11.1.1a) combines a 90° electrostatic sector with a 90° magnetic sector. The slit between the analysers transmits only a narrow range of ion energies (as determined by the electrostatic field) to the magnetic analyser. Ions of the desired m/z are brought to the collector slit by appropriate setting of the magnetic field and recorded electrically. This configuration, which is capable of highest mass resolution, is favoured for SIMS (Chapter 15) and organic mass spectrometry (Chapter 16).

With the Mattauch–Herzog design (Fig. 11.1.1b), owing to the way in which the electrostatic and magnetic fields are configured, the energy dispersion of the magnetic field is compensated by equal but opposite dispersion in the preceding electrostatic sector. A wider range of ion energies can therefore be accepted by the magnetic sector without loss of resolution, and this design thus offers higher ion transmission (allowing the measurement of feebler ion beams). It also has the capability to focus all masses in one plane (e.g. a photographic plate), and is therefore the only design used for SSMS. For electrical detection, the photoplate is replaced by a slit which admits the required m/z to the detector, or by a multi-ion-counting (MIC) system consisting of 20 separate **channeltrons**.

KPJ, NS and RG

From optical density to concentration

An automated microdensitometer is used to read the **optical density** D or **transparency** τ of the mass lines on the photoplates (Fig. 11.1). D and τ are defined in terms of light intensities transmitted by the plate:

$$D = 1 - \tau = (I_0 - I)/I_0 \qquad [11.1]$$

where I_0 is the light intensity on an uncoated region of the plate and I the light intensity transmitted through the densest part of a spectral line.

D and τ values for a particular line are functions of (a) concentration of isotope and (b) exposure. However, this relationship is rather complicated because of the non-linearity inherent in the transparency curves (Fig. 11.2).

The concentration C_x of an element x can be calculated from

$$C_x = C_s(Q_i/Q_j)\,(X_{j,s}/X_{i,x})\,(M_x/M_s)\,(1/f_x) \qquad [11.2]$$

Figure 11.1.1 Alternative geometries of double-focusing mass spectrometers. ϕ_e and ϕ_m represent the angles of deflection of ion trajectories in the electrostatic and magnetic fields respectively; (b) also shows the configuration of the spark source in SSMS.

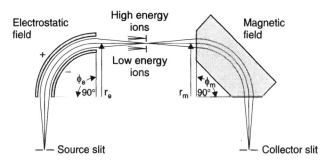

(a) Nier–Johnson geometry

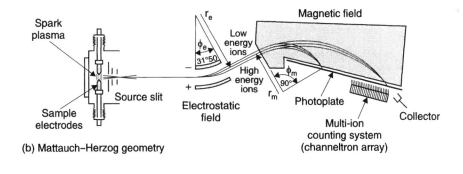

(b) Mattauch–Herzog geometry

where C_s is the concentration of an internal standard element s; Q_i/Q_j is the measured ratio of the ion abundances of isotope i of element x and of isotope j of element s; $X_{i,x}$, $X_{j,s}$ are the isotopic abundances of isotopes i and j, respectively; M_x and M_s are the relative atomic masses of the elements x and s; and f_x is an empirical 'relative sensitivity factor' (RSF).

Spectral interference

There are four potential sources of interference:

(1) **isobaric interference** from isotopes of neighbouring elements;
(2) multiply-charged ions of heavier elements;
(3) cluster ions of the type C_2^+, C_3^+, ... formed from the graphite electrodes in the discharge;
(4) molecular fragments, oxides and carbides, such as $CaAlC_2^+$, BaO^+, BaC_2^+.

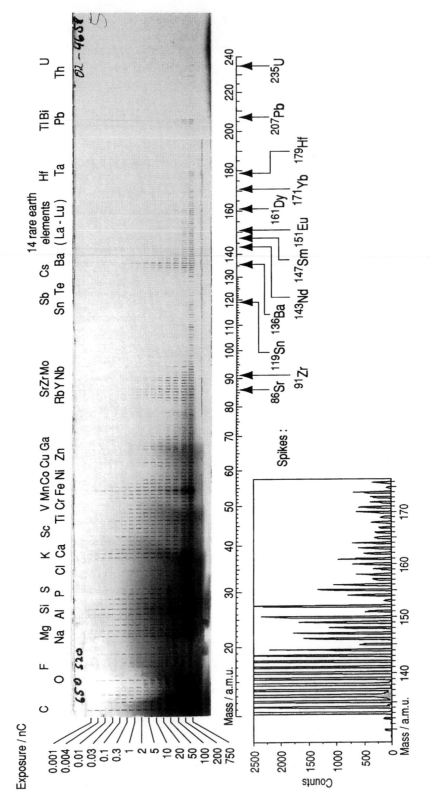

Figure 11.1   Mass lines on a photoplate obtained from the analysis of the geologic standard rock BCR–1. The plate records a graded series of 15 exposures (left-hand scale). The fogged area in the region of the major matrix lines is caused by positive secondary ions produced by the impact of primary ions on the photoplate. Isotopically enriched **spikes** have been added to the sample (lower horizontal scale) for quantitative analysis by multi-element **isotope dilution**.

The inset shows the mass spectra of Cs, Ba and the rare earth elements in the spiked standard sample JR–2, which were obtained by ion counting using a total ion charge of 1 nC for one measurement.

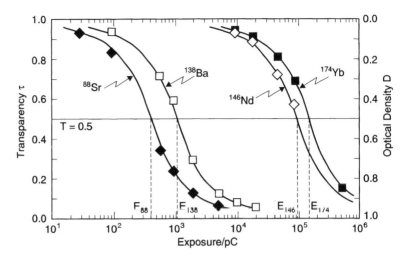

Figure 11.2   Transparency curves for four mass lines. Successive points represent measurements on different exposures in Fig. 11.1. The curves are fitted to the data by empirical non-linear functions (see Ahearn, 1972). To obtain relative ion abundances, hypothetical exposure values E_i are read off from the curve at the same transparency value (e.g. horizontal line at 0.5); ion abundance is inversely proportional to this exposure value, for example $Q_{^{174}Yb}/Q_{^{88}Sr} = E_{88}/E_{174}$.

Relative sensitivity factors (RSFs)

Differences in ion formation, transmission and detection of the various elements are corrected by using empirical RSFs. The RSF f_x is defined as

$$f_x = C_x(\text{meas})/C_x(\text{true}) \qquad\qquad [11.3]$$

where C_x (meas) is the apparent concentration of element x obtained by reference to the internal standard element and C_x(true) is the 'true' concentration in a **standard** sample. The principal factors which influence the RSFs are element specific (Fig. 11.4). The determination of RSFs by analysing standard rocks is the most commonly used technique.

The RSFs have been established to a high degree of precision (Fig. 11.4). If the measuring conditions are kept constant, the RSFs remain constant – with the possible exception of volatile elements, like Rb and Cs – over a period of many years. They are independent of trace element compound, sample mineralogy and concentration (over five orders of magnitude).

Multi-element isotope dilution

Isotope dilution (ID) (Box 8.3) has gained particular importance in SSMS. The ID–SSMS technique yields the most precise and accurate trace element analyses because isotope ratios can generally be measured in one exposure and no sensitivity factors are needed. It offers the possibility of simultaneous multi-element ID analyses (Figs 11.1 and 11.3), with high sensitivity, and a minimum of sample pretreatment (Jochum *et al.*, 1988). A 'dry' technique has been developed in which the spike solutions are added to ultrapure graphite and dried. This 'spiked' graphite can then be mixed with the undissolved sample powder for analysis.

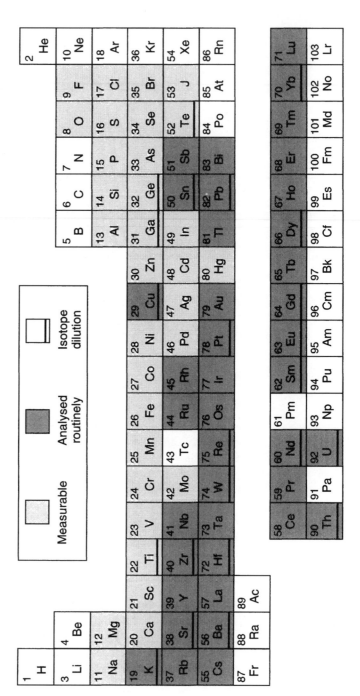

Figure 11.3   Major, minor and trace elements which are measurable in many terrestrial and extraterrestrial samples by spark source mass spectrometry (SSMS); 40 trace elements can be analysed routinely by SSMS laboratories with high precision and accuracy (Taylor and Gorton, 1977, Jochum *et al.*, 1988).

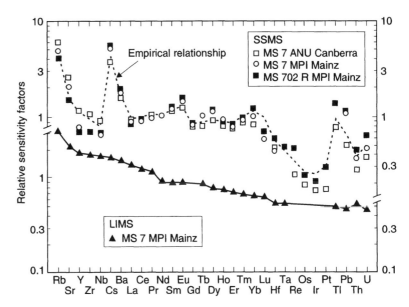

Figure 11.4   Relative sensitivity factors (RSFs) (compared by setting Nd = 1) of trace elements determined by different laboratories. Sensitivity decreases as a function of mass by variations of linewidth and mass response of the photoplate emulsion. In SSMS, the experimentally determined RSFs agree well. They are similar to the model values calculated by Taylor and Gorton (1977) from boiling point, first ionisation potential and relative atomic mass. In LIMS, nearly all elements are ionised with about the same efficiency at a power density of 10^{14} W m^{-2}.

Performance of SSMS

Elements

Nearly all stable and long-lived elements have been analysed by SSMS (Fig. 11.3). However, the detection of elements can be impaired by the interferences of spectral lines. For example, plagioclase-rich samples invariably yield complex molecular species, and one of these ($^{28}Si_2{}^{27}Al^{12}C_2$) interferes on ^{107}Ag (Taylor and Gorton, 1977).

Detection limits

The concentration ranges of 63 elements are plotted in Fig. 11.5. For elements having atomic numbers $Z > 36$ (Rb–U), the detection limits generally vary between 0.001 and 0.1 ppm ($\mu g\,g^{-1}$). In the lower-mass range, detection limits are much higher because of interferences and high background caused by major matrix lines.

   SSMS is one of the most sensitive methods for analysing solid materials. This is especially true for some trace elements (e.g. Nb, Zr, Y, Sn, Sb, Th) where reliable analyses in the ppb ($ng\,g^{-1}$) range can be performed (Jochum et al., 1990).

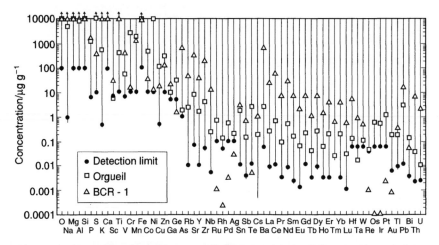

Figure 11.5   Concentration ranges and detection limits (3s of photoplate background) of 63 elements. To demonstrate the analytical capabilities of SSMS, the element abundances of two typical samples (terrestrial basalt standard BCR-1 (Govindaraju, 1994); stony meteorite Orgueil (data of Anders and Grevesse, 1989)) are plotted. Nearly all elements can be measured by SSMS in these samples except those with extremely low concentrations (e.g. noble metals in BCR-1).

Precision

Precision of the analyses depends on several factors, such as the homogeneity of the sample–graphite–internal standard mixtures, and the precision of isotope ratio determinations. Optimal spiking, with isotope ratios close to unity, and optimal selection of internal standard elements yield the best results. Overall, **reproducibility** is better than 5% (1 sigma) for most elements. Significantly better results (1–3%) are obtainable using the isotope dilution technique and/or the multi-ion counting technique (Jochum *et al.*, 1994).

Accuracy

The accuracy of SSMS analyses can be estimated by using standard reference materials. Figure 11.6 shows a comparison of SSMS analyses with reference values. Most data agree with the recommended values to within 5%.

Disadvantages

There are only a few laboratories that use SSMS. The main reasons why SSMS has not become a routine method are the complex and expensive instrumentation and the low sample throughput (about one analysis per day). However, the high sensitivity of the recently developed multi-ion counting system (Box 11.1) offers much shorter measuring times of 1 minute to 1 hour.

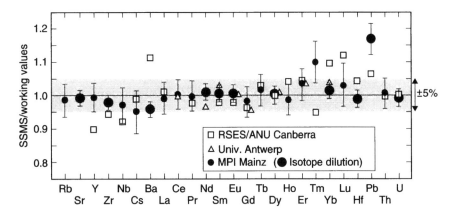

Figure 11.6 Comparison of SSMS analyses of standard rock BCR-1 from several laboratories with the **working values** compiled by Govindaraju (1994). Error bars ±1σ (Mainz data only).

Applications

Typical applications of SSMS are found in (a) the determination of all the elements within an element group (e.g. REEs, noble metals), (b) element correlations, (c) determining certain elements at extremely low concentrations which cannot be done by other techniques (e.g. Nb, Y, Zr, Rh, Th, Sn), and (d) multi–element analysis where only small amounts of sample are available (e.g. basalt glasses, minerals, meteorites, and lunar samples). Additionally, laser plasma ionisation mass spectrometry (LIMS), a new development from SSMS, allows *in situ* microanalyses of rock samples and minerals.

Chemistry and evolution of the Earth's mantle

In investigations of mantle chemistry and its time evolution, comprehensive element coverage is a particularly important requirement that SSMS fulfils to a high degree. Figure 11.7 shows the average composition of a worldwide sampling of normal type mid ocean ridge basalt (N-MORB) glasses. The average composition of the continental crust is also plotted for comparison. The average analyses, surprisingly, appear to form complementary patterns. This relationship can be understood in terms of a simple, two–stage model for extracting first continental, and then oceanic, crust from the mantle (Hofmann, 1988). This model satisfactorily reproduces the characteristic concentration maximum observed in MORB, and yields quantitative constraints on the effective aggregate melt fractions during both stages: about 1.5% for the continental and 8–10% for the oceanic crust.

Certain element ratios (e.g. Nb/U, Ce/Pb, K/U, Sn/Sm mostly established by SSMS) have become extremely important in constraining the composition of the Earth and its evolution. Hofmann *et al.* (1986) demonstrated that Nb/U ratios are remarkably uniform in both MORB and oceanic island basalts (OIB) (Fig. 11.7 inset). The average Nb/U ratio of 47 for these rocks is higher than the primitive mantle ratio of about 30, as deduced from

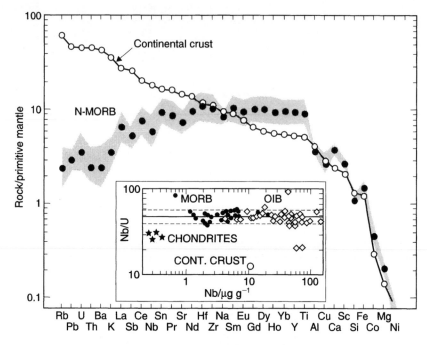

Figure 11.7   Average compositions of 26 normal-type mid-ocean ridge basalt (N-MORB) glasses obtained by the
SSMS lab in Mainz (except major elements, Sc, Co and Ni), normalised to primitive mantle (Hofmann, 1988). The stippled band shows population variability ($\pm 1\sigma$). Elements are plotted in order
of decreasing concentration in average continental crust, which is also shown (Taylor and McLennan,
1985). The inset shows that the Nb/U ratios in MORB and oceanic island basalts (OIB) differ significantly from measured or estimated values for chondrites and for continental crust (Hofmann *et al.*,
1986).

SSMS analyses of carbonaceous chondrites; it is also greater than the respective ratio in the
continental crust (about 10). The explanation put forward by these authors was that segregation of continental crust from the mantle during the Archean and early Proterozoic
occurred through a process that enriched the crust in U more strongly than it did Nb. The
residual mantle, relatively depleted in U, was homogenised during or after this segregation
process, and subsequently differentiated into the chemically and isotopically distinct sources
of present MORB and OIB. SSMS analyses of Archean to Tertiary rocks (Jochum *et al.*,
1991) show a secular increase in Nb/Th ratio. This variation may indicate a change in
upper mantle composition resulting from the progressive withdrawal of continental crust
over the past 3 Ga.

Microanalysis using laser plasma ionisation mass spectrometry (LIMS)

One promising new development in the area of SSMS has been the replacement of the
entire spark source assembly by a laser ion source (e.g. Matus *et al.*, 1994). The resulting
instrument offers further advantages compared with conventional SSMS. LIMS allows
the direct analysis of rock surfaces without any sample pretreatment and has already been
used in the microanalysis of individual mineral phases without separation. The high power

density (10^{14} W m^{-2}) results in more uniform ionisation efficiency (Fig. 11.4), so that only small corrections are necessary for quantitative analysis. The number of measurable elements and their detection limits are similar to conventional SSMS. It is expected that LIMS will find wide geochemical application in the future.

Chapter 12

Accelerator mass spectrometry

Louis Brown

Accelerator mass spectrometry employs techniques of nuclear physics to extend the applicability of the mass spectrometer to unprecedentedly low analyte concentrations. The motivation behind this development has been the need to analyse vanishingly small concentrations of cosmogenic nuclides such as ^{10}Be and ^{14}C in the presence of enormously greater amounts of the common isotopes of the same elements.

Cosmogenic nuclides

Cosmogenic nuclides are radionuclides produced by the interaction of cosmic rays either with interplanetary matter or with elements present in the Earth's atmosphere or surface rock. The most important is indisputably ^{14}C, which decays rapidly enough to allow its application to the dating of a wide variety of natural samples ('radiocarbon dating'). The traditional method for determining ^{14}C is to count its emitted radiation, but the decay rate is so low that very large samples and long counting times are required. Five other cosmogenic isotopes (^{10}Be, ^{26}Al, ^{36}Cl, ^{41}Ca and ^{129}I) have half-lives much too long for decay counting techniques to be of use.

Conventional mass spectrometry has limitations that preclude its application to cosmogenic nuclides because the ratio of the cosmogenic isotope to the stable isotopes of the same element is so small (as little as 10^{-12}): the cosmogenic isotope is at such a low level that its measurement by mass spectrometry requires the counting of individual ions, sometimes as slowly as a few per minute, whereas the ion current derived from the attendant stable reference isotope is intense enough to be measured with a common laboratory multimeter. Conventional mass spectrometry offers no adequate means of coping with this extreme disparity in abundance.

The cause of these limitations is the arrival at the detector of ions that are not the ones desired and against which the detector is unable to discriminate. Such interfering ions have three origins: (a) **isobars** of another element and (b) molecular ions, both of which have obeyed all the laws of ion optics on their way from ion source to detector, thereby causing them to enter it at very nearly the same mass setting as the desired isotope; (c) ions from the inevitably intense reference beam (another isotope of the same element) that should be excluded on ion optical grounds but which reach the detector through being scattered by the residual gas of the vacuum or from the solid parts of the machine. An instrument with very high mass resolution could distinguish the first two sources of interference but the third always appears at some level.

By 1980 instrument designers had developed techniques to overcome these difficulties to a remarkable degree by making spectrometers capable of eliminating molecular ions completely and discriminating against scattered ions to an extraordinarily high degree. The interference of isobars, which could have wrecked the scheme, was overcome through luck and ingenuity. Unfortunately, these benefits depend on the use of a nuclear particle accelerator, substantially increasing the expense and complexity of the equipment required.

For information on the production of cosmogenic isotopes the reader is referred to Lal and Peters (1967); for applications with emphasis on the accelerator method see Brown (1984), Faure (1986) and Elmore and Phillips (1987); for applications in determining the particularly useful nuclide ^{10}Be, see Brown (1987) and Morris (1991). The accelerator method has complemented rather than replaced gas proportional and liquid scintillation counting (Box 6.3) for radiocarbon research; for comparisons, evaluations and applications see Long (1989).

Technique fundamentals

A nuclear particle accelerator consists of the same basic elements as a mass spectrometer: an ion source (which produces copious amounts of unwanted as well as desired ions), an accelerating system for achieving the particle energies needed for the experiment, and some form of mass analysis. The ways of achieving these ends are many and varied, and at times the basic functions are almost unrecognisable amid a confusion of equipment.

Fortunately, accelerator mass spectrometry is now done with only one type of particle accelerator – the tandem electrostatic accelerator – the principles of which are easy to grasp (Fig. 12.1). Its design offers four advantages over other accelerators: (a) the ion source is readily accessible for changing samples; (b) the use of negative ions allows efficient ionisation of a large fraction of the periodic table; (c) the necessary conversion of negative to positive ions completely eliminates interference from molecular ions; and (d) the high ion energies allow numerous instrumental techniques of nuclear physics to be used to identify the particles entering the detector assembly. For an extensive bibliography of technique papers the reader is referred to Elmore and Phillips (1987).

Using the simplified diagram of a tandem accelerator shown in Fig. 12.1, let us trace the course of particles through the machine. Negatively charged ions of the analyte are produced in a caesium **sputter** source: the specimen is bombarded by ions of Cs with energies of a few keV, sufficient to eject atoms or molecules from the specimen surface, a process called *sputtering*. The ion source contains Cs vapour at low pressure and, because the specimen is held at a lower temperature than the rest of the ion source, it acquires a thin coating of Cs. Caesium has the lowest **ionisation energy** of all natural elements and, as they pass through the Cs film, the sputtered atoms or molecules capture an extra electron from the Cs film, escape with a negative charge and are accelerated out of the ion source and into the main accelerator. Negative-ion currents of microamperes – several orders of magnitude greater than those encountered in thermal source mass spectrometers – are obtained with sample use efficiencies usually of a few per cent.

Ions leaving the source pass through a magnetic sector mass spectrometer prior to their introduction into the main accelerator. This is necessary in order not to overload the accel-

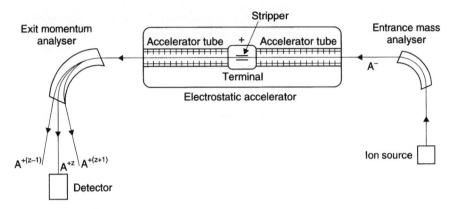

Figure 12.1.   Schematic of an accelerator mass spectrometer. Negative ions (atomic or molecular) are produced in
          the ion source, generally with a wide variety of mass and species. The beam having the isotope of inter-
          est is selected by the entrance mass analyser for acceleration to the high-voltage terminal where elec-
          trons are stripped away, leaving a distribution of positive charge states, illustrated as $z - 1$, z and
          $z + 1$. Generally the most intense of these beams is selected for the detector. The ion optics external
          to the accelerator and the detector schemes vary greatly according to the isotope being measured and
          are much more complicated than shown. The accelerator tube provides a uniform electric field gradi-
          ent between earth and the potential of the positive terminal.

erator with unwanted ions, which otherwise give rise to electrical, radiation and mass
resolution problems. On entering the accelerating tube, negative ions are attracted to the
central, positively charged terminal where on arrival they suffer collisions with low-pressure
gas or a thin carbon foil or both. This interaction destroys molecular ions completely and
gives the particles leaving the 'stripper' a distribution of positive charge states that depends
on their atomic number and velocity. The positive ions so produced undergo a second
stage of acceleration from the terminal back to **earth** potential.

The kinetic energy acquired by the negative ions is Uq, where U is the potential of the
terminal relative to **earth** and q is the negative charge on the ion. Terminal potentials vary
according to the nuclide under study and the requirement of the accelerator, the lowest
being 2 MV for ^{14}C and the highest being about 8 MV for ^{36}Cl. If the negative ion is molecular, its kinetic energy on break-up is distributed to the constituent atoms in proportion
to their masses. The additional energy a positive atomic ion picks up in passing back to
ground at the other end of the accelerator tube is again the product of its charge and the
terminal voltage.

On leaving the accelerator the ions pass through another magnetic sector (a very large one
in view of the high ion energies) whose function is to select the most abundant of the charge
states formed at the positive terminal. The beam emerging from the second magnetic analyser may have significant impurities despite the mass and momentum selection to which the
ions have been subjected. Isobars follow to this point perfectly, if they have formed negative
ions. Only a few elements are unable to form negative ions, and it is a bit of good luck that,
of the six nuclides for which the accelerator method is employed, four have isobars in elements that do not form negative ions. Beams of high intensity relative to the beams of interest may enter the accelerator as molecular ions and produce, through details that need not
trouble us, spurious ions for the detector. Add to these the ions that come from scattering
off the vacuum chamber walls or from other subtle causes, and one has a substantial number
of problem ions for the detector to sort out.

Each of the six cosmogenic isotopes for which the accelerator method is used has unique requirements for detection, and the numerous interfering ions that enter the detector are eliminated from the data by means that vary according to isotope and experimenter.

Nuclide-specific techniques

Six cosmogenic isotopes have been analysed successfully by the accelerator method, the two lightest of which, ^{10}Be and ^{14}C, have found wide application. Beryllium is introduced as an oxide from which the BeO$^-$ ion is produced; unfortunately so is BO$^-$, and ^{10}B therefore follows ^{10}Be through the entire apparatus. Boron is a fearful contaminant and its presence would make detection of ^{10}Be impossible except that it loses energy in matter at a much faster rate than does beryllium, owing to its higher atomic number. A gas absorber placed in the detector preceding the counter stops it completely.

Carbon is introduced either as graphite or CO_2, depending on the kind of ion source. Isobaric interference from ^{14}N is prevented because nitrogen does not form negative ions.

The four heavier nuclides, ^{26}Al, ^{36}Cl, ^{41}Ca and ^{129}I, have limited but expanding uses. Aluminium is introduced as the oxide and runs as Al$^-$. Its natural isobaric interference would be ^{26}Mg, but this also forms no negative ions. Calcium is loaded as CaH$_2$ from which CaH$_3^-$ is formed, a trick to avoid an otherwise fatal interference of ^{41}K. Chlorine is loaded as AgCl and runs as Cl$^-$; ^{36}Ar cannot interfere because it does not form negative ions, but ^{36}S, though a very rare isotope, forces the experimenters to resort to extreme measures. Iodine is loaded as AgI, runs as I$^-$ and escapes the interference of ^{129}Xe through the absence of a negative Xe ion.

Detector details

A simple form of atomic particle detector can distinguish ions of different atomic number because an ion loses energy at a lower rate in passing through matter than does another with the same total energy but higher atomic number. Detectors that make use of this yield two electronic pulses for each ion that enters, one proportional to the ion's total energy, the other proportional to the rate at which it loses energy. The pulses are recorded simultaneously in a two-dimensional computer array whose x and y co-ordinates are proportional to the two pulse heights. The counts attributable to the cosmogenic isotope stand out as an island in an x–y plot from the counts of the many ions that do not satisfy the desired criteria; Fig. 12.2 shows a graphical representation of such data. One variant of this detector produces four to six rate-of-energy-loss pulses as the particle slows down, yielding a distinctive profile for each element. Though the energy loss detector is very effective for light isotopes, it is difficult to use for ^{129}I, but iodine ions are slow enough that the time elapsed on passing two detectors having picosecond resolution allows identification. Detectors are generally employed with supplementary magnetic and electric beam analysis configured to the isotope being measured and in ways not shown in Fig. 12.1.

The detectors are formed of combinations of conventional charged-particle counters. Semiconductor (or 'surface-barrier') counters are often combined with gas-filled counters or ionisation chambers.

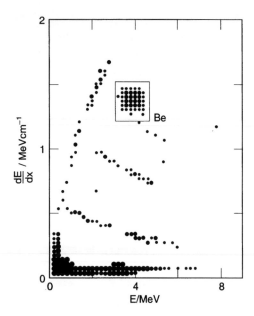

Figure 12.2   A two–dimensional display of data taken in the measurement of ^{10}Be. Detector pulses are sorted into bins of a two–dimensional computer array. The x co-ordinate is proportional to the total energy of the particle entering the detector; the y co-ordinate is proportional to the rate at which the particle loses energy in the detector. The number of counts in a bin is illustrated by a circle with radius proportional to the logarithm of the number of counts. Most bins have no counts whatsoever. The 'island' of ^{10}Be stands out. The computer integrates the number of counts within the 'island' to determine the number of ^{10}Be atoms detected.

Applications

Accelerator mass spectrometry must be employed at a laboratory specialising in the technique. Converting a tandem van de Graaff accelerator for this purpose is costly in time, labour and money, and has been accomplished by very few. Investigators requiring such analyses must deal with a laboratory possessing the equipment and requisite knowledge; costs depend on the nature of the operator's support and research goals. Some laboratories are organised as user laboratories, even preparing the ion source loads from raw sample material. Knowledge of the details is of little use to a prospective user, but some understanding of the overall technique is essential.

^{10}Be has provided a geochemical answer to a question that arose almost immediately after it became accepted that the Earth's structure is made up of dynamic tectonic plates: is any of the sediment carried down by subduction at island arcs incorporated into the magma of the arc volcanoes? The answer has come from numerous measurements of ^{10}Be in lava at levels ranging from 10^4 to 10^7 atoms g^{-1} and is an unequivocal 'yes'. ^{10}Be has too short a half–life (only 1.5 million years) for any to remain inside the Earth from the time of accretion; it must be a component of ocean–floor sediment into which it was incorporated by atmospheric fall–out. The analysis involves the standard mass spectrometric technique of **isotope dilution** (see Box 8.3), wherein about 1 mg of ^{9}Be spike is added to about 5 g of basalt sample; this typically produces ^{10}Be/^{9}Be ratios of 10^{-12}. For recent developments the reader is referred to Morris (1991) and references contained therein.

Industrial civilisation has added CO_2 to the atmosphere in unprecedented amounts, whose climatic effects cannot be confidently predicted. How much of the present increase can be attributed to fossil fuel burning? Radiocarbon can be used in tracing the origin of polycyclic aromatic hydrocarbons, soot and polymeric carbon. Samples originating in fossil carbon will be devoid of ^{14}C whereas those originating from burning living vegetation will have modern carbon. In principle one can collect samples sufficiently large for conventional counters to be used, but the much smaller ones that the accelerator method uses allows a much wider range of investigation. For details of this application, see Currie (1992).

Chapter 13

Which method should I use?

Robin Gill

It is easy for the reader to be overwhelmed by the wide range of bulk–sample inorganic analysis methods described in Chapters 4 to 12. Many of the instrumental techniques in use today offer an impressive multi-analyte capability spanning a wide concentration range, so that for routine elemental analyses of the commoner elements a number of alternative methods may seem, on the face of it, to cover the same parts of the periodic table and to perform equally well. How does the analyst (or the client selecting a commercial analysis service) decide which of these techniques best serves their immediate need? In what instrumentation will an expanding analytical service company invest its shareholders' capital in order to meet most effectively the expected demands of its clients?

This chapter attempts to provide a rationale for making such decisions, concentrating on bulk-analysis methods for determining inorganic constituents in essentially homogeneous samples. The microbeam techniques used for *in situ* microanalysis of inhomogeneous samples or the analysis of very small samples are considered in Chapters 14 and 15 (and in Box 6.8), and the analytical methods used to determine organic components are the subject of Chapter 16. The applications of these techniques speak for themselves.

One of the prime factors in deciding which method to use for a specific analytical task is the facilities already available in the analyst's own laboratory, but little consideration will be given to this question in the present chapter. Deciding whether to 'make do' with an in-house method or to seek superior performance elsewhere at additional cost, for example, is a subject about which it is virtually impossible to generalise, since so many local factors come into play: the terms and expense of access to other laboratories, staff workload, the required turnaround time, the volume of work to be carried out and so on, not to mention more arcane issues of company policy and finance. The skills deployed in answering such questions are as much managerial as analytical.

Whatever the circumstances, the analyst's primary objectives must be:

- to define with the user the data quality required to support the immediate (and foreseeable) uses of the data (Chapter 2) – the notion of **fitness for purpose**;
- to acquire data of the required quality at the lowest 'cost' (though how 'cost' is defined will vary from one organisation to another);
- to secure adequate quality control as part of the analytical process.

In considering how these needs can be met, one must give some thought not only to the instrumental method but also to what is the most appropriate form of sample preparation to use; these two aspects of analysis go hand in hand.

For the purpose of this chapter it is conceptually helpful to draw a distinction between the analysis *producer* and the analysis *user*, even though these two functions may in some cases be performed by the same person (e.g. a research student carrying out his or her own analyses in a university laboratory).

Data quality

The first decisions that affect the choice of method should be made not by the analyst, but by the user of the analyses. In a well-planned investigation, analyses are commissioned to fulfil a specific purpose: to establish whether the level of pollution in a river exceeds some statutory limit, or to construct a quantitative model of crystallisation processes in a sub-volcanic magma chamber, or to locate an ore deposit in an area of granite moorland, to take three simple examples. These applications make quite different demands in terms of data quality, and it is the responsibility of the analysis user to define, in addition to the elements to be determined, what standards of data 'quality' will be required for the success of the investigation. Together with the user, the analyst must ensure that 'measurement' variances (Fig. 2.3) will not obscure the real geochemical variation being investigated. Selecting the analytical method to ensure that the analyses produced **fit the purpose** for which they are being carried out (as defined by the user) is primarily the analyst's responsibility, whether the analyst works in academe or in industry.

Let us briefly examine each of the above examples. The primary consideration, when comparing the concentration of pollutants in river water with a statutory definition of the maximum acceptable concentration, is that the measured values should not be subject to **systematic error** or **bias** (Chapter 1). We can overcome the effect of **random error** (poor **precision**) by replicating the analysis (Box 2.1), but if the values we obtain are consistently different from the true analyte concentration, then our conclusion as to whether or not the concentration exceeds the permitted value may be false. For a utility facing prosecution for not maintaining adequate water quality, for example, such an error could prove costly. In work of this kind, steps must always be taken to determine the **bias** of the analyses carried out, by analysing appropriate CRMs at the same time as the unknowns (Chapter 2). The analyst may also be required to follow a prescribed quality control protocol (such as that laid down by the US Environmental Protection Agency – see Box 13.1) designed to limit the bias to an acceptably low level, but the analyst cannot rely on the protocol alone.

It is important to note that **detection limit** may also be a decisive factor in determining which analytical method to use. Many elements in natural waters are present at levels below or very close to the ICP–AES detection limit (see Cd and Pb in Table 13.1), for example, and therefore it may well be necessary to resort to ICP–MS instead.

The compositions of successive lavas erupted from a volcano commonly vary as a consequence of fractional crystallisation taking place in the magma 'chamber' supplying the eruptions. Geochemists can quantify the parameters of this process (e.g. how much of mineral X has crystallised?) by statistical analysis of the observed changes in composition. Because the method relies upon measuring quite small *differences* in composition between sample A and sample B in the same data set, successful modelling demands analyses of high **precision**; it matters less if the data are subject to bias, provided the bias is common to all of the analyses being used (including the mineral analyses used in the calculations[1]). The user should be able to estimate the standard deviation that can be tolerated, and the analyst must tailor the method accordingly, using longer counting times for XRFS analysis of Sc, V, Cr, Co, Ni, Rb and Sr (Chapter 6), for example, or exploiting the high reproducibility of ID–TIMS (Chapter 9) for rare earth elements instead of using INAA.

[1] Bias may have to be taken into account if *different methods* are used for the bulk rock and mineral analyses (e.g. XRFS and electron microprobe).

Box 13.1 Analytical quality control and quality assurance

In many industrialised countries, environmental legisla-
tion imposes ever higher standards of water, air, soil
and food quality, and places tighter restriction on dis-
charges from industrial plant, utilities, agriculture and
waste disposal facilities. The enforcement of such standards, and an industry's need to monitor its own emissions, have stimulated the growth of a commercial
environmental testing sector selling a range of analytical
services. To enable clients (who may possess no specialist analytical expertise) to identify reliable laboratories,
the industry relies upon voluntary national accreditation
of analytical service companies, usually against codes of
practice defined by authoritative international and
national bodies[1]. Though accreditation is awarded for
specific analytical procedures (e.g. for 'organochlorine
pesticides in soil'), the process tends to raise awareness
of general good practice in everything the laboratory
does.

To qualify for accreditation (see Broderick, 1991), a
laboratory regularly undergoes rigorous 'third-party'
audit, usually by a national organisation (in the UK by
the National Measurement Accreditation Service
(NAMAS[2]); in the USA, by various organisations such
as the federal Food and Drug Administration). The
laboratory must demonstrate on-going adherence to
ISO Guide 25 through internal *quality control* and *quality
assurance* practices (see Garfield, 1991, Prichard, 1995).
In the context of an analytical laboratory these concepts
may be defined as follows:

Quality control   a planned, systematic programme of
measures to verify and document the
precision, accuracy and reliability of
all analyses produced in the laboratory, introducing **anonymously** into
the work stream appropriate replicates and **reference materials**, and

keeping for immediate and future
reference a continuous record of
results obtained.

Quality assurance   'planned activities designed to
ensure that a **quality control** pro-
gramme is being properly implemen-
ted' (after Garfield, 1991).

In a well-run testing laboratory, all samples received for
analysis will on arrival be documented in a **chain of custody** record, indicating the type of sample, the preparation required, the analytes to be determined, the
personnel through whose hands the sample will pass
for analysis, and the end destination of the samples; this
record is updated each time the samples change hands.
For each type of analysis it offers, the laboratory will
maintain on file a written protocol detailing how it should
be carried out, stipulating reagent grade, the standards to
be used, the replicates and reference materials to be run
(thus ensuring **traceability**), and listing the staff trained
to use the method; commonly the method will comply
with a protocol laid down by a recognised organisation,
such as the US Environmental Protection Agency. The
lab will also keep maintenance and performance records
of all analytical equipment, and will maintain a quality
manual detailing the laboratory's quality assurance policy
and procedures.

Practice in academic analytical laboratories has varied
widely. Though in some the standard of data scrutiny and
validation has been high, sample and data administration
is often left to individuals to organise, and few academic
research laboratories have undertaken the challenge of
formal accreditation, as research journals as a rule have
not imposed the most stringent quality standards. The
increasing need for academic laboratories to earn commercial income, however, may soon lead to formal
accreditation being more widely sought.

---

[1] Relevant organisations and standards include:
  • Organization for Economic Co-operation and Development (OECD Good Laboratory Practice standard).
  • International Organization for Standardization (ISO 9000 quality standards and ISO Guides 25 and 49 on lab practice).
  • European Community for Standardisation, CEN (EN 29000 and EN45000 standards).
  • British Standards Institution (BS 5750 Quality Standard, now engagingly known as 'BS EN ISO 9000').
[2] Administered by the UK National Physical Laboratory, Teddington.

The standards of accuracy and precision required in mineral exploration, on the other
hand, are often far less demanding. The reason is that an ore body usually contains concentrations of the elements of interest that are several orders of magnitude greater than in surrounding unmineralised rocks. Thus in attempting to locate the ore body by analysing
sediments in streams draining the target area (in order to identify the streams in whose
catchment the ore body lies), we are looking for relatively large anomalies in media that are
inherently somewhat heterogeneous, and to purchase high-precision analyses or to adopt
stringent quality control measures for such a project would simply be a waste of money.

Table 13.1   Limits of determination ($6s_{background}$ in ng g^{-1}) of selected elements by various solution methods. Data from Potts (1987) and K. Jarvis (personal communication)

Element	AAS flame	AAS graphite furnace	ICP–AES	ICP–MS	Public water standard[†]
Li	3	0.03	45	7	–
Be	3	0.006	0.54	0.3	–
B	2100	45	9.6	20	5000
Na	1	0.009	58	2.3	200,000
Mg	0.6	0.0006	60	1	–
Al	60	0.03	40	0.6	–
Ca	3	0.03	20	16[‡]	–
Fe	15	0.03	12.4	7	300
Zn	3	0.006	8	2	5000
Cd	3	0.0006	7	0.05	5
Tl	60	0.03	80	0.04	–
Pb	60	0.021	84	0.06	10
Machine time per sample[*]	1 h	5 h	1 min	4 min	

[*]Estimated instrument time per sample to measure 15 elements.
[†]Interim Canadian Environmental Quality Criteria for Contaminated Sites, drinking water standard, 1991.
[‡]Interferences require use of less abundant isotope.

The emphasis in many geochemical exploration programmes is on high sample throughput using rapid methods of sample preparation and analysis, which may even be carried out in a mobile trailer deployed in the field area.

Multi-element instrumental techniques

Considerations of data quality are not the only criteria used in selecting the best analytical method for a given investigation. There are many aspects of practicality that have to be weighed up, among the most important of which is the physical state of the material to be analysed.

Liquid media

Hundreds of thousands of analyses are carried out each year in the UK alone to test the potability of domestic water supplies. For analysing aqueous solutions, it is natural to restrict consideration to techniques that accept the sample directly in liquid form, as this minimises sample processing. These are primarily AAS, ICP–AES and ICP–MS, whose performance for a range of trace elements (by comparison with one internationally recognised water quality standard to indicate the concentration range of interest) is shown in Table 13.1. Note the following points:

- *AAS* is in a somewhat different category from the other two methods: each element requires the introduction of a separate hollow-cathode lamp, and this requires that elements are analysed sequentially rather than simultaneously (although some recent AAS instruments offer a limited multi-element capability). These factors make flame AAS at least an order of magnitude slower for multi-element analysis than the plasma methods. If the investigation is limited to one or two analytes, however, this may be no disadvantage.
- For water analysis flame AAS has few advantages other than its modest capital cost, but in *graphite furnace* or *cold vapour* (hydride) mode AAS offers extremely low detection limits for certain elements (albeit at some penalty in analysis time), an advantage that leads to quite widespread use in hydrogeological and other environmental analysis.
- *ICP–AES* generally offers detection limits similar to flame AAS, for many elements barely low enough to be applicable to natural waters without preconcentration (e.g. by evaporation).
- Detection limits (expressed in element terms) are, with certain exceptions, 10–100 times lower for *ICP–MS* than ICP–AES.

These techniques may be applied directly to fresh waters, but solutions of higher ionic strength such as sea water would need to be diluted to bring the total dissolved solids to below 10,000 μg ml^{-1} (below 2000 μg ml^{-1} for ICP–MS), and standards would need to be matched to unknowns to minimise matrix effects.

None of these techniques is successful at determining anions in solution, and the analyst must turn to colorimetric methods or ion chromatography (Chapter 5) to analyse the halides, nitrate and sulphate, or (in the case of fluoride) the ion-selective electrode.

Solid media

For analysing rock powders, soils and sediments the analyst faces a choice between: (a) dissolving the samples to analyse them as solutions (AAS, ICP–AES, ICP–MS), and

Table 13.2   Advantages of analysing samples in solution and in solid form

Advantages of solution method:	• easier standardisation
	• easy to match matrices to minimise matrix effects
	• straightforward separation and preconcentration
	• selective analysis of readily leached components may be more relevant in some applications than total content
	• option to dissolve large sample and preconcentrate if analyte is inhomogeneously distributed (e.g. Au)
Advantages of analysing solids directly:	• minimal sample preparation (e.g. INAA)
	• less dilution of trace elements
	• long-term stability of sample preparations (e.g. reference materials used for QC)
	• no loss of insoluble residues (e.g. zircon)
	• no loss of volatile products (e.g. SiF$_4$)

(b) employing a method that analyses solid samples directly (XRF, INAA, SSMS). The advantages of each approach are summarised in Table 13.2.

When a sample is in solution, it is easy to prepare standards that cover the concentration range and to eliminate or buffer differences in composition that give rise to matrix effects. Solution methods offer the opportunity to dilute the solution or preconcentrate the analyte to bring it into the optimum concentration range for analysis. Moreover, for some environmental applications the relevant question is not 'how much of element X is present in the sample?' but 'how much is sufficiently mobile to be taken up by plants or otherwise find its way into the ecosystem or the food chain?', and the latter question can be more appropriately addressed by acid-leaching digestions (Chapter 3) and analysis of the filtrate than by bulk analysis of the entire sample. There are other cases where the selective removal of a major consituent (e.g. Si as SiF_4) can be advantageous for trace element analysis, although it brings obvious disadvantages for major element analysis.

The direct analysis of a solid sample, because it avoids the dissolution step, results in less dilution of trace elements and may save considerable preparation time (as in the case of INAA, for example). XRFS, however, requires the sample to be fused or pelletised (Chapter 6) and SSMS requires it to be mixed with a conducting matrix and formed into appropriately shaped electrodes (Chapter 11), so any gain in preparation time relative to solution methods may be slight. Analysis in the solid state avoids the loss of volatile fluorides that occurs during HF digestion in open beakers, and it can be advantageous for the trace element geochemist as it overcomes the problem that sometimes arises with highly resistant minerals like zircon, which may require time–consuming special steps to deliver their trace element constituents quantitatively into solution (Chapter 3). Some forms of solid–state analysis also have the advantage of being non–destructive.

Table 13.3 Comparison of multi-analyte methods, showing applicability to three concentration ranges with regard to sample preparation, sample form, machine time, and potential for enhancing precision and accuracy by isotope dilution

Method	Analysed as	Major elements	Trace elements 1–1000 ppm	Trace elements < 1 ppm	Analysis time*	Isotope dilution?
XRFS (WD†)	Powder	Fusion disk	Powder pellet	–	2 h	No
INAA	Solid/powder	–	Sealed capsule	Sealed capsule	1–2 h count time	No
SSMS (photoplate)	Powder	–	Formed electrode	Formed electrode	6h	Possible
ICP–AES	Solution	Acid digest or fusion‡	Acid digest or fusion‡	–	1 min	No
ICP–MS	Solution or slurry	–	Acid digest or slurry	Acid digest or slurry	4 min	Possible
Flame AAS	Solution	Fusion or closed HF digestion	Acid digest	–	1.5 h	No
TIMS	Solution	–	Acid digest + ion exchange	Acid digest + ion exchange	1 day +	Essential

*Machine time per sample per 20 elements.
†Wavelength–dispersive spectrometer assumed. Energy dispersive analysis is quicker but less widely used.
‡Fusion may be required to take resistant phases into solution.

Table 13.3 shows a comparison of the concentration capabilities of multi-element methods applicable to solid samples. ICP–AES, XRF and flame AAS potentially cover both major elements and trace elements above the ppm level, but cannot determine lower-level trace elements without some form of preconcentration. It must be remembered that X-ray methods are not applicable to light elements ($Z < 9$) owing to the poor excitation and strong absorption of soft X-rays. Unlike AAS and ICP–AES, XRFS and INAA are applicable to electronegative elements as well as to electropositive ones.

For trace element concentrations near or below 1 ppm, one must resort either to the more sensitive methods like SSMS and ICP–MS, or to preconcentration. Detection limits for INAA, being dependent on nuclear-capture cross-sections, are highly variable (Table 7.1).

Particulates

One application of environmental interest is the analysis of filters used for collecting airborne particulates. Though acid digestion followed by a solution analysis method is possible (essential for isotopic analysis), the most convenient and widely used approach is to analyse the filter directly by XRFS or by instrumental neutron activation analysis (Chapter 7). If the filter size is chosen correctly it can fit directly into an XRFS sample holder, or into an INAA sample vial.

Rare earth elements in rocks

The trade-off between precision and accuracy on the one hand and analyst time, instrument time and cost on the other is illustrated by the range of methods used for analysing REEs (lanthanides) in igneous rocks. Because chemical behaviour and **normalised** abundance (Box 13.2) change only gradually across this series of 14 elements, it is only necessary to analyse a representative subset. The essentials of a rare earth analysis are:

- La or Ce and possibly Nd representing the 'light' REEs;
- Sm, Eu and preferably Gd to highlight anomalous behaviour of Eu (which may behave differently from other REEs in geological processes on account of its M^{2+} state); and
- Yb or Lu, representing the 'heavy' REEs, with a selection of intervening elements to confirm a smooth pattern and indicate the degree of curvature.

A wide variety of techniques has been used in recent years for carrying out such analyses (Table 13.4). INAA and SSMS are sensitive enough to accomplish REE analysis on solid samples without preconcentration. Analysis in solid form is advantageous because acid-resistant minerals in the sample (some of which host REEs) are necessarily included in the analysis. SSMS requires considerable operator involvement, whereas INAA requires little. Both are capable of ±5–10% precision, adequate for determining the general pattern of REE abundance but insufficiently precise for quantitative modelling of geological processes. It is common in interpreting REE data to calculate element ratios (La/Lu, Ce/Yb, La/Sm, Gd/Yb, etc.), in which the relatively poor precision characteristic of these methods will be amplified.

Box 13.2. Rare earth elements

The abundances of individual rare earth elements vary in a sawtooth manner in all cosmic and terrestrial matter, even-Z elements being about 10 times more abundant than odd-Z elements (Fig. 13.2.1). Early members of the series ('light rare earth elements' – LREEs – like lanthanum and cerium) are also noticeably more abundant than heavier members ('heavy rare earth elements' – HREEs – such as ytterbium). To filter out these circumstantial factors, it is conventional to divide the individual REE concentrations in any terrestrial material by the corresponding element concentration in 'average' chondrite meteorite. Such 'chondrite-**normalised**' concentrations, when plotted against Z, give smooth rather than jagged patterns (Fig. 13.2.2) that are much easier to interpret.

The more abundant light REEs are readily analysed by a number of methods. X-ray fluorescence spectrometry, for example, can detect La, Ce and Nd in most geological materials without preconcentration. Greater analytical sophistication is, however, required to determine HREE concentrations, because of their much lower abundance.

Figure 13.2.1 The composite solar system abundance curve showing alternating high–low abundances for even- and odd-Z rare earth elements. In accordance with normal cosmochemical practice, abundances are expressed in terms of numbers of atoms per 10^6 atoms of silicon. (From Gill, 1995, Fig. 10.2.)

Figure 13.2.2 Chondrite-normalised rare earth patterns for two lunar basalts. (From Gill, 1995, Fig. 9.10, using data from Taylor, 1982. Reproduced with permission.)

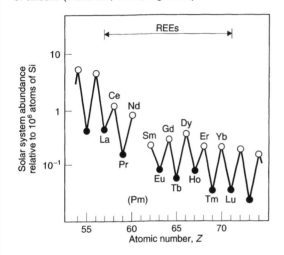

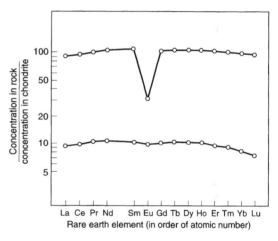

Table 13.4 Methods for the determination of rare earth elements in bulk geological samples. 'ID' denotes isotope dilution (Box 8.3)

		Chapter
Techniques sensitive enough often to require no separation	INAA	7
	SSMS ± ID	12
Techniques requiring solution but no preconcentration:	ICP–MS	10
Techniques involving ion exchange separation/preconcentration:	XRFS	6
	ICP–AES	4
	ID–TIMS	8

In many rocks the concentrations of the less abundant rare earths are at or below the ppm level, and therefore XRFS and ICP–AES both require ion exchange concentration in order to be able to measure them (Eby, 1972, Walsh *et al.*, 1981). The precision obtainable is comparable with INAA, but significantly greater specimen preparation time is required.

Isotope dilution TIMS (Box 8.3) stands out as the method of choice for REEs to be used in quantitative modelling studies, for example to shed light on the source domains in the mantle or crust from which igneous melts are derived. The time commitment on the part of the analyst (in terms of dissolving the rock powder, spiking and separating the REEs) and the mass spectrometer hours required are considerable, but precision of the order of 1% **RSD** or better can be obtained, and for certain REE *ratios* the precision can be as low as 0.1% RSD (Box 8.3).

In all REE methods, significant inter-element interferences occur and corrections must usually be applied. This introduces error, especially into determinations of the less abundant heavy rare earth elements (HREEs). Note that in isotope dilution analysis one is confined to analysing the polyisotopic HREEs Dy, Er and Yb (which, having even atomic numbers, are also the more abundant HREEs), whereas in INAA analysis it is advantageous to analyse the monoisotopic odd-Z elements Tb, Ho and Tm instead (usually with Yb or Lu), in spite of their lower abundance, because their higher neutron–capture cross-sections result in substantially lower detection limits.

Chapter 14

Electron beam methods

Robin Gill

The finely focused high-energy electron beam used to create the image in a scanning electron microscope (SEM) happens also to generate characteristic **X-rays** in the parts of the specimen exposed to it, and can therefore be used as a chemical 'probe'. An SEM equipped with a suitable X-ray analyser provides a versatile, non-destructive means of analysing chemical variation *on the micrometre scale* in the specimen surface. 'Electron probe microanalysis' (EPMA) has found innumerable applications in earth science, metallurgy, solid-state physics, biology and environmental science. General reviews of the subject are given by Potts (1987), Goldstein *et al.* (1992), Reed (1993, 1995) and Champness (1995).

An electron probe microanalysis system consists of two parts, often originating from separate manufacturers:

(1) the electron optics 'column' (usually a commercially available SEM) providing the electron beam, the specimen-handling and imaging facilities, and the vacuum system;
(2) the X-ray spectrometer or 'analyser' which unravels each X-ray spectrum into individual element peaks; the associated computer translates peak intensities into element concentrations or element distribution patterns.

The electron probe technique, as developed during the 1950s and 1960s, relied upon elaborate X-ray diffraction spectrometers (see Fig. 14.7) that separated element peaks according to wavelength. Though such 'wavelength-dispersive' spectrometers are still widely used, the development in the 1970s of cheaper semiconductor-based 'energy-dispersive' X-ray analysers has made X-ray analysis attainable at relatively low cost even on the most basic electron microscope. Such systems commonly offer software options capable of driving various functions of the electron microscope under computer control, allowing the instrument as a whole to perform a predetermined analytical programme unattended.

The general layout of an electron microprobe is shown in Fig. 14.1. To minimise electron scattering and X-ray absorption, the specimen chamber and the X-ray spectrometer/detector system are pumped to a high vacuum of around 10^{-3} Pa (10^{-8} atmospheres), usually by an oil diffusion pump 'backed' by a rotary pump (Appendix A). Traces of diffusion pump oil may break down under electron bombardment and contaminate the specimen surface, and turbomolecular pumps (Appendix A) are increasingly popular as they avoid this problem.

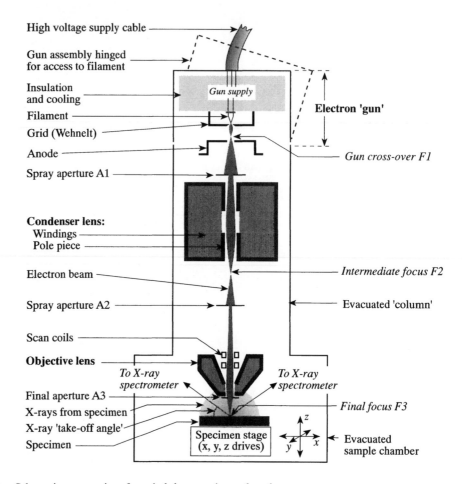

High voltage supply cable

Gun assembly hinged
for access to filament

Insulation
and cooling

Filament

Grid (Wehnelt)

Anode

Spray aperture A1

Condenser lens:
Windings
Pole piece

Electron beam

Spray aperture A2

Scan coils

Objective lens

Final aperture A3

X-rays from specimen

X-ray 'take-off angle'

Specimen

Gun supply

Electron 'gun'

Gun cross-over F1

Intermediate focus F2

Evacuated 'column'

*To X-ray
spectrometer*

*To X-ray
spectrometer*

Final focus F3

Specimen stage
(x, y, z drives)

Evacuated
sample chamber

Figure 14.1   Schematic cross-section of a typical electron microprobe column.

The electron microscope

The electron 'gun'

The device that 'fires' the electron beam towards the specimen is called the electron *gun* (Fig. 14.2a). Free electrons are generated by **thermionic emission** from the tip of a hairpin-shaped tungsten *filament* operated at about 2400°C. The electron cloud is converted into a high-energy beam by two cylindrical metal cups, the *grid* (or 'Wehnelt cylinder') and the *anode*, each having a central hole through which electrons can pass. The high voltage difference between them creates a strong electric field that accelerates the electrons. The cylindrical field gradient created by the hole in the grid (see the equipotential surfaces in Fig. 14.2a) also acts as an electrostatic 'lens': it draws the beam together into a crude focus or *cross-over* about 50 μm in diameter. The beam focus formed on the specimen surface is a 'demagnified' image of this cross-over.

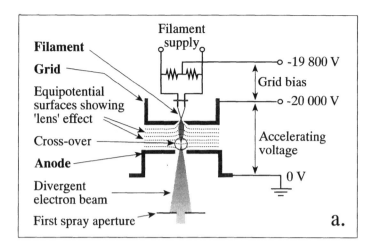

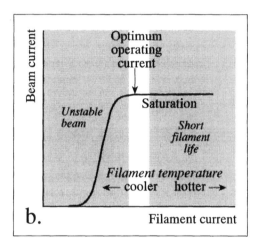

Figure 14.2   (a) The conventional electron 'gun'. Voltages shown are illustrative only. (b) Gun emission as a function of filament current. 'Beam current' refers to the total current emerging from the anode. The window indicates the optimum operating conditions.

For efficient X-ray excitation, the kinetic energy of the electrons bombarding the specimen needs to be in the range 15–30 keV. This is accomplished by applying a positive potential difference of between 15 and 30 kV (typically 20 kV) between grid and anode as shown in Fig. 14.2(a), using a stabilised high-voltage generator.

The beam current emerging through the anode hole rises with increasing filament current until a plateau is reached ('saturation' in Fig. 14.2b). The plateau provides the optimum operating condition, where electron emission from the gun is insensitive to fluctuations in filament current. Setting the current at the low end of the plateau maximises filament life. As filaments have a limited life (<100 hours), gun design must allow for isolation from the column vacuum and easy access for changing the filament (Fig. 14.1), which is mounted on a removable assembly to minimise replacement down-time.

Advanced gun designs developed for high-resolution electron microscopy (LaB$_6$ and field-emission guns) offer no advantage in quantitative microanalysis.

The electron optics column: 'condenser' and 'objective' lenses

Forming a convergent electron beam that attains its narrowest diameter at the specimen surface is analogous to focusing the Sun's rays into a small area to set paper alight. In place of a glass lens, the electron microscope uses a series of cylindrical electromagnets ('magnetic lenses') to focus the divergent electron beam from the gun to form successively smaller images of the gun cross-over.

Modern high-resolution SEMs may use three or even four magnetic lenses, but two are sufficient to produce a beam focus less than 1 μm in diameter, adequate for conventional microprobe analysis. The first, the *condenser lens*, produces an intermediate demagnified image of the cross-over (point F2 in Fig. 14.1). The beam then passes through a smaller *objective lens* which produces the final image ('F3') focused at the specimen surface. Each lens consists of copper wire wound symmetrically around the axis of symmetry of the lens (shown in cross-section in Figs 14.1 and 14.3) and encased in an iron shell with a gap in the inner cylindrical surface, across which the magnetic field is concentrated.

Circular apertures made of Pt, Mo or Ta are positioned at various points along the beam path to restrict beam divergence, primarily to minimise a form of optical distortion called **spherical aberration**, to which magnetic lenses are particularly susceptible.

The focal length of a magnetic lens (unlike a glass lens) can be varied by altering the current passing through the windings: the greater the current, the stronger the magnetic field, the more strongly the beam converges and therefore the shorter the focal length of the lens. This property of magnetic lenses is exploited to control two important parameters of the electron beam. The condenser current regulates the proportion of the beam that passes through the second aperture (A2) and therefore determines the **probe current** reaching the specimen (Fig. 14.3a and b). Selecting an appropriate probe current is an important factor in accurate quantitative analysis. Regulating the current supplied to the objective lens, on the other hand, provides the means for focusing the beam on the specimen surface (Fig. 14.3c).

The X-ray intensity obtained from each chemical element in the specimen depends on the intensity of the electron beam (the **probe current**) and on the kinetic energy of the elec-

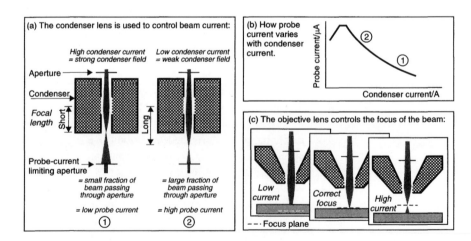

Figure 14.3   Electron–optical controls.

trons in the beam as well as on element concentration. For quantitative analysis, therefore, it is essential to regulate these beam parameters within narrow limits. The electron energy is solely a function of the gun accelerating voltage, which is electronically stabilised to 1 part in 10^5 in modern instruments. The probe current fluctuates in response to mechanical as well as electrical factors, in particular the position of the filament in relation to the grid, and distortion of the filament as it ages may cause the current to vary. The probe current must therefore be monitored regularly during quantitative analysis. Any drift may be corrected by adjusting the condenser current manually, or by electronic feedback from the current meter to the condenser control circuitry.

The microprobe operator routinely adjusts the beam trajectory when starting work (and periodically thereafter) to ensure that it is directed along the axis of each lens and is sharply focused in the specimen plane. The means for optimising beam alignment are summarised in Box 14.1.

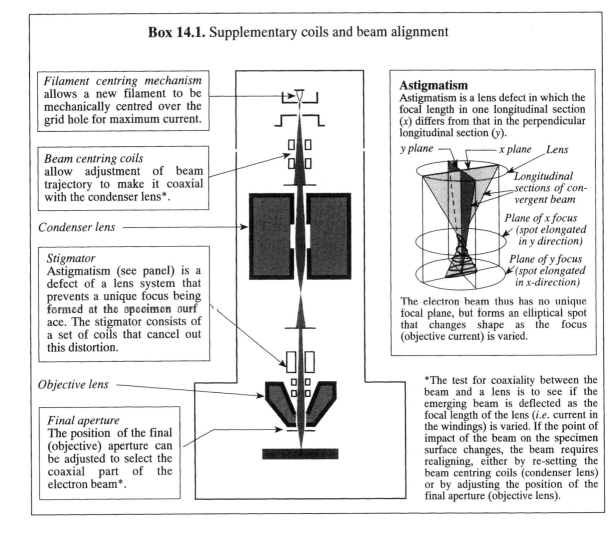

Box 14.1. Supplementary coils and beam alignment

Filament centring mechanism allows a new filament to be mechanically centred over the grid hole for maximum current.

Beam centring coils allow adjustment of beam trajectory to make it coaxial with the condenser lens*.

Condenser lens

Stigmator Astigmatism (see panel) is a defect of a lens system that prevents a unique focus being formed at the specimen surface. The stigmator consists of a set of coils that cancel out this distortion.

Objective lens

Final aperture The position of the final (objective) aperture can be adjusted to select the coaxial part of the electron beam*.

Astigmatism
Astigmatism is a lens defect in which the focal length in one longitudinal section (x) differs from that in the perpendicular longitudinal section (y).

y plane ———   ——— *x plane*   *Lens*

Longitudinal sections of convergent beam

Plane of x focus (spot elongated in y direction)

Plane of y focus (spot elongated in x-direction)

The electron beam thus has no unique focal plane, but forms an elliptical spot that changes shape as the focus (objective current) is varied.

*The test for coaxiality between the beam and a lens is to see if the emerging beam is deflected as the focal length of the lens (*i.e.* current in the windings) is varied. If the point of impact of the beam on the specimen surface changes, the beam requires realigning, either by re-setting the beam centring coils (condenser lens) or by adjusting the position of the final aperture (objective lens).

Scan coils and electron imaging

An electron image of the specimen surface is essential for locating features of interest and displaying elemental distribution patterns ('element mapping'). To this end, an electron microprobe is equipped with a set of *scan coils* near to the objective lens (Fig. 14.1), which scan the beam from side to side. The standard form of scan is the two–dimensional *raster* (Fig. 14.4a) in which an image of an area of the specimen surface is built up of successive line scans like a TV picture. A synchronised raster timebase applied to the display screen allows the image to be displayed simultaneously on screen (Fig. 14.4b).

In an SEM image, the brightness on the screen represents the intensity of electrons emitted from the point of impact of the primary electron beam. The most energetic of these, called *back-scattered electrons* (BSEs), are simply incident electrons that re-emerge as a result of elastic scattering in the specimen. Electrons emerging with much lower energies (≤ 50 eV), resulting from beam-induced ionisations within the specimen, are termed *secondary electrons*. Electron detectors (scintillators or diodes) mounted around the objective lens allow either type to be used to form an electron image.

The efficiency with which primary electrons are back-scattered increases with the *mean atomic number*, $\bar{Z}$ (the mean of the atomic numbers of all elements present, weighted according to their concentration), of the substance being irradiated by the beam. A high-$\bar{Z}$ phase therefore appears as a bright area in a BSE image (Fig. 14.4b). Such compositional contrasts are usually sufficient to distinguish between different types of crystal in a rock specimen, for example aiding the location of areas of analytical interest. Back-scattering is also influenced by surface topography, and therefore the atomic number effect is most apparent in a polished specimen. Topographic imaging (the primary purpose of an SEM) is usually accomplished using secondary electrons. Scanning facilities can also be used to investigate the distribution of individual elements by means of X-ray maps (Fig. 14.4c).

Most microprobes are equipped with a high-power optical microscope that allows the specimen to be viewed in reflected and sometimes transmitted light.

Area of raster on specimen surface

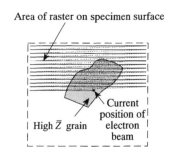  

(*a*) Beam raster on the specimen surface. Dotted lines represent the rapid 'flyback' of the beam to the *x* baseline.

(*b*) Back-scattered electron image generated by a synchronous raster on display screen showing higher back-scatter efficiency of high-$\bar{Z}$ grain

(*c*) Fe Kα X-ray map on display screen showing concentration of iron in the same grain.

Figure 14.4.   Imaging the specimen surface with a rectangular raster of the electron beam. The relation between beam raster and the synchronous screen raster is illustrated by 'freezing' the beam in mid raster.

Specimen chamber and stage

As in an optical microscope, the specimen in a microprobe is mounted on a *stage* (Fig. 14.1). This has to provide:

- a secure, vibration-free mounting for one or several specimens, together with appropriate standards;
- electrical connection between specimen surface and earth (ground) to prevent charge accumulation;
- illumination and light path for an optical microscope if fitted;
- the capability of driving the specimen in x and y directions (Fig. 14.1) rapidly and reproducibly (± 1 μm) under manual or computer control to bring preselected parts of the specimen(s) under the beam for analysis;
- vertical (z) adjustment of specimen position to correct for variation in specimen thickness, particularly for wavelength-dispersive analysis.

The stage is mounted inside a *specimen chamber* that maintains high vacuum (typically $\sim 10^{-3}$ Pa) around the specimen during analysis, but which allows access for specimen changing. An airlock mechanism allows the specimen to be changed without admitting air to the chamber itself; the specimen is mounted on a shuttle that is manipulated between airlock and stage by means of a push rod.

Specimen preparation

Qualitative analyses may be carried out on specimens of any shape, but for reasons discussed later the specimen must have a flat polished surface if reliable quantitative analyses are to be obtained. Non-conducting specimens must be vacuum coated with a thin film (typically 20 nm thick) of carbon or another conducting medium to avoid charge accumulation on the specimen surface; as the coating diminishes X-ray intensities by a few per cent, its thickness must be carefully controlled. The conducting coating is earthed to the specimen holder using proprietary colloidal graphite paint ('Aquadag').

X-ray spectrometers

Characteristic X-rays emerge in all directions from the small area of specimen surface excited by the electron beam (Fig. 14.1). Their spectral content is analysed by one or more X-ray spectrometers positioned around the specimen chamber. The spectrometers 'view' the X-ray source from a predetermined *take-off angle*, a design compromise between the desire to minimise X-ray path length and absorption in the specimen (favouring a high take-off angle – see Figs 14.1 and 14.8c) and the geometrical restriction that the design of the final lens imposes on the X-ray path. Most instruments offer a take-off angle of 35–40°.

X-ray spectrometers separate the components of the X-ray spectrum either on the basis of photon energy ('energy-dispersive' or ED spectrometry), or by separating wavelengths using a crystal lattice as a diffraction grating ('wavelength-dispersive' or WD spectrometry) as in Chapter 6.

The energy-dispersive (ED) X-ray spectrometer

The detector   The basic physics of the ED X-ray detector has been introduced in Box 7.1. For X-ray microanalysis the Si(Li) crystal is typically 3 mm thick, has an active area of about 10 mm^2 and is mounted, together with a low-noise **FET** preamplifier chip, at the end of a long tube extending into the evacuated specimen chamber (Fig. 14.5), pointing directly at the X-ray source. Both devices are cooled to suppress electronic noise by coupling the Si(Li)–FET assembly to a copper rod connected to a liquid nitrogen reservoir.

The Si(Li) crystal is normally protected by an 8 μm thick beryllium window, which (a) keeps the detector sealed under a high static vacuum (10^{-5} Pa), even when air is admitted to the specimen chamber, (b) prevents contaminants condensing onto the cold detector surface, and (c) prevents secondary electrons from the specimen (as well as stray light and infrared radiation) generating spurious signals in the Si(Li) crystal.

Pulse processing   The output from the FET is transformed into a series of electrical pulses, whose height is approximately proportional to the quantum energy of the photon that generated them. These pulses pass into an **ADC** and are digitally sorted by pulse height into several thousand elements or *channels* of a computer array . Each channel represents a narrow interval of pulse height (proxying for photon energy, Box 7.1), and at any instant contains the number of pulses so far registered that had amplitudes falling within this interval. Plotting this value against the corresponding channel number reveals a series of peaks representing the various characteristic lines present in the X-ray spectrum (Fig. 7.1.2). This 'histogram' is displayed on a VDU in real time, and peaks can be seen 'growing' as the spectrum is accumulated during the preselected acquisition time.

Two types of counting error arise when two or more photons are detected in rapid succession:

(1)   *Dead time* If a pulse arrives while the amplifier is still 'busy' dealing with the previous one, the second pulse escapes detection. This **dead time** is at least 10 times longer than that for proportional and scintillation counter circuits (Chapter 6) because, to achieve high signal to noise performance and high energy resolution, ED systems must work with pulses of relatively long duration (≤ 50 μs versus ~ 1 μs for a WD system). This restricts the maximum count rate (X-ray intensity) that an ED spectrometer can reliably measure.

ED system software routinely calculates the accumulated dead time for a measured counting rate in real time, and compensates for it by extending the counting period until the *live time* (elapsed time minus calculated dead time) has reached the predetermined nominal counting time (say 200 s). Even with such correction, it is normal in quantitative ED analysis to restrict the probe current to a few nanoamperes in order to keep count rates below 10 kHz and dead time below 30%. As well as making intensity measurement less reliable, high count rates may result in a loss of energy resolution

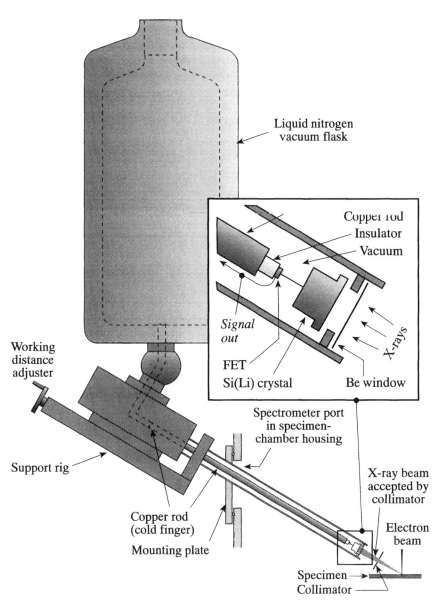

Figure 14.5   Construction of an energy–dispersive X-ray detector.

and thereby increase peak overlap. The analyst must therefore use relatively low probe currents (a few nanoamperes).

(2) *Pulse pile-up* If a second pulse arrives during the initial rise-time of the preceding one, the amplifier may simply stack one on top of the other, outputting a single pulse whose amplitude is the *sum* of the individual pulse heights, which contributes to an spurious *pulse pile-up* peak at a hypothetical 'energy' corresponding to the sum of the individual photon energies. Such artefacts are removed in commercial systems by an electronic fix called a *pulse pile-up rejector*.

The ED spectrum   A typical ED spectrum (Fig. 14.6b) consists of a series of element peaks attributable to the characteristic X-ray lines of the elements present, and a significant background signal between the peaks which represents the X-ray *continuum* spectrum.

ED element peaks are broader than those obtained by WD spectrometry (Fig. 14.6a) owing to statistical fluctuation in the number of ionisations in the detector associated with photons of a particular energy; instead of pulses of uniform amplitude, a spread of pulse heights is produced yielding a broad peak of **Gaussian** shape in pulse–amplitude space (Box 7.1; Fig. 14.6b). Preamplifier noise widens the peak further, but this contribution is minimised by liquid nitrogen cooling. The energy **resolution** of an ED detector is quoted as the width in eV of the 5.89 keV Mn Kα peak at half the maximum peak height, typically around 138 eV for a high-performance detector.

Because the count rate representing a particular element in Fig. 14.6(b) is spread over a number of channels, the best estimate of that element's X-ray intensity is obtained by inte-

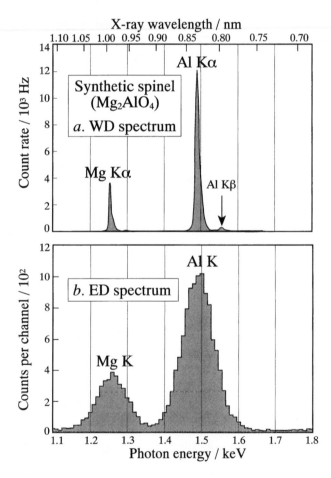

Figure 14.6   (a) Part of a wavelength-dispersed X-ray spectrum for a synthetic spinel crystal. The spectrum has been plotted against photon energy (rather than wavelength or 2θ) to allow direct comparison with the ED spectrum (below); an inverse wavelength scale is shown at the top. Note that the Al Kβ peak, for example, is clearly resolved from the corresponding Kα peak. (b) The energy-dispersive X-ray spectrum of the same specimen for a 200 s acquisition time. Note the broader peaks and poorer peak to background ratio. As the ED spectrometer cannot resolve the Kα and Kβ energies, the peaks are labelled simply as 'Mg K' and 'Al K'. Spectra reproduced by kind permission of Dr. N. Charnley, University of Oxford.

grating the peak *area*. Quantitative interpretation of ED spectra requires (a) removal of the background signal (attributable largely to the X-ray continuum), and (b) 'deconvolution' (unravelling) of the remaining multi-peak spectrum into a series of individual peak profiles that can be integrated separately.

Details of the advanced numerical techniques used may be found in such books as Reed (1993). These calculations are carried out automatically by proprietary software provided with most ED systems.

The wavelength-dispersive (WD) X-ray spectrometer

The WD spectrometer fitted to an electron microprobe differs in design from an X-ray fluorescence spectrometer (Chapter 6), which is designed to analyse a parallel beam of X-rays of relatively high power emanating from a finite area of sample (~ 10 cm^2). The electron microprobe works with much lower total X-ray power emanating from a point source on the sample surface. If a flat crystal were used, the angle of incidence of this divergent beam would vary across the crystal, and the Bragg condition (Chapter 6) could be satisfied only over a small part of its area. The microprobe spectrometer therefore uses a curved crystal, which allows more of the divergent beam to satisfy the Bragg condition and *focuses* the beam into the detector slit to make more efficient use of the limited X-ray intensity available (Fig. 14.7).

In a focusing spectrometer, the X-ray source, the crystal and the detector lie on the circumference of a hypothetical circle of constant radius (typically 15 to 25 cm), called the 'Rowland circle' (Fig. 14.7). Spectrometers are usually mounted with the Rowland circle vertical. To maintain optimum geometry across a range of wavelength, the sample–crystal

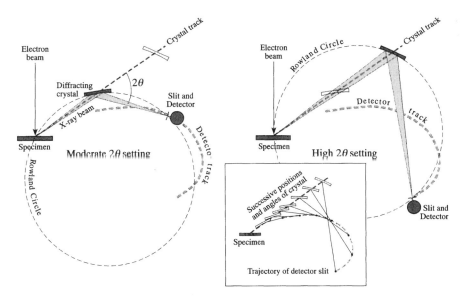

Figure 14.7   The geometry of a focusing X-ray spectrometer used in wavelength-dispersive electron probe micro-analysis, shown at two different wavelength settings. Note that the Rowland circle moves as the wavelength setting (which may be graduated in 2θ or l, the specimen–crystal distance) is altered. The inset shows how crystal and detector slit move in relation to each other.

and crystal–detector distances are varied as shown in Fig. 14.7. The complicated mechanism required accounts for the relatively high cost of a WD spectrometer. Modern spectrometers are computer driven by means of high-precision stepping motors.

A WD spectrometer is highly sensitive to movement of the X-ray source in the direction of the beam (z direction); a positioning error of even a few micrometres is sufficient to cause unacceptable peak shift and intensity loss (Reed, 1993, page 77). This can be avoided by using a built-in high-power optical microscope to 'focus' the specimen surface at the optimum plane.

The diffracting crystal of a focusing spectrometer is fixed to a curved former in such a way that its lattice has a radius of curvature *twice* the radius of the Rowland circle (the crystal must be thin). The crystals used in microprobe analysis are of the same types as for X-ray fluorescence (Chapter 6). Three crystals (typically TlAP, PET and LiF) are needed to cover the routine wavelength range. Spectrometers commonly incorporate a crystal-change mechanism, but fitting 3–4 spectrometers makes time-consuming crystal changing unnecessary.

A WD spectrometer incorporates a gas flow proportional detector (Box 6.3) mounted inside the spectrometer vacuum, producing pulses whose height is crudely proportional to photon energy. Its energy resolution is poorer than an ED detector, but second-order reflections (with $n = 2$ in the Bragg equation) and **escape peaks** can be filtered out by an electronic pulse-height discriminator that transmits pulses falling within a preset pulse-height 'window', but rejects those lying outside. The signal from each spectrometer may be recorded on a digital display, or passed directly to computer memory.

Spectrum processing is much simpler than for ED spectrometry. Analyte intensity is adequately represented by peak height (the count rate measured by the detector at the peak maximum). For quantitative analysis, peak intensities must be corrected for dead time, for background (by subtracting a background signal measured adjacent to the peak) and for any spectral interference. All of these corrections are much smaller than for ED.

X-ray generation

Figure 14.8(a) shows a numerical simulation of the fate of incoming electrons as they are scattered by the atoms of the specimen. *Elastic collisions* (involving no loss of energy) merely deflect electrons, causing the spidery electron paths in the upper diagram. *Inelastic collisions*, however, transfer energy from electrons to atoms, and the resulting inner-shell ionisation causes the atoms to emit characteristic X-rays. The simulated paths terminate in two ways:

(1) At the specimen surface, through which some electrons escape as *back-scattered electrons* that do not contribute to X-ray generation. The proportion of electrons that escape in this way increases with the mean atomic number $\overline{Z}$ of the specimen. Displaying the variation in the intensity of BSE emission as the beam rasters the specimen therefore maps the distribution of high-$\overline{Z}$ domains in the specimen surface.

(2) At the point where, after successive inelastic collisions, the electron's remaining energy is insufficient to excite further X-rays. Such terminations delineate the extent of the

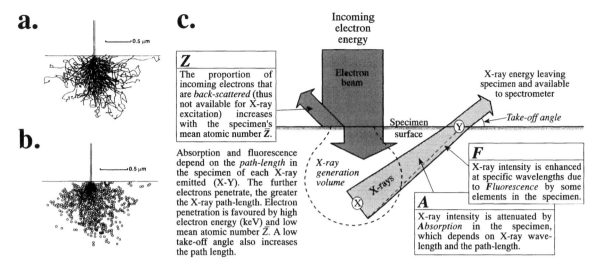

Figure 14.8 (a) Computer simulation of electron scattering in a specimen, illustrated by Fe metal. (b) Corresponding distribution of X-ray generation. Beam energy 20 keV. Parts (a) and (b) reproduced by permission of Plenum Press from Goldstein *et al.* (1992) *Scanning Electron Microscopy and X-ray Microanalysis.* (c) The effects of electron back-scattering, X-ray absorption and fluorescence on the measured X-ray intensity. Arrow width represents electron or X-ray beam intensity.

X-ray generation volume (Fig. 14.8b), which extends several micrometres beneath the specimen surface.

The figure shows that the depth profile of X-ray excitation peaks at a finite depth *within the specimen*.

Qualitative analysis and element distribution

Some information about element distribution is available from a back-scattered electron image of the specimen, which shows the distribution of high- and low-$\overline{Z}$ (mean atomic number) phases such as pyroxene and feldspar (Fig. 14.3b). In most microanalytical applications, however, the user wants to identify or map individual elements by means of their X-ray spectra.

Commercial ED software systems offer convenient element-labelling routines for identifying unknown peaks in a spectrum. A second important application is *element mapping*, in which the distribution of a specified element is imaged by modulating the brightness of the SEM display screen with detector pulses from a WD spectrometer tuned to the X-ray line of the element in question (or with pulses from the appropriate pulse-height window of an ED spectrum). When the electron beam excites emission from the element at a particular point in the raster, the X-ray photons detected appear as bright dots at corresponding points on the display screen (Fig. 14.3c). Most ED analysis systems offer powerful element-mapping packages that allow the distributions of several elements to be displayed on screen simultaneously in different colours.

Quantitative analysis

Matrix ('ZAF') corrections

If the characteristic X-rays measured in microprobe analysis were generated at the very surface of the specimen, the intensity of the analyte spectrum would depend on just two parameters: (a) the concentration of the analyte in the surface layer of the specimen, and (b) the probe current and electron energy employed. If the same conditions (b) were employed for measuring both the sample and standard, the ratio of intensities obtained would be the same as the concentration ratio:

$$[I_i]_u / [I_i]_s = [C_i]_u / [C_i]_s \qquad\qquad [14.1]$$

where $[I_i]_u$ and $[I_i]_s$ refer to the measured intensity of the analyte peak (corrected for dead time, background and peak overlap) obtained from unknown and standard respectively and $[C_i]_u$ and $[C_i]_s$ refer to the analyte concentration in each material.

In reality, however, the X-rays are generated a few micrometres *beneath* the specimen surface (Fig. 14.8b). Electrons and X-rays interact with the specimen (Fig. 14.8c) in a composition-dependent manner that leads to a more complicated relationship between intensity and analyte concentration. Electron back-scattering, X-ray absorption and fluorescence (each dependent on the concentrations of *all* major elements present) affect the **sensitivity** of the analyte, and a systematic error will arise if the specimen composition differs from that of the standard. Accurate quantitative analysis thus requires apparent concentrations (Equation 14.1) to be corrected for these effects. The most popular correction procedure is referred to as 'ZAF correction' as it consists of $\overline{Z}$-dependent (back-scattering and electron penetration), absorption and fluorescence factors. Because the specimen composition (initially unknown) is used in calculating the correction, iteration must be used. For details see Goldstein *et al.* (1992) and Reed (1993), who also describe recent advances. Instrument manufacturers offer on-line ZAF correction as standard, so that each analysis can be inspected in fully corrected form before the next spot is analysed.

The absorption and fluorescence corrections are critically dependent on X-ray path length, and for rough surfaces path length uncertainty degrades the analysis. Surface features may also mask radiation from low points on the surface. For these reasons a highly polished surface is essential for quantitative analysis.

Choice of standards

Quantitative analysis demands the use of standards of accurately known composition. It is normally sufficient to use one standard per element being analysed; in some cases one standard may serve for two elements (e.g. wollastonite, $CaSiO_3$, for Si and Ca). A microprobe standard needs to be:

- of accurately known chemical composition;
- chemically homogeneous on the micrometre scale;
- hard enough to be easily polished and to resist abrasion;
- available as a crystal or glass that is chemically stable in the atmosphere, under vacuum and under electron bombardment;

- preferably similar in analyte valency and bonding to the unknown to avoid **chemical shift** in wavelength, particularly for lighter elements.

Standards, even if conducting, must be carbon coated to the same thickness as the unknowns.

When analysing metallic specimens it is normal to use elemental standards (metals or homogeneous binary alloys). The choice of standards for analysing silicate minerals or other complex materials involves compromise. Though it seems desirable to minimise uncertainties in ZAF corrections by using a standard whose composition closely matches that of the unknown, this places undue reliance on the quality of analysis and uniformity of a chemically complex material. Mineral analysts prefer to use simple compounds such as pure stoichiometric oxides (e.g. synthetic MgO), whose compositions can be determined by calculation. Carefully analysed silicate minerals may be used to standardise elements for which no such simple material is available or stable, such as jadeite ($NaAlSi_2O_6$) for Na. For heavier elements like Mn and Fe, metals or oxides are equally acceptable (Table 14.1).

In WD analysis it is usual to measure all of the relevant standards once or twice during every analysis session. ED systems are inherently so stable, however, that frequent reference to primary standards is unnecessary. Primary standard spectra, together with that of a pure metal standard (usually cobalt), may be measured carefully perhaps once a year and stored digitally as reference spectra. In routine analysis (using the same instrumental settings, e.g. kV) the analyst simply measures the Co standard periodically. Expressing these Co spectra as a ratio the reference Co spectrum allows each unknown to be calibrated against the appropriate reference spectra for the elements concerned.

Analytical resolution

Modern scanning electron microscopes and electron microprobes can focus an electron beam to a diameter of a few nanometres. Nevertheless, except for ultra-thin specimens of the kind used in transmission electron microscopy, scattering within the specimen causes the incoming electrons to spread out beneath the surface of the specimen (Fig. 14.8a), and the *analytical* resolution (the diameter of the X-ray generation volume, Fig. 14.8b and c) is therefore no better than a few micrometres.

Table 14.1   Commonly used standards for silicate analysis

Element	Standard (mineral/*synthetic*)	Ideal composition
Na	Jadeite	$NaAlSi_2O_6$
Mg	*MgO*	MgO
Al	Corundum	Al_2O_3
Si	Wollastonite	$CaSiO_3$
K	Orthoclase	$KAlSi_3O_8$
Ca	Wollastonite	$CaSiO_3$
Ti	*TiO_2*	TiO_2
Cr	*Cr metal*	Cr
Mn	Rhodonite	$MnSiO_3$
Fe	Hematite	Fe_2O_3

When analysing close to the boundary between two crystals of different composition, the X-ray generation volume or X-ray path may include the margin of the neighbouring crystal (particularly where a low-angle boundary is present), and the resulting composite spectrum will give misleading results. Such 'edge effects' also have to be considered when analysing crystals smaller than 20 μm.

On instruments equipped with an optical microscope, it is easy to forget that the beam may not impact on the specimen exactly at the spot indicated by the microscope crosshairs, and the analysis obtained may not represent the intended target position. The analyst must record where the 'beam spot' lies relative to the microscope crosswire, using a **cathodoluminescent** material (such as MgO) or by comparing a high-magnification electron image with the microscope field of view.

Analysing light elements

Elements having atomic numbers below 10 (e.g. O, N and B) present several challenges for X-ray analysis:

- The 'soft' X-rays from these elements are strongly absorbed in the specimen or detector window, so count rates are low and susceptible to surface contamination.
- Some elements lie beyond the wavelength range of conventional diffracting crystals.
- Spectral interference from L and M lines (Box 6.1) of heavier elements may be troublesome, and rising continuum intensity at long wavelengths leads to low peak to background ratios.
- **Chemical shifts** due to differences in bonding are more noticeable than for heavier elements.
- Accuracy is subject to greater uncertainty in mass absorption coefficients (Reed, 1993).

Recent years have seen important technological advances. ED detectors are available in which the beryllium window can be replaced (once the specimen chamber has been evacuated) with an 'ultra-thin window' of aluminised polymer, or which can be used in a completely windowless mode. Long-wavelength WD spectrometry has kept pace through the development of *multi-layer dispersion elements* (MLDEs), made by evaporating alternating thin films of a high-Z element (e.g. W) and a lighter element (e.g. Si). MLDEs with $2d$ spacings of 6–15 nm are commercially available to cover the wavelength range 2.4 nm (oxygen) to 11.4 nm (beryllium). Their high diffraction efficiency greatly improves the detection limits for light elements (Fig. 14.9).

ED and WD spectrometry compared

Energy-dispersive analysis offers three major advantages:

(1) an ED detector's high collection efficiency (enabling low probe currents to be used, minimising beam damage to delicate specimens) and insensitivity to specimen position;

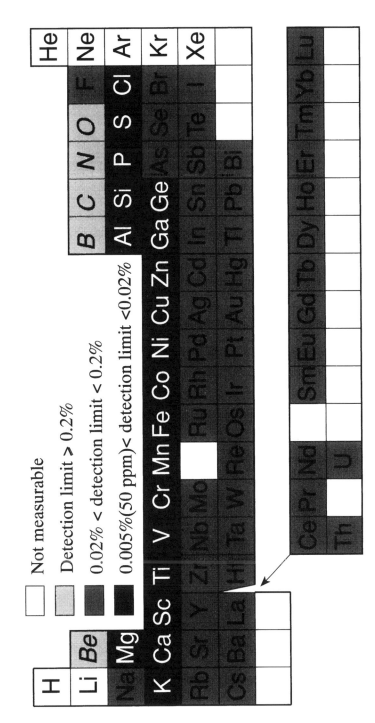

Figure 14.9 Ranges of detection limit (3σ) for wavelength–dispersive electron probe microanalysis. Detector: limits of energy-dispersive analysis are generally about 10× higher. The detection limits refer to analysis under optimum conditions in a light-element matrix, and will be higher in the presence of peak overlap or other interference. Light elements shown in italics require non-routine conditions (see text).

(2) simultaneous acquisition of the entire spectrum, providing rapid multi-element capability;

(3) lower capital cost and ease of use.

These advantages are overwhelming for qualitative work such as element mapping, and apply to many quantitative applications too.

Wavelength-dispersive spectrometers offer advantages for more demanding quantitative applications:

- tolerance of higher count rates (shorter dead time);
- much narrower peaks and higher spectral resolution, giving better peak separation in complex spectra (see inset, Fig. 14.6);
- higher peak to background ratios and therefore lower detection limits.

Optimum versatility is gained by having both techniques available on the same instrument.

Applications

Environmental biomonitoring

Electron probe microanalysis has found many environmental applications, from the purely inorganic (e.g. characterisation of asbestos fibres) to the analysis of biological tissue. Many marine organisms accumulate heavy metals in response to coastal pollution. To ensure the viability of coastal fisheries, it is essential to understand how this uptake is manifested higher in the food chain. Sediment-feeding molluscs such as the tower shell (*Cerithium vulgatum*) may accumulate high levels of toxic metals such as Zn and Cd in certain tissues (Walker *et al.*, 1975, Ray and McLeese, 1987). Electron probe microanalysis shows such metals are concentrated in the digestive gland, where they are bound in phosphate granules that limit their bioavailability (Nott, 1993): carnivores feeding on such creatures are evidently unable to metabolise the metalliferous granules, which pass through the gut into the faeces. X-ray microanalysis provides a means of investigating the mechanisms by which organisms 'detoxify' heavy-metal concentrations in polluted waters, and offers the prospect of quantitative biomonitoring of coastal water pollution.

Geothermobarometry

The minerals biotite and garnet co-exist in many **metamorphic** rocks. Both are solid solutions in which Fe and Mg compete for the same crystallographic sites, and their Mg/Fe ratios are found to vary according to the temperature at which they have equilibrated. Figure 14.10(a) shows how the ratio $K_{gb} = (Mg/Fe)_{garnet}/(Mg/Fe)_{biotite}$ varies with temperature in laboratory experiments. These experiments provide the calibration for a widely used *geothermometer*: accurate electron microprobe analyses of co-existing biotite and garnet crystals in a natural metamorphic rock can be related to the laboratory data and used to esti-

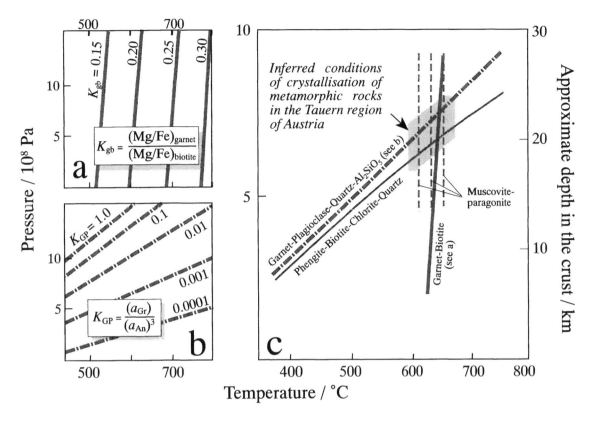

Figure 14.10   (a) Isopleths for Mg/Fe distribution (from laboratory experiment and microprobe analysis) between garnet and biotite (Ferry and Spear, 1978). (b) Isopleths for the assemblage garnet–quartz–plagioclase–Al$_2$SiO$_5$ as a function of mineral composition. The numbers represent values of the equilibrium constant K_{GP}; a_{Gr} is the activity of the grossular (Mg$_3$Al$_2$Si$_3$O$_{12}$) component in garnet, and a_{An} is the activity of the anorthite (CaAl$_2$Si$_2$O$_8$) component in plagioclase. Activities are thermodynamic concentration parameters calculated from microprobe analyses of the minerals. (c) Simplified P–T diagram showing the best estimate (stippled polygon) of the conditions of formation of selected metamorphic rocks in the Tauern Window, Austria, adapted from Droop (1985, Figs. 5 and 6). The size of the polygon reflects the aggregate uncertainty associated with each set of isopleths.

mate the temperature at which the rock (and its constituent minerals) crystallised, assuming their compositions have not changed subsequently.

Figure 14.10(b) shows the behaviour in pressure–temperature (P–T) space of another metamorphic mineral assemblage (garnet + quartz + plagioclase + *either* kyanite *or* sillimanite (alternative crystal forms of Al$_2$SiO$_5$)). Garnet and plagioclase are solid-solution minerals whose compositions vary according to the P–T conditions under which they attained chemical equilibrium. Each radiating line indicates a restricted range of P–T conditions under which all four minerals can co-exist in mutual equilibrium for a different value of the equilibrium constant K_{GP} (defined in the caption). The lines are based on laboratory experiment and thermodynamic calculation. Microprobe analyses of garnet and plagioclase crystals in a garnet–quartz–plagioclase–Al$_2$SiO$_5$-bearing metamorphic rock can be used to calculate a value for K_{GP} for the rock, which identifies the P–T contour ('isopleth') upon which the conditions of formation of the rock lay. Unlike the garnet–biotite case, this equilibrium is more sensitive to pressure than temperature and can be used as a *geobarometer* to estimate the pressure (i.e. depth in the Earth) at which the rock was formed.

Neither example indicates a unique pressure or temperature of formation, because the isopleths are all oblique (i.e. vary with temperature *and* pressure). The isopleths for the two assemblages differ markedly in gradient, however, so their intersection for a single rock, or a group of closely related rocks, pinpoints the *P–T* conditions under which the rock has apparently formed. Metamorphic petrologists increase the confidence of such estimates by using rock suites whose mineral asemblages allow four or five isopleths to be drawn. An example is shown in Fig. 14.10(c); the stippled polygon indicates the best estimate of the conditions under which the rocks concerned originally crystallised. An introduction to the science of *geothermobarometry* can be found in Yardley (1989, pages 51–58).

Acknowledgements

I am grateful to Dr Norman Charnley for allowing me to reproduce the spectra in Fig. 14.6 and to Dr Elizabeth Andrews for leads on environmental applications.

Chapter 15

Principles of SIMS and modern ion microprobes

N. Shimizu

Introduction

When a beam of ions with energy of the order of 10 keV strikes the surface of a solid, kinetic energy and momentum are transferred from incident ions to atoms in the solid. With the inter-atomic bonding energy up to 100 kcal mol^{-1} (approximately 4–5 eV) in rock-forming minerals (e.g. Si–O bonding), the ion beam bombardment sets many atoms in motion and cascades of collisions occur. As a result, some atoms on or near the surface could acquire energy and momentum sufficient to leave the surface. Some of these particles (atoms and clusters of atoms) leave the surface in charged states (secondary ions). This phenomenon is called **sputtering**, and secondary ion mass spectrometry (SIMS) is a technique for mass spectrometric analysis of secondary ions formed by sputtering. The ion microprobe is an anaytical instrument based on the principles of SIMS, and uses a well-collimated ion beam so that analysis of micrometre-scale areas is possible. Since it is fundamentally mass spectrometry, isotopic compositions of elements can be determined *in situ*, and quantitative chemical analysis can also be made *in situ* by converting secondary ion intensities to concentrations.

Sputtering produces diverse ion species (ions with either polarity, monatomic ions of various charge states, multi-atomic ions such as dimers, trimers and more complex molecules). Sputtered ion formation is also a selective process, and a population of secondary ions is chemically and isotopically fractionated relative to the true composition of sample. It was essentially because of this complexity of SIMS spectra that it took more than a decade of effort before ion microprobe analysis proved useful in geochemistry. Technical difficulties caused by the complexity of SIMS spectra have, however, motivated and led instrumentalists to improved machine design and performance. The part played by the earth and planetary sciences community in recent technical development has been essential.

W. Compston and colleagues (e.g. Clement *et al.*, 1977) at the Australian National University were the first to design and build a large-radius high-resolution ion microprobe ('SHRIMP': Sensitive High-Resolution Ion MicroProbe). They foresaw the necessity of the type of instrument for U–Pb dating of zircons. Experience accumulated over the years with small-radius low-resolution (and relatively low-transmission) instruments such as IMS 3f (and 4f, 5f, 6f) has finally convinced users and manufacturers alike that the SHRIMP-type large-radius instruments are the only alternative to meet the technical demands of complex geologic samples and research needs in geochemistry. As of late 1994, two instruments are commercially available: SHRIMP-II (ANUtech, Australia) and IMS 1270 (Cameca, France). These instruments share many design principles, and their mass spectrometer capabilities are very similar. A brief description of the essential components of a modern ion microprobe follows, using the IMS 1270 as an example (Fig. 15.1).

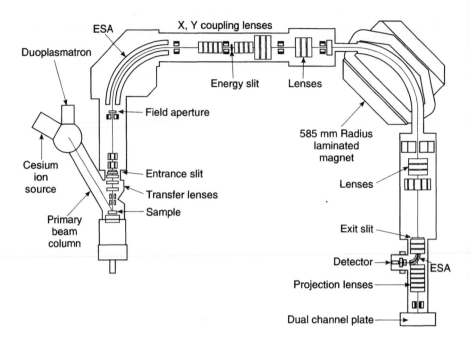

Figure 15.1   An overhead view of Cameca IMS 1270, showing the basic components in a modern large–radius, high–resolution ion microprobe. The radius of curvature for both electrostatic analyser (ESA) and magnet is 58.5 cm and can be used as a scale.

Ion beams

The primary ion beam most frequently used in geochemical analysis is either negatively or positively charged oxygen ions $(O^-, O_2^-$ or $O_2^+)$. These beams are produced with a duoplasmatron ion source (Fig. 15.1), in which a plasma is formed by arcing oxygen gas at the tip of a cold hollow cathode (made of Ni). Either positively charged (concentrated at the centre of the plasma) or negatively charged (concentrated at the periphery of the plasma) ions are extracted, accelerated to an energy of the order of 10 keV and collimated to form a beam with a circular cross-section of uniform brightness (or uniform current density). Since the beam drills a crater on the sample surface and secondary ions come from the crater, the uniformity of the beam cross-section with respect to current density is important. In order to ensure uniformity, SHRIMP-II employs the Koehler illumination principles, in which the beam is an apertured uniform portion of a larger beam. The IMS 1270 can use either the Koehler or the critical illumination principles, in which the sample surface is bombarded with a high–density cross-over of the ion beam.

The usefulness of an oxygen ion beam in geochemical analysis is based on high secondary ion yield and stable emission of secondary ions compared with primary ion beams of Ar^+ or Xe^+ that were more commonly used in early studies of sputtering dynamics in pure metals (e.g. Sigmund, 1969). Another species useful for geochemical analysis is a Cs^+ ion beam produced in a thermal ionisation source. The Cs^+ ion beam is effective (see p. 201) in producing negatively charged secondary ions of electronegative elements (e.g. P, Se) and also of H, O, S, C, F, Cl, etc. Figure 15.1 shows that IMS 1270 is equipped with a Cs ion source built together with a duoplasmatron with a Y–shaped dual–entry system that allows a quick switchover from one source to the other without lengthy mechanical work breaking the vacuum of the system.

The Cs^+ beam, when used on insulators such as typical rock-forming minerals, is known to charge up the sample; the excess positive charge should then be compensated by electrons supplied to the surface. The IMS 1270 is equipped with an electron gun (Slodzian *et al.*, 1986) which supplies a beam of electrons with energy such that charge compensation is most effective.

Samples are usually polished sections (or polished thin sections) prepared with the same techniques required for electron probe microanalysis. The surface is coated with a film ($\sim$300 Å thick) of conductive material such as carbon or gold. The sample position in the instrument is controlled by stepping motors with a precision and backlash better than ± 1 μm; the sample is viewed via a TV camera and a display on a CRT.

Mass spectrometer design

The mass spectrometer design for the analysis of secondary ions is similar in the two modern instruments with a forward double-focusing principle (Box 11.1). Since secondary ions leave the sample with various initial velocities, an electrostatic analyser (ESA) is required for 'sorting out' secondary ion velocities in order to carry out high-resolution mass analysis by a magnetic sector (Box 11.1). A slit (energy slit) is placed at a cross-over between the ESA and magnet, where the 'energy spectrum' is formed by the ESA, and it can be adjusted to select a specific portion of the secondary ion energy spectrum (energy filtering). An ultimate mass resolution of 100,000 is attainable with the modern instruments. Mass-analysed secondary ions are collected and electronically detected/measured. Simultaneous collection/detection of ions with different masses can be achieved by an array of collectors/detectors placed on the focal plane of the secondary ion optics after the magnet. Simultaneous (as opposed to sequential) collection improves the efficiency of analysis by reducing sample consumption, and is essential for the high-precision analysis of large isotopic ratios such as $^{16}O/^{18}O$ ($\sim$500) and $^{12}C/^{13}C$ ($\sim$100); with a single collector, signal fluctuations during the long count time of the minor isotope (^{18}O and ^{13}C) introduce an error that cannot be corrected for and this limits the precision of the analysis. When a pulsed laser beam is used for ionisation in resonance ionisation mass spectrometry (RIMS) and/or the sputtered–neutrals mass spectrometry (SNMS) mode, secondary ion signals are also pulsed and the simultaneous collection of mass-analysed ions becomes imperative.

An additional ESA after the magnet can be important in order to achieve a level of abundance sensitivity (Box 8.5) required for the accurate determination of extreme isotope ratios (such as $^{232}Th/^{230}Th$ of $\sim$10^6). The abundance sensitivity is defined as the ratio of ion intensities of two neighbouring atomic mass units and is used as a measure of 'tailing' of a very intense peak onto the neighbouring mass position due essentially to the scattering of ions by collision with residual gas molecules in the flight tube (cf. Chapter 12). Since these scattered ions have energy levels different from the ones to be measured, they can be reduced effectively by an additional ESA. Indeed, the abundance sensitivity of 7×10^{-9} can be achieved at mass 237.

Complexity of SIMS spectra

A classical example of the complexity of SIMS spectra is isobaric interferences by molecular ion species. Sputtering produces various ion species, and in the case of silicate samples under oxygen ion bombardment the complexity is manifested even more, with abundant oxide, double-oxide and hydroxide species. For instance, when determination of Sr abundance in a basalt glass is attempted by measuring the secondary ion intensity of ^{88}Sr, interferences of univalent ion species such as $^{44}Ca_2$, $^{27}Al_2{}^{16}O_2{}^1H_2$, $^{44}Ca^{28}Si^{16}O$, $^{28}Si_2{}^{16}O_2$, $^{40}Ca^{16}O_3$, $^{56}Fe^{16}O_2$, $^{40}Ca^{24}Mg_2$, among others of the same nominal mass number, could be serious. When the Pb isotopic composition of zircon is attempted, HfSi, Zr_2O, HfO_2, etc., must be dealt with. Although with early low-resolution instruments most of the interferences were unresolvable mass spectrometrically, the modern instruments mentioned above, with a 'routine' mass resolution in excess of 5000, can resolve them easily. Atomic mass units (amu) and required mass resolution (mr) relative to ^{88}Sr (amu = 87.905625) for the molecular ion species listed above are as follows: $^{44}Ca_2$ (amu = 87.91096; mr = 16480), $^{27}Al_2{}^{16}O_2{}^1H_2$ (87.96859; 1400), $^{44}Ca^{28}Si^{16}O$ (87.92732; 4050), $^{28}Si_2{}^{16}O_2$ (87.94368; 2310), $^{40}Ca^{16}O_3$ (87.94732; 2110), $^{56}Fe^{16}O_2$ (87.92482; 4580), $^{40}Ca^{24}Mg_2$ (87.93269; 3250). A mass resolution of 3920 is required to resolve $^{180}Hf^{28}Si$ (207.92349) from ^{208}Pb (207.97664).

Methods of quantitative geochemical analysis

With molecular ion interferences resolved, the next step towards quantitative SIMS analysis is a conversion of secondary ion intensity to concentration. The complexity of SIMS spectra plays a role in this process, because sputtering is a selective process and a secondary ion population is chemically fractionated relative to the sample. This phenomenon is generally termed 'matrix effects', and results from interaction between atoms during sputtered ion formation (e.g. Shimizu, 1986). Practically, the conversion can be made adequately using empirical relationships between intensities and concentrations obtained for standards. Figure 15.2 shows examples of such relationships (working curves) for rare earth elements (REEs) in pyroxene and garnet, and water in silicate glasses. It is noticeable that the REE working curves pass through the origin and are identical between these pyroxene and garnet standards, demonstrating that matrix effects are similar between these minerals under the specific analytical conditions employed.

The quality of ion microprobe data is critically dependent on counting error, the documentation of standards, the quality of data with which working curves are established, and the short- and long-term stability of working curves. Experience at the Woods Hole Laboratory (IMS 3f) is that, once established, working curves are reproducible to within counting statistics for a period of several years or more, indicating that drifts of any type, if present, are of negligible magnitude. Counting error appears to be the most significant uncertainty associated with data, and routine determinations of REEs in many rock-forming minerals with a primary beam spot size of $\sim$20 μm and with energy filtering (voltage offset -60 ± 10 volts) are reproducible to approximately 2–5% for light REEs (LREEs; La, for instance) and 5–7% for heavy REEs (HREEs; Yb, for instance) at concentration levels of $\sim$10 times C1 chondrite. Since the documentation of standards relies on bulk analysis techniques, the homogeneity of standards must be checked very carefully before use.

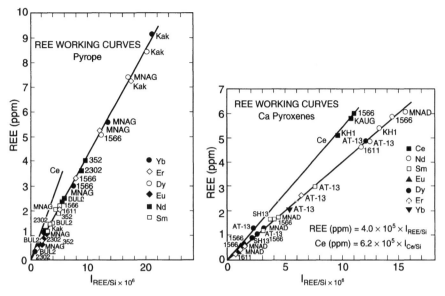

Figure 15.2(a) Relationships between secondary ion intensities and concentrations (working curves) for rare earth elements (REEs). Clinopyroxene and garnet are shown separately for clarity, but the working curves for these mineral groups are identical. Samples used for establishing the curves are as follows. Kak: garnet megacryst from Kakanui, New Zealand (Mason and Allen, 1973); KAUG: clinopyroxene megacryst from Kakanui, New Zealand (Mason and Allen, 1973); MNAG: garnet megacryst from Monastery, South Africa (Shimizu, unpub. data); MNAD: clinopyroxene megacryst from Monastery, South Africa (Shimizu, unpub. data); 1566: garnet and clinopyroxene from garnet lherzolite PHN 1566 from Thaba Putsoa, Lesotho (Shimizu, 1975); 352: garnet from garnet lherzolite JJG 352 from Bultfontein, South Africa (Shimizu, 1975); 2302: garnet from garnet lherzolite PHN 2302 from Liqhobong, Lesotho (Shimizu, 1975); 1611: garnet (core) and clinopyroxene from garnet lherzolite PHN 1611 from Thaba Putsoa, Lesotho (Shimizu, 1975); BUL 2: garnet from garnet lherzolite BUL 2 from Bultfontein, South Africa (Shimizu, unpub. data); KH 1: clinopyroxene megacryst from Kilbourne Hole, New Mexico (Irving and Frey, 1984); AT-13: clinopyroxene from harzburgite from Romanche Fracture Zone, Atlantic (Shimizu, unpub. data); SH 13: clinopyroxene megacryst from Shaeffer, Colorado (Shimizu, unpub. data).

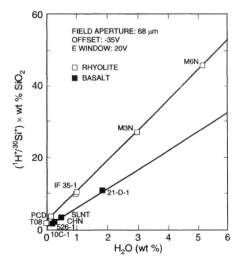

Figure 15.2(b) Working curves for water in rhyolite and basalt glasses (Layne and Stix, 1990). Note that residual $^1H^+$ intensities at zero concentration are indicated as positive intercepts of the working curves.

Detection limits are strongly a function of ion yield. For a given element, the detection limits associated with ion microprobe analysis are a function of primary beam species and current, and other operating conditions. For instance, for REEs with the conditions mentioned above, detection limits (3σ) are approximately 1 ppb (or 0.005 times chondrite for La) and 10 ppb (or 0.06 times chondrite for Yb). Indeed, Lundberg *et al.* (1990) reported a successful analysis of a pyroxene with an La content of less than 1 ppb. Detection limits with the modern instruments are yet to be determined, but are expected to be much better than those mentioned above because of their better transmission.

The sputtered ion formation process enhances the ionisation of light isotopes relative to heavy ones (isotope fractionation), and it is necessary to determine correction factors for specific purposes. The magnitude of isotopic fractionation is a complex function of secondary ion energy and matrix. In pure metals, Shimizu and Hart (1982a) showed that the magnitude is of the order of relative mass difference (in per cent) between two isotopes. For instance, fractionation for ^{10}B and ^{11}B is approximately 10%, whereas that for ^{206}Pb and ^{207}Pb is about 0.5%. As discussed by Slodzian *et al.* (1980), however, the matrix exerts very significant effects, and the observations made on pure metals are not directly applicable to silicates. Shimizu (1986) reported that the silicon-induced enhancement of sputtered ion formation in silicate glasses is mirrored by a decrease in the magnitude of isotopic fractionation and argued that the decrease in isotopic fractionation of Mg in the order pure Mg > olivine > pyroxene with increasing silicon abundance may be explained in the same way. The matrix dependence can be very significant for S isotopes, as Eldridge *et al.* (1987) and Macfarlane and Shimizu (1991) showed that fractionation effects are in the order of barite > pyrite = pyrrhotite > chalcopyrite > galena, with galena displaying the smallest dependence on the secondary ion energy.

A brief summary of geochemical applications

An explosive increase in the use of ion microprobes in geochemistry is evident in recent geochemistry literature, and it is impossible to present a comprehensive picture in the limited space of this chapter. This section summarises only some examples of various types of geochemical applications.

As an *in situ* technique, the ion microprobe eliminates the lengthy, labour-intensive steps of mineral separation, chemistry, irradiation, counting, mass spectrometry, etc., associated with bulk-analysis techniques, and therefore its data throughput is a considerable advantage. For instance, at the ion microprobe laboratory at Woods Hole Oceanographic Institution (Cameca IMS 3f) a routine determination of an REE abundance pattern in basalt glass can be made in approximately 30 minutes. This enables us to carry out the most detailed geochemical documentation of a large number of specimens from individual sample suites. As most geochemical processes produce chemical compositions of minerals variable over length scales ranging from micrometres to kilometres, this capability is crucial for establishing an unbiased comprehensive data set. The ion microprobe also provides quantitative data for a wide range of elements; there are no restrictions, in principle, in terms of elements, from the determination of water content in volcanic glasses with secondary ^{1}H$^+$ intensities as pioneered by R. Hervig and colleagues (e.g. Hervig and Dunbar, 1992), to U–Th–Pb in zircons. Concentrations of many elements can be determined simultaneously (e.g. REEs, HFSEs and alkaline earths on one spot). This is an additional advantage, because conventional bulk-analysis techniques are often limited individually to certain groups of elements.

Another very important aspect of ion microprobe techniques is the capability of directly connecting petrography with trace element/isotope chemistry. Particularly when combined with major element analysis using the electron probe, hypotheses based on petrography can be tested directly and thoroughly.

Determinations of trace element abundances in rock-forming minerals and glasses using the ion microprobe have advanced our knowledge of their behaviour in geochemical processes and re-established their usefulness as geochemical tracers. A key aspect in this area is the determination of the partitioning of trace elements in mineral–melt and mineral–mineral pairs in natural and experimental systems. The sensitivity of ion microprobe techniques generally enables working at natural abundance levels. Analyses of laboratory-experiment-run products have produced a crucial database for the partitioning of many trace elements between clinopyroxene and silicate melt (e.g. Hart and Dunn, 1993). Ion probe data of trace elements in mantle peridotite minerals have been combined with the clinopyroxene–melt data providing a set of partition coefficients for mantle mineral–melt pairs (e.g. Kelemen *et al.*, 1992) that can be used in melting models.

Trace element data obtained in clinopyroxenes in abyssal peridotites (Johnson *et al.*, 1990, Johnson and Dick, 1992) showed extreme depletions and fractionations of incompatible elements and strongly suggested that small-degree melting was followed by efficient extraction of melt fractions beneath mid ocean ridge spreading centres. Sobolev and Shimizu (1993) confirmed the suggestion, showing that a glass inclusion in an olivine phenocryst in a basalt from the Mid Atlantic Ridge possesses trace element characteristics expected for a fractional melt at an advanced stage of the process.

Ion-probe-based trace element data have also advanced our understanding of other geochemical processes in the upper mantle. For instance, REE patterns obtained for garnet inclusions in diamonds (Shimizu and Richardson, 1987, Shimizu *et al.*, 1989, Moore *et al.*, 1991) place important constraints on mechanisms of diamond formation and on the chemical characteristics of materials involved in the process. The interaction between the mantle and silicate melts that pass through it has also been studied in alpine peridotite massifs, using trace element data of pyroxenes and model calculations involving chromatographic effects (e.g. Kelemen *et al.*, 1992, Takazawa *et al.*, 1992).

The surface analysis capability of the ion microprobe is particularly useful in studies of diffusion in solids. By rastering a primary ion beam over a controlled area, the rate of sputtering (or the rate of sample excavation) can be regulated. Continuous data acquisition during excavation (depth profiling) provides a relationship between intensity and time, or concentration vs depth (or isotope ratio vs depth). Using this technique on crystals whose surface had exchanged isotopes with surrounding hydrothermal solution in the laboratory, B. Giletti and co-workers have determined diffusion coefficients of oxygen, strontium and other elements in some rock-forming minerals (e.g. Giletti *et al.*, 1978, Giletti and Yund, 1984, Farver and Giletti, 1985). With a depth resolution of approximately 10 nm, this is a powerful technique in determining very low rates of diffusive transport.

One of the most spectacular contributions of the ion microprobe to geology is undoubtedly the U–Pb age determination of zircons, pioneered by W. Compston and colleagues at the Australian National University using SHRIMP (and recently SHRIMP-II). Based on an empirical approach for converting relative secondary ion intensities of U and Pb to U/Pb atomic ratios and on the measured isotopic composition of Pb, they determined U–Pb ages *in situ* on various parts of individual zircon crystals. Using zircons as an indicator of granitic crust, they not only discovered the oldest rocks on Earth (Froude *et al.*, 1983), but established a new way to investigate early geologic evolution of the Earth (e.g. Bowring *et al.*, 1989 among others). Combined with isotopic geochemistry and tectonic studies, early Earth processes and evolution are coming alive and this field is developing rapidly.

Ion microprobe data of excess ^{26}Mg due to *in situ* decay of extinct ^{26}Al in Al-rich, Mg-poor minerals such as anorthite and hibonite in chondrites (both carbonaceous and ordinary chondrites) contributed to constraining the time scale between the last phase of nucleosynthesis and the formation of the solar system and supporting the possibility that ^{26}Al could have contributed significantly to the thermal history of early planetary bodies.

Significant discoveries were made, however, more recently on anomalous isotopic compositions of Ca, Ti and C in small mineral grains in carbonaceous chondrites (particularly Murchison), and E. Zinner and colleagues at Washington University, St Louis, spearheaded the effort.

The anomalous isotopic compositions of Ca and Ti were observed as large excesses and deficiencies of high-mass isotopes such as ^{48}Ca and ^{50}Ti amounting to the order of several per cent in hibonite, one of the early condensates from the solar nebula (e.g. Zinner *et al.*, 1986, Ireland *et al.*, 1985, Fahey *et al.*, 1985, among others). The presence of isotopically anomalous Ca and Ti indicates that the nucleosynthetic processes resulted in isotopically inhomogeneous presolar materials, which were subsequently incorporated in the condensation of hibonite. The large isotopic variations observed among hibonite grains from a single meteorite show that isotopic equilibrium was not attained at any time since nucleosynthesis. The existence of high-neutron isotopes of Ca and Ti places important constraints on models of nucleosynthetic mechanisms, and their survival throughout the period since then is crucial in understanding the thermal history of the early solar system.

Most of the geochemical studies summarised here used early low-resolution instruments. With the modern instruments installed (or to be installed in the near future) at geochemistry laboratories around the world (Cambridge University (UK); University of Nancy (France); NORDSIM, the Nordic Consortium, Stockholm (Sweden); University of California, Los Angeles (USA); Curtin Technical University, Perth (Australia); Australian National University, Canberra (Australia); Geological Survey of Canada, Ottawa (Canada); Geological Survey of Japan, Tsukuba (Japan); Hiroshima University, Hiroshima (Japan); Woods Hole Oceanographic Institution, Woods Hole (USA)), geochemical and cosmochemical applications of ion microprobe analysis will steadily expand into the 21st century.

Chapter 16

Analytical techniques in organic geochemistry

C. Anthony Lewis

Introduction

Organic geochemistry is the study of the organic chemistry of the geosphere. It is an inter-disciplinary subject encompassing many of the traditional sub-divisions of the physical sciences, for example biochemistry, organic chemistry and sedimentary geology, although a working knowledge of many other subjects (e.g. microbiology, physical chemistry and aquatic chemistry) is also helpful. Organic matter, that is reduced carbon, occurs in many guises in the geosphere and it is convenient to distinguish between fractions (e.g. **lipid**, **asphaltene** and **kerogen**) which are typically operationally defined by their solubility, or otherwise, in a particular solvent system and, consequently, are analysed using different techniques. These gross fractions are usually further sub-divided, as exemplified by the lipids.

Lipids isolated from the geosphere are usually a mixture of different types of organic compound, having a molecular weight ranging up to about 1000 amu. For example, the lipids isolated from a contemporary marine sediment may be a mixture of **aliphatic** and **aromatic hydrocarbons**, alcohols, carboxylic acids, ketones and aldehydes, and **pigments** (a term encompassing **carotenoids** and **tetrapyrroles**, both of which are themselves general terms encompassing functionalised as well as unfunctionalised compounds). The lipids isolated from an ancient sediment or a crude oil, on the other hand, are typically a mixture of aliphatic and aromatic hydrocarbons, and **NSO compounds**. Although these sub-fractions often share a chemical attribute (e.g. alcohols possess a hydroxyl group) their fractionation is also operationally defined, in this instance by their chromatographic **polarity**.

As a result of the chemical and physical diversity of geological organic matter, there is no single analytical scheme that is applicable to all samples. Figure 16.1 illustrates a generalised flow chart for the processes that are described in this chapter.

Organic geochemistry is a relatively new discipline and, because of the complexity of geological organic matter, it has made use of sophisticated and often innovative instrumentation and analytical methods. Together, these are capable of separating the complex mixtures of components observed in the geosphere as well as detecting, quantifying and, most importantly, identifying minute amounts of individual compounds.

The first authoritative book discussing organic geochemistry, published over 25 years ago, was edited by Eglinton and Murphy (1969). Since then, a number of important organic geochemistry books have been published, most notably those written by Hunt (1979), Tissot and Welte (1984), Peters and Moldowan (1993), Killops and Killops (1993) and Tyson (1995), and those edited by Bordenave (1993) and Engel and Macko (1993). The interested reader is directed towards these books for a detailed discourse of most aspects of the subject.

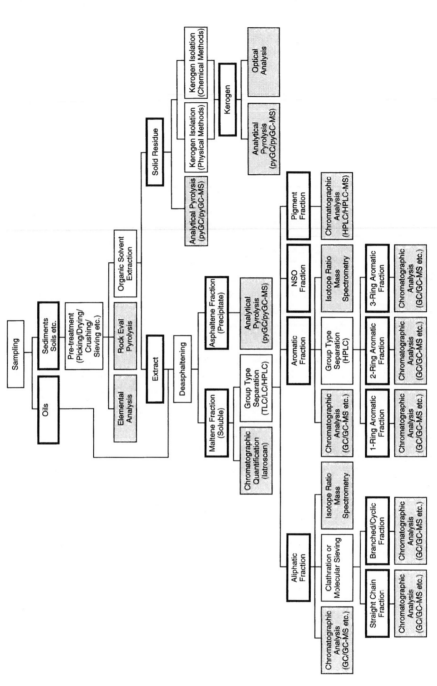

Figure 16.1   A flow chart illustrating the overall analytical scheme to which petroleum and sedimentary organic matter may be subjected. Not all the work-up and analytical processes illustrated will necessarily be performed on the same sample and neither is the scheme intended to be exhaustive with respect to all the possible experimental procedures. **Boxes outlined** indicate the type of fractions obtained whilst the shaded boxes indicate the analytical methods used.

Sampling and storage

Some aspects of sampling strategy have already been covered in Chapter 2. Samples of a surface exposure or of a core or cuttings from drilling operations may be available. Those from a surface exposure are subject to weathering and this may affect the composition of the organic matter to a depth of about 1 m. Drill cuttings are often the only samples available and they may become contaminated with either solid material and/or fluids. Solid contamination includes material added, intentionally or otherwise, to the drilling mud as well as admixture with cuttings from other lithologies (e.g. cavings from above). Fluid contamination may arise from lubricants, drilling mud additives or petroleum products (e.g. diesel) added to the mud.

It is necessary to ensure that the sample is not contaminated during the sampling procedure, during transport to the laboratory and storage, or during subsequent handling and analysis. The sample must be stored in a container before and after analysis (the type depending, to some extent, on the sample itself), which should neither introduce any contamination to nor cause any change in the distribution of the analytes of interest (e.g. loss of volatile components). Glass or metal vessels are typically the most appropriate; however, they should be cleaned of any traces of organic matter. Large pieces of rock may be wrapped in aluminium foil (and newspaper during transportation). For samples with a very low level of organic carbon, PTFE-lined foil bags are sometimes used, although these are rather expensive. Samples that are likely to be affected by microbial activity (e.g. water samples or the retentate from a filtered volume of water) may be sterilised with a biocide (e.g. $HgCl_2$).

The choice of glass or metal depends largely on the sampling strategy employed and their availability. It may be convenient for an academic research team to collect estuarine mud samples in glass jars but this is unlikely to be the case for a geochemist collecting drill cuttings from the shale shaker on an operating rig or oil samples from a completed well head. Plastic containers should, on the whole, be avoided as they usually contain a significant amount of **phthalate esters** which may leach into the sample. This may not be problematical for the analysis of a crude oil, since they may be easily identified and have not been observed to occur in significant amount naturally, but determination of the level of anthropogenic pollutants, including phthalate esters, in a contemporary sediment, for example, would be compromised.

As with storage, samples should be handled so as to reduce the possibility of contamination. Ideally, low-level work should be carried out in a clean room appropriate for organic analysis (unlike a clean room designed for inorganic analysis, plastic surfaces and utensils should be avoided and replaced with the equivalent in metal and glass). Most samples may be manipulated in the open laboratory as long as precautions are taken to minimise contamination by dust and any other hydrocarbon analyses being performed. Benches should not be polished since the wax used may be a source of hydrocarbon contamination. Aluminium foil is used extensively as a temporary covering for samples and also as a cap liner; 'labfilms' should be avoided owing to their waxy composition.

Bulk samples (i.e. exposure, core and cuttings and crude oil) may be stored in a cool, dry place, although the appropriate precautions should be taken if a significant quantity of crude oil is stored. Unconsolidated sediments and other samples susceptible to microbial alteration should be stored frozen at $-20°C$ or, if the analysis of pigments (i.e. carotenoids and tetrapyrroles) is intended, at $-70°C$ and in the dark. Alternatively, unconsolidated sediments may be freeze dried and subsequently stored at ambient conditions (not recommended if pigments are to be analysed). Isolated organic matter and fractions thereof should be stored in a freezer or refrigerator and analysed promptly.

Contamination and its prevention

Contamination of a sample with components that interfere with, or are similar to, the analytes of interest is a matter of concern to all analytical geochemists. The sample should not be contaminated by other carbon compounds and since many of the procedures employed in the analysis of sedimentary organic matter include the addition of *organic* reagents to the sample it follows that they must be of adequate purity. Particular attention should be paid to solvent purity since a relatively large amount is used during isolation and separation and, as it is typically removed again before analysis, any contaminant will be concentrated in the final sample. Glass distilled solvents are usually employed, although it is good laboratory practice to test each batch of solvent received by evaporation of an aliquot and analysis using conditions similar to those typically employed for real samples. Inorganic reagents, chromatographic adsorbents and drying agents should be extracted with an organic solvent (e.g. dichloromethane) prior to use since even the highest-purity grade (e.g. AristaR®) is only certified with respect to inorganic and not organic impurities. Other consumable materials, for example cotton or glass wool and aluminium foil, should be similarly treated to remove natural or other lipids or the rolling oil used during manufacture.

All storage vessels and laboratory apparatus should be cleaned of organic matter which may be achieved by soaking in aqueous detergent, rinsing with tap and double-distilled or de-ionised water and oven drying. Items may then be wrapped or covered in foil until immediately prior to use when it is usual to rinse with dichloromethane. Some workers soak glass apparatus in chromic acid instead of the detergent bath. However, chromic acid is a strong oxidant and should be used with caution and it is debatable whether it offers any benefit over modern laboratory detergents. Handling of samples and apparatus with bare hands should be kept to a minimum, especially in the case of low-level work.

Even when care is exercised, contamination can occur which may be out of the control of the analyst. For example, clean practices may not be rigorously followed during sample collection by a third party. The ease with which contamination can be identified depends, to a large part, on the type of contamination. For example, contamination of a rock sample by a light lubricating oil would normally be readily apparent (e.g. by gas chromatography), whereas contamination of a crude oil by a similar oil might be much more difficult to detect.

It is important that procedural blanks of the entire analytical methodology are performed and that these data are both taken into account when discussing the samples and fully reported. A compendium of most of the contaminants likely to be encountered by the organic geochemist or organic environmental chemist (i.e. trace organic analyst) has been published (Middleditch, 1989).

Isolation of sedimentary organic matter

Consolidated sediments should be broken into small pieces with a hammer or, in exceptional cases, a diamond-tipped circular saw. In the latter case it is important that a non-organic lubricant/coolant is used that will not contaminate the sample. The pieces are ground to a fine powder (e.g. to pass 200 mesh) in a disc mill. Unconsolidated, wet sediments may be separated from excess water and extracted without further treatment. Alternatively, they may be freeze dried and, if necessary, ground to a fine powder, or mixed with a drying agent such as anhydrous sodium sulphate prior to extraction with an organic solvent. Low-temperature oven drying is not recommended.

Soluble organic matter

The lipid or **bitumen fraction** may be isolated from a finely powdered sediment by extraction with an organic solvent. Toluene/methanol (1:1), dichloromethane/methanol (1:1) or dichloromethane are commonly used solvent systems. A propan-2-ol/n-hexane (4:1) solvent system may be used for wet, unconsolidated sediments. Since organic solvents are used in relatively large amounts they usually represent the major health and safety hazard in the organic geochemistry laboratory and proper safety precautions should be taken to minimise exposure to workers.

Various means of solvent extraction are regularly employed, for example mechanical vortex mixing, Soxhlet extraction (Fig. 16.2) and ultrasonic agitation. Although Soxhlet extraction is slightly more efficient than ultrasonic extraction both are widely used and each method has its merits. Soxhlet extraction is somewhat less labour intensive than ultrasonic extraction, since it does not need continuous attention and a bank of six extractors may be used, but an efficient extraction can take at least 24 h. In addition, Soxhlet extraction consumes a large amount of solvent, adding to consumable and disposal costs, and initial glassware expenditure is high.

Super-critical fluid and microwave-assisted extraction of organic compounds have been used in environmental analysis and since they are as efficient as other methods, reduce the amount of solvent used and are rapid, they are likely to become increasingly popular. However, care should be exercised in order to avoid the risk of explosion and the initial capital cost of commercial equipment that ameliorates the problem is substantial.

Whether the sample is a crude oil or sediment extract, it is often desirable to remove the asphaltene fraction. The usual method of isolation is to dissolve the oil or extract in a minimum of a polar solvent (e.g. dichloromethane) with subsequent addition of an excess (40 fold is common) of an apolar solvent (n-pentane, n-hexane or n-heptane). After precipitation overnight the asphaltenes are collected by filtration. The filtrate has been called the **maltene fraction**. Redissolution, precipitation and filtration a further two times will afford an asphaltene fraction free from occluded soluble components. The choice of n-**alkane** solvent is often based on convenience and which fraction is of primary interest: if the asphaltene fraction is sought, n-heptane is typically chosen, whereas n-pentane is often employed if the maltene fraction is required (since it has a lower boiling point and is easily removed).

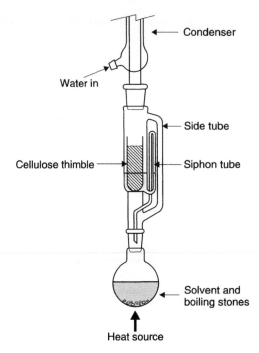

Figure 16.2   A Soxhlet extractor. The sample to be extracted is placed in a precleaned thick cellulose thimble. Solvent evaporated from the heated flask passes through the outer wide-bore side tube and condenses onto the sample. When the solution of solvent and extract reaches the top of the siphon tube, it siphons back into the flask and the extraction is repeated with fresh solvent evaporated from the flask (modified from Harwood and Moody, (1989) *Experimental Organic Chemistry: Principles and Practice*. Blackwell Scientific).

Insoluble organic matter

The sedimentary organic matter remaining after solvent extraction is called kerogen ('proto-kerogen' or **humin** in unconsolidated sediments and soils). Since kerogen is insoluble, the methods devised to separate it from the inorganic mineral matrix rely on the chemical solubilisation of the matrix and/or the difference in density between the kerogen and the matrix. Whether a chemical or a physical method is chosen depends on the type of analysis to be conducted: chemical analysis necessitates the use of chemical separatory methods, whilst petrographic analysis requires physical separatory methods since these tend to cause less damage to the texture of the kerogen.

A significant proportion of most sedimentary rocks may be dissolved in hydrochloric or hydrofluoric acid or mixtures thereof. Extreme care should be exercised when using hydrofluoric acid (Box 3.1). Procedures for the isolation of kerogen are discussed by Durand and Nicaise (1980). Typically, the powdered sediment (previously extracted with an organic solvent) is treated with HCl, washed and then treated with an HCl/HF mixture and, after another wash, with HCl again. This procedure will remove the carbonate and a majority of the silicate minerals. Thorough washing with distilled water and maintenance of a temperature of about 65°C to ensure removal of the fluorides formed affords a kerogen concentrate. Resistant minerals, such as pyrite, zircon, titanium dioxide and tourmaline, may remain. Barite and tungsten carbide have also been found in kerogen concentrates and probably derive from the drilling mud and the drill bit, respectively.

Pyrite is quantitatively the most important resistant mineral associated with the kerogen. Both reducing (e.g. $LiAlH_4$) and oxidising (e.g. HNO_3) agents have been used to remove it but they may alter the kerogen itself: the former reduces carbonyl and carboxyl groups (i.e. $>C=O$), whereas the latter can cause significant carbon loss and an *increase* in the amount of carbonyl and carboxyl groups. A milder oxidant, ferric sulphate, has been used with some success.

The most common physical method for isolating kerogen concentrates depends on the difference in density between the kerogen and the mineral matrix. Centrifugation of a ground sediment in a dense liquid (e.g. aqueous $ZnBr_2$) will result in a crude separation of the kerogen, which will tend to float, from the minerals, which will tend to sink. To avoid excessive damage to the texture of the organic matter the sediment should only be coarsely ground. However, since the organic and mineral phases are often intimately associated the separation is never complete and this is especially true in the case of pyrite.

Group-type separation

The lipid or bitumen fraction is typically a complex mixture of different classes of organic compounds and it is usual to separate this into a number of sub-fractions, each of which ideally comprises one compound class. This 'group-type' separation may be achieved by thin-layer chromatography, open-column liquid chromatography or preparative high-performance liquid chromatography initially into polar and non-polar fractions and then into sub-fractions of these classes. Further separation of selected compound classes may be achieved by **clathration** or **molecular sieving**.

Thin-layer chromatography (TLC)

TLC is a commonly used and relatively quick and inexpensive method of separation. Briefly, a thin layer of adsorbent (i.e. the **stationary phase**) is spread on a flat solid support and dried. The sample is then applied close to one edge of the 'plate' which is 'developed' in a solvent (i.e. the **mobile phase**). A separation is achieved because the solutes partition to a different extent between the stationary and mobile phases as the latter migrates upwards by capillary action. The separation is governed by the polarity of the solutes and the stationary and mobile phases. The history, theory and practice of TLC are discussed in detail by Stahl (1969) and Fried and Sherma (1986).

Commercial plates are available and the solid support is usually glass, plastic or heavy aluminium foil. These are relatively expensive and, if a spreader is available, it is not difficult to prepare TLC plates using 20 cm × 20 cm pieces of glass. The most common stationary phase is silica gel, which usually contains a binder (10–15% gypsum) to maintain integrity of the layer during use. TLC plates coated with silica gel impregnated with silver ions (5–10% $AgNO_3$) are also widely used to separate saturated and **unsaturated** analytes (e.g. alkanes and alkenes, respectively). Other stationary phases, for example aluminium oxide, are used occasionally. Some stationary phases are mixed with a fluorescent indicator to aid with visualisation of the developed plate by ultraviolet light.

The mobile phase used to develop the TLC plate depends on the sample being separated and the type of separation required; the choice is almost limitless. However, two or three are used for the majority of separations performed in the organic geochemistry laboratory. The extracts from recent and contemporary sediments may be initially separated into hydrocarbons, ketones, alcohols and more polar components by development with dichloromethane. Further separation of the hydrocarbon fraction so obtained, as well as crude oil and the extract from ancient sediments, may be achieved by development with n-hexane. Thus, fractions corresponding to aliphatic hydrocarbons, aromatic hydrocarbons and NSO compounds are obtained. Mobile phases of intermediate polarity are useful for separating the more polar fractions (e.g. acidic fraction) and these typically comprise a mixture of predominantly n-hexane and a minor amount of polar solvent (e.g. ethyl acetate, 9:1; diethyl ether, 96:4).

To remove contaminants from the stationary phase, it is common to predevelop a TLC plate in a polar solvent such as ethyl acetate. Contaminants will move with the solvent front which, after allowing the solvent to evaporate, may be removed by scraping off and discarding the top 3–4 cm of the stationary phase. The plate is then activated (2–4 h, 100–120°C) prior to use. After cooling, the sample to be separated is dissolved in a minimum of solvent (e.g. dichloromethane) and carefully applied to the stationary phase as a line or series of dots with a dedicated spreader or drawn-out capillary tube. To assist the identification of bands after development, it is normal to apply a small amount of one or more reference compounds (e.g. n-heptadecane and phenanthrene might be used as a representative alkane and aromatic hydrocarbon, respectively) to the TLC plate alongside the sample. Scoring a line in the stationary phase between sample and reference compound fields prevents cross-contamination. For the best resolution, the sample and reference compounds should not be applied right up to the edge of the stationary phase and the surface of the stationary phase should not be damaged.

The TLC plate must be developed in an atmosphere saturated with the mobile phase. This may be achieved by lining a tank with filter paper and adding sufficient mobile phase so that there is a layer of solvent in the bottom of the tank after the filter paper has been wetted. However, it is imperative that the sample and reference compounds are applied to the plate so that they are above the level of the solvent. After 10–15 min, the TLC plate may be carefully placed vertically in the tank so that the mobile phase moves evenly up the stationary phase. The plate is left in the tank until the solvent front reaches the top of the stationary phase; further development will only broaden the bands and decrease the resolution.

There are numerous procedures for detecting the sample bands on the developed TLC plate. The stationary phase may already contain a fluorescent indicator; if it does not then the developed plate may be sprayed with a fluorescent dye (e.g. rhodamine 6G). The bands are then easily observed as bright (e.g. aliphatic hydrocarbons) or variously coloured (e.g. aromatic hydrocarbons) regions under an ultraviolet lamp. Fractions may be recovered by scraping the stationary phase from the support and extraction with a solvent (e.g. dichloromethane). Figure 16.3(a) shows a sketch of a prepared TLC plate before development, whilst Fig. 16.3(b) illustrates a typical separation of a crude oil or sediment extract using silica gel and n-hexane as stationary and mobile phases, respectively, and visualised with rhodamine 6G.

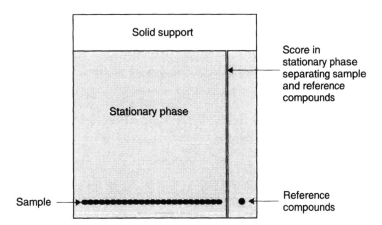

Figure 16.3(a)   An undeveloped TLC plate. The sample to be separated is applied to the stationary phase as a line or series of dots and a small amount of a solution of reference compounds is applied at the same level. It is important that the sample and reference compounds are not in direct contact with the solvent used as the mobile phase.

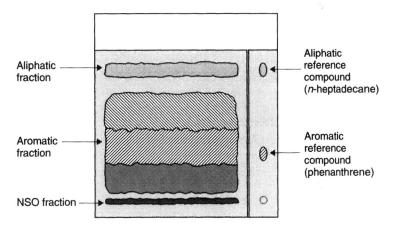

Figure 16.3 (b)   The TLC plate after development and visualisation (e.g. with rhodamine 6G). A separation of a crude oil or sediment extract using silica and *n*-hexane as the stationary and mobile phases, respectively, is illustrated. The least polar components (i.e. aliphatic hydrocarbons) will migrate with or slightly behind the solvent front, whilst the more polar components (i.e. aromatic hydrocarbons and NSO components) will be retained more strongly by the stationary phase. Compared with the reference compounds, the complex mixtures of hydrocarbons present in a crude oil or sediment extract produce rather broad bands on the TLC plate.

Liquid chromatography

TLC is capable of handling only relatively small amounts of material (typically about 50 mg) and offers limited chromatographic resolution. These limitations may be overcome by using open–column liquid chromatographic (LC) or high–performance liquid chromatography (HPLC), in which the stationary phase is held in a tube and the mobile phase passes through it by gravity (open–column) or applied pressure (HPLC).

Open-column LC is a simple and commonly used separatory technique. The chromatography column itself is a glass tube with a stopcock (PTFE is recommended) at one end and some means for attachment of a solvent reservoir at the other. A coarse glass frit may also be fitted just above the stopcock to retain the stationary phase. The internal diameter and length of the column are typically 10–20 mm and 50–100 cm, respectively; however, the size of the column should be chosen to suit the application and larger columns are used occasionally. The stationary phase (typically $\times 50$ amount of sample) may be poured in dry or added as a slurry in a suitable solvent. The latter technique usually results in columns exhibiting better chromatography.

The most commonly used stationary phase is silica gel, although alumina or a layer of alumina on top of silica gel are also used. Traditionally, a relatively coarse grade (i.e. 60–120 mesh) has been used, because finer grades reduce the flow rate of mobile phase through the column under the action of gravity, thereby increasing the time taken to effect the chromatographic separation. However, a finer grade of stationary phase (i.e. 200–400 mesh) may be used by applying low pressure from a compressed-gas cylinder (nitrogen is typically used). This technique, known as flash chromatography, is quicker than traditional open-column LC and affords similar resolution.

The mobile phase used for the separation depends on the sample and the type of separation required and, as with TLC, the choice is almost limitless. Usually a low-polarity solvent is used initially and it is then increased to elute progressively more polar components. This may be achieved by an initial elution with n-pentane to elute aliphatic hydrocarbons, followed by elution with a mixture of n-pentane and a more polar solvent (e.g. diethyl ether, 5:95) to elute aromatic hydrocarbons and finally a still more polar solvent (e.g. dichloromethane or methanol) to elute polar components (e.g. **PAH** and functionalised and NSO components). This general procedure will normally produce aliphatic and aromatic fractions amenable to instrumental analysis; however, the more polar fraction normally requires further simplification.

Although a larger sample may be separated by open-column LC than by TLC, the chromatographic resolution of these two methods is comparable. Preparative, as opposed to analytical, HLPC has been found to exhibit a higher resolution. Stainless steel tubes (10–25 mm i.d. $\times$ 10–25 cm) are used in place of glass and a pressure up to about 1500–2000 psi is applied. In addition, a wide range of stationary phases is available and includes so-called normal as well as reverse phases. Practical details may be found in McMaster (1994).

'Normal' stationary phase packings are those already mentioned for TLC and LC, that is silica and alumina, substances with polar surface groups that may be used with either a non-polar mobile phase or one of successively increasing polarity. 'Reverse' stationary phase packings are those that have non-polar surface groups; they are used with a polar mobile phase or one of successively decreasing polarity. The most common reverse phase packing is formed by derivatising the surface hydroxyl groups of silica with a dimethyloctadecylsilyl group (i.e. $-\mathrm{Si(CH_3)_2C_{18}H_{37}}$) and is often called an ODS phase. Both normal and reverse phase HPLC are used in the organic geochemistry laboratory; however, for group-type separation, amino-bonded (i.e. $-\mathrm{NH_2}$) silica has been found to be particularly useful.

Modern HPLC instruments are capable of controlling both the composition of the mobile phase and its flow rate, and therefore considerable selectivity may be achieved. Consequently, not only is it possible to separate a de-asphaltened crude oil or sediment extract into compound classes (i.e. aliphatic and aromatic hydrocarbon and polar fractions), but it is also possible to separate the aromatic hydrocarbons according to the number of aromatic rings. Although the former separation may also be performed by simple LC, with lower initial capital cost, the HPLC method is ideally suited to automation which is advantageous for routine analysis. Separation of a de-asphaltened crude oil or sediment extract into aliphatic hydrocarbon, mono-, di-, tri- and polyaromatic hydrocarbon and polar fractions is difficult to achieve except by HPLC methods and yields fractions consisting of a simpler distribution of hydrocarbons.

Clathration and molecular sieving

The fractions obtained from the chromatographic procedures discussed above are often quite complex and it is advantageous to sub-divide them further. This is especially true for aliphatic hydrocarbons since the components of most interest, the polycyclic sterane and triterpane **biological markers**, are typically present in relatively low amounts. In this instance, it is useful to remove the n-alkanes which, owing to their relatively high concentration, may interfere with the quantification of polycyclic biological markers.

Clathration is the process by which molecules of one compound become trapped within the growing crystal lattice of another. The clathrates (often called adducts) of **urea** are particularly useful to organic geochemists. When urea crystallises in the presence of n-alkanes (or other long-chain compounds), a clathrate of the urea and n-alkanes is formed. These crystals may be isolated by filtration and after dissolution in water the n-alkanes may be obtained by extraction with an organic solvent. The filtrate, or urea non-adduct, comprises compounds that do not form clathrates, that is branched and cyclic compounds.

Whereas the utility of urea clathration depends on the presence of the guest molecule at the time of crystallisation, molecular sieves possess pores of uniform diameter into which small molecules may be adsorbed. Only a restricted number of the available molecular sieves are used by geochemists, the most common being called Type 5A. It absorbs molecules up to about 0.5 nm (i.e. 5 Å) in diameter and may also be used to remove n-alkanes from the aliphatic hydrocarbon fraction but will not adsorb branched alkanes or cycloalkanes with rings of more than three carbon atoms. Type 5A will also remove simple molecules (e.g. water, carbon dioxide, hydrogen sulphide and hydrocarbons with less than three carbon atoms) from gases and is used for this purpose to purify the gas supply to sensitive analytical instrumentation (e.g. the carrier gas used for gas chromatography).

Other types (e.g. Types 10X and 13X) adsorb larger molecules having a diameter up to about 0.9 nm (i.e. 9 Å) and 1 nm (i.e. 10 Å), respectively. Both these larger sieves were used in the API Research Project 6 during the 1950s and 1960s to isolate pure aromatic hydrocarbons from crude oil (e.g. Rossini *et al.*, 1953). Type 13X and Ultrastable Y molecular sieves may be used to isolate fractions enriched in polycyclic terpenoids.

Bulk analytical techniques

Elemental analysis

Of the elements present in organic matter, the amounts of carbon, hydrogen and nitrogen are routinely measured whilst sulphur and oxygen are only sometimes measured.

A number of 'elemental (or CHN) analysers' are commercially available and they are typically sophisticated instruments, with both computer control and collection of data, which are able to quantify carbon, hydrogen, nitrogen and sulphur in one analysis and, after slight modification, oxygen. Before analysis, it is usual to treat sediments with an acid to remove carbonate minerals that would otherwise decompose to CO_2 and therefore be measured as organic carbon. An amount (about 0.5–2 mg) of the remaining sediment is then weighed into small tin capsules and loaded into the combustion furnace. For the measurement of carbon, hydrogen, nitrogen and sulphur the sample is oxidised at $1800°C$ in the presence of oxygen. The oxidation products (i.e. CO_2, H_2O, nitrous oxides and SO_2) are flushed by the helium carrier gas over copper (see lower part of Fig. 9.6), where the nitrous oxides are reduced to N_2, and, after the gases are separated chromatographically, they are quantified by thermal conductivity detection (Box 16.1). Oxygen may be determined by reduction, over nickel-coated carbon, to CO at $1060°C$ and similar separation and quantification.

The data are typically presented as a weight per cent of each element relative to the total sample. The amount of carbon, often called the total organic carbon (TOC), is used as a measure of the organic richness of the sediment and it is common to normalise the amount of other fractions of organic matter or of individual biological markers to the amount of TOC. The atomic H/C and O/C ratios may also be calculated and these data are often plotted against each other on a van Krevelen diagram (see Tissot and Welte, 1984).

Rock Eval® pyrolysis

Rock Eval® is a widely used proprietary pyrolysis technique[1] that provides a rapid assessment of the quality, quantity and type of organic matter present in a potential fossil fuel source rock and is often used to screen samples before choosing those that will be subjected to further geochemical analysis.

The finely ground sediment is weighted into a crucible and placed on a carousel with other samples and calibration standards. Crucibles are loaded automatically one by one into the oven which is initially held at about $50°C$ and is then increased to about $300°C$ at which temperature it is held for around 3 min. This process ensures that all the thermally distillable organic matter (assumed to consist only of hydrocarbons) is removed from the sample by the flow of helium carrier gas (the S_1 peak). Increasing the temperature to $550°C$ pyrolyses the remaining organic matter (i.e. kerogen). The material released is split into two components, the first (the S_2 peak) represents the quantity of pyrolytically produced hydrocarbon, whilst the second (the S_3 peak) represents the quantity of oxygen in the kerogen. Both the S_1 and S_2 peaks are detected by a flame ionisation detector (Box 16.1) and are expressed as mg hydrocarbon g^{-1} rock, whilst the S_3 peak is detected by a thermal conductivity detector (Box 16.1) and expressed as mg CO_2 g^{-1} rock. Contribution to the S_3 peak by carbon dioxide formed by the degradation of carbonate minerals is minimised by heating to only $550°C$. The temperature at which the S_2 peak reaches a maximum (T_{max}) is also measured and is used as an approximate measure of the **thermal maturity** of the sediment. Whilst the S_1 peak is used as a measure of the free (solvent extractable or thermally distillable) hydrocarbons in the sediment, the S_2 peak is used both to assess the

---

[1]   Vinci Technologies (14, rue Auguste-Neveu, BP 91, 92503 Rueil Malmaison, France).

Box 16.1 GC detectors and their characteristics

Type	Specificity	LOD[†] (ng)	Linear-ity	Stability[‡]	Carrier gas
FID	None[*]	0.1	10^7	Excellent	H_2 He N_2
FPD	P S	0.1	Non-linear	Fair	H_2 He N_2
TID (NPD)	N P S halogens	0.01	10^4	Fair	He N_2
ECD	Halogens	0.001	10^2	Fair	He N_2
TCD	None	10	10^5	Good	He H_2

[*]None   responds to all organic compounds except CS_2, COS and formic acid.

[†]LOD   limit of detection (to some extent this will be governed by the compound class and this is particularly true for the ECD).

[‡]Stability   stability to variation of the operating conditions, e.g. temperature, carrier gas flow rate.

The flame ionisation detector (FID) works by combustion of the analyte in a hydrogen–air diffusion flame. The mechanism of ion formation is not well understood, but it is believed that the dominant ion produced is CHO^+. The increase in the number of charged species causes an increase in the current flowing between the flame tip and another electrode of opposite potential. The current is detected by a sensitive electrometer and subsequently amplified electronically.

The flame photometric detector (FPD) may be used to analyse compounds containing sulphur or phosphorus which are decomposed in a hydrogen diffusion flame to yield S_2 and PO, respectively. These species are excited to a higher electronic state by the flame and therefore emit electomagnetic radiation on returning to the ground state. The emission is detected by a photomultiplier tube through either a 394 nm or a 526 nm filter for sulphur or phosphorus, respectively. In the case of sulphur, the emission originates from S_2, and the detector shows a square root dependence on the mass of the component. The signal may be quenched by co-eluting components and these difficulties make the quantitative use of FPD problematical.

The thermionic ionisation detector (TID), also known as the nitrogen–phosphorus detector (NPD), is selective for compounds containing nitrogen and phosphorus. It consists of an electrically heated (600–800°C) bead of

an alkali metal salt (e.g. rubidium silicate) in the region of which a plasma is sustained by the flow of hydrogen and air. At a low hydrogen flow rate (1–5 ml min^{-1}) a flame is not supported and the detector operates in the nitrogen–phosphorus mode. At a higher flow rate (about 30 ml min^{-1}) the sensitivity to nitrogen is reduced ($\geq \times 10$) and the detector operates in the phosphorus selective mode. The precise mechanism of detection is poorly understood. The TID is difficult to use and special attention must be paid to many of the operating variables (e.g. gas flow rate and temperature).

Compounds containing an electronegative element (e.g. the halogens) may be very sensitively determined using the electron-capture detector (ECD). The detector contains a β-radiation emitting source (e.g. ^{63}Ni) which sets up a standing current by ionisation of the carrier gas. When an analyte passes through the detector, electrons will be captured by any electronegative atoms present causing a decrease in the standing current.

The thermal conductivity detector (TCD) measures the resistance of a filament, through which a current is passing. In the absence of an analyte, the temperature of the filament, and therefore its resistance, is constant. When an analyte passes through the detector the thermal conductivity of the gas stream is reduced, causing a concomitant increase in both the temperature and the resistance of the wire, which is measured by a Wheatstone bridge circuit.

The response of these detectors depends on the particular operating conditions employed (e.g. carrier gas type and flow rate) as well as the type of analyte (e.g. aliphatic or aromatic hydrocarbon). Consequently, if a quantitative rather than a qualitative analysis is required then the detector must be calibrated. The most successful method for the absolute quantification of individual analytes is to use an internal standard. To use this method, a known amount of an appropriate internal standard (i.e. one that is readily available, stable and structurally similar to, but does not co-chromatograph with, the analytes of interest) is added to the sample and the mixture analysed. By comparing the peak areas of the analytes and the internal standard, the amount of each analyte may be calculated; however, to accomplish this successfully it is necessary to know the relative response factor of each analyte to the internal standard.

quantity of hydrocarbons that could be generated from the sediment and to calculate its hydrogen index (HI; S_2/TOC). The S_3 peak is used to calculate the oxygen index (OI; S_3/TOC) of the kerogen. The HI and OI may be used instead of the atomic H/C and O/C ratios to assess the source of the organic matter.

Iatroscan® TLC–FID analysis

The utility of TLC as a technique for the preparative separation of lipids and crude oil into compound class fractions has already been discussed. Although the technique may be used to quantify the amount of each fraction in the original mixture, it is both time consuming and costly in consumable materials to analyse a large number of samples. The Iatroscan®[2] TLC–FID analyser combines the ease of use of TLC with the sensitivity of flame ionisation detection (Box 16.1) and enables rapid, quantitative analyses to be performed. This technique is discussed in detail by Ranný (1987).

Separation is performed on reusable quartz rods coated with a thin layer (75 μm) of stationary phase, 10 of which are held in a metal rack. Both silica-gel- and alumina-coated rods are available, and they may be modified by impregnation with, for example, silver nitrate. The crude oil or sediment extract to be separated is dissolved in a suitable solvent and a small amount (1–5 μl), corresponding to 10–30 μg of sample (about 2–5 μg of each compound class), is carefully applied to one end of the rod. Up to 10 samples may be analysed at a time, although it is more common to use one rod for authentic reference compounds (cf. planar TLC) and perform duplicate or triplicate analyses of each sample (i.e. three or four samples analysed at a time). The rods are then placed in the mobile phase and developed in a manner similar to planar TLC. Once developed, the rods are removed from the mobile phase, dried and analysed by scanning them with the flame ionisation detector (Box 16.1) and the signal displayed on a chart recorder or, more commonly, acquired by a computer. To obtain quantitative data, a calibration curve must be constructed for each class separated since the response of this type of detector depends on the type of compound analysed.

Instrumental analytical techniques

The previous sections have attempted to outline the most common procedures employed to obtain fractions amenable to detailed instrumental analysis. They are used both to convert a *very complex* mixture (i.e. the original crude oil or sediment extract) into a number of *slightly less complex* fractions and to remove components that may be detrimental to the analytical techniques used.

Gas chromatography (GC)

Gas chromatography is arguably the most commonly employed analytical technique in the organic geochemistry laboratory and may be applied to a wide range of analytes. Separation is achieved by partition of the analyte between a gaseous mobile phase (i.e. the carrier gas) and a liquid, stationary phase. The polarity of the analytes and the stationary phase and the temperature programme which is employed all affect the separation. Many books have

---

[2]   Iatron Laboratories, Inc., 1-11-4 Higashi-Kanda, Chiyoda-Ku, Tokyo 101, Japan.

been published describing the theory and practice of GC and, for example, those by Poole and Schuette (1984) and Jennings (1987) are useful.

Overview of a gas chromatograph   Although chromatographic separation occurs in a capillary column it is necessary firstly to introduce the sample into the column, which is achieved via a syringe and injector. Secondly, after passage through the column, the separated components must be detected. Thirdly, it is necessary to display and manipulate the data. This may be accomplished using a chart recorder or, more commonly, the analogue detector signal is digitised and the data collected on a dedicated integrator or computer. A schematic diagram of a gas chromatograph is shown in Fig. 16.4.

The injector   The most widely used methods of introducing the sample into the column are the vaporising injector and the on-column injector. In the former, a solution of the sample is introduced by syringe into the heated injector volume where the sample and solvent are immediately vaporised and subsequently transferred into the column by the carrier gas (Fig. 16.4b). The sample is introduced into the closed injector volume and is directed entirely onto the column. However, the large excess of solvent vapour results in a severely tailing solvent peak and to overcome this problem the injector is backflushed with carrier gas 30–60 s after injection to prevent the remaining solvent from entering the column. Although this injection method works well in many instances, it discriminates against less volatile higher-relative-molecular-mass components and, for samples comprising analytes with a wide range of volatility, on-column injection is recommended.

In the on-column injector the sample is introduced directly into the column as a liquid. This is achieved by inserting the fine needle of a syringe *inside* the column and introducing a solution of the sample as a contiguous plug of liquid. Whilst injecting the sample, the temperature of the beginning of the column must be maintained at least 10°C below the boiling point of the solvent used. After withdrawing the syringe the oven temperature may be increased, at which point the plug of solvent in the column will begin to evaporate and finally the sample will be concentrated as a narrow band. Although on-column injection has a higher precision and suffers from less discrimination than vaporising injection and is, therefore, the method of choice for the majority of trace analytical applications, it does suffer from some disadvantages. For example, since the entire sample is introduced into the column any non-volatile or polar components will concentrate at the beginning of the column and will, most probably, be detrimental to the quality of subsequent chromatography. This may be overcome by ensuring that the samples analysed are 'clean' and do not contain components that are soluble in the injection solvent but that will not chromatograph. In addition, it is considerably more difficult to automate on-column injection than vaporising injection and therefore it is difficult to use this method with an autoinjector. These difficulties notwithstanding, on-column injection is often the method of choice.

The column and stationary phase   The majority of GC columns are capillaries (about 0.2–0.5 mm i.d.) made from fused silica, coated on the outside with a thermostable polymer to strengthen the brittle capillary. For high-temperature GC (>370°C), the polymer is replaced by aluminium. Glass or occasionally stainless steel capillaries are also used. Stationary phases are viscous liquids or gums and are coated on the inside of the capillary tube. By careful choice of stationary phase and **derivatisation** the range of compounds amenable to analysis by GC is impressive.

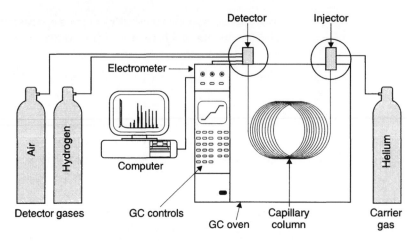

Figure 16.4(a)   Schematic of a gas chromatograph. A solution of the sample is introduced by syringe into the injector. The mobile phase (i.e. carrier gas) sweeps the sample onto the capillary column where the individual analytes are separated. On elution from the column an analyte passes into the detector where a signal is generated which is then amplified and converted into a digital signal which is usually stored by a computer. Alternatively, the analogue signal may be sent directly to a chart recorder (not shown).

Figure 16.4(b)   A vaporising injector.

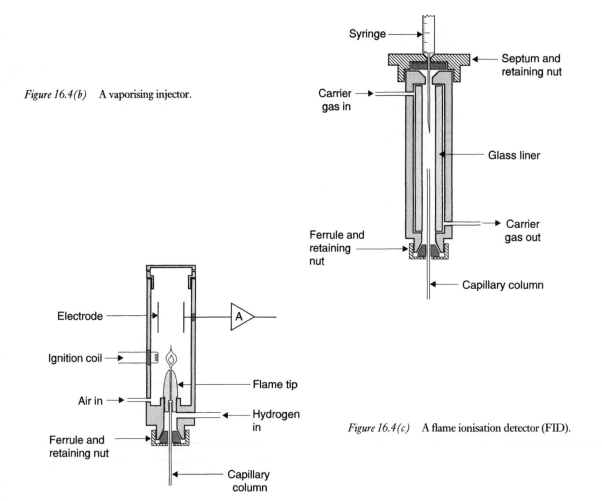

Figure 16.4(c)   A flame ionisation detector (FID).

Most of the non-polar stationary phases in use today are siloxane polymers which, by covalent bonding between the polymer itself and between the polymer and the solid support (i.e. glass or silica), exhibit good thermal stability. GC columns coated with these phases may be routinely used to about 300°C with brief periods at about 350°C; however, high-temperature columns are also available that may be similarly used to about 400°C and about 450°C, respectively. These high-temperature GC columns enable the analysis of *n*-alkanes having in excess of 100 carbon atoms.

The most commonly used phases for organic geochemical work are poly(dimethylsilox-ane) and poly(dimethyldiphenylsiloxane), where 5% of the methyl groups are replaced with phenyl groups, the former being the most non-polar. For the analysis of aromatic hydrocarbons, including biological markers, the latter gives slightly better resolution. Polar stationary phases are obtained by progressively replacing the methyl groups with more polar substituents (e.g. cyanopropyl or trifluoropropyl) or are based on poly(ethylene glycol) or poly(diethylene glycol), either as the pure polymer or as the adipate or succinate esters.

The mobile phase   The choice of mobile phase or carrier gas is governed by a number of variables, not the least of which are that it should be unreactive, safe, inexpensive and, preferably, monoatomic and have a low molecular weight. The only gas that meets all these criteria is helium and this is in fact a widely used carrier gas. Both nitrogen and hydrogen meet some of the criteria and they too are often used. As a result of the physical properties of these gases, only helium and hydrogen are employed as carrier gases for capillary GC. Owing to the possibility of explosion some laboratories do not use hydrogen, although with the correct electrical safety equipment and good laboratory practice this risk can be minimised. Whichever carrier gas is chosen, it is important that a high-purity grade is used so as to minimise contamination and that an oxy-gen trap and a molecular sieve trap, to remove water and hydrocarbons, are included as part of the manifold so that the gas reaching the GC injector is as clean as possible.

The detector   Many types of detector have been developed for use with GC and those most commonly used are described in Box 16.1. The FID (Fig. 16.4c) is particularly common because it responds to virtually all organic compounds, has a reasonably low limit of detection and has a linear range of up to seven orders of magnitude (Box 16.1). Consequently, all organic compounds of geochemical interest may be detected and components having widely differ-ent relative abundance may be quantified. The FPD is used by organic geochemists to detect sulphur-containing components, often in combination with an FID to produce a dual FID/FPD trace. The TID is occasionally used by organic geochemists to analyse, for example, amino acids or heterocyclic compounds containing nitrogen. However, the detec-tor is difficult to operate and it is perhaps for this reason that the TID has been neglected by many laboratories.

The ECD and TCD are rarely used by organic geochemists because the former is insen-sitive to organic compounds that do not contain electronegative atoms and the latter is con-siderably less sensitive than other detectors. However, ECD is widely used in environmental analysis and the TCD is used in elemental analysers and the Rock Eval® instrument. The mass spectrometer may also be used as a detector for GC; such *combined* methods are discussed later.

GC performance   *Mass range* The molecular mass range that is amenable to analysis by GC is demonstrated by the successful separation of light hydrocarbons (e.g. C_3–C_7) and of derivatised

porphyrins which have a molecular mass of <100 amu and 800–900 amu, respectively. As might be expected, different analytical conditions are often necessary and, for example, the former require a relatively thick stationary phase (e.g. 0.5–1.0 μm) whereas the latter require a thermostable stationary phase.

Structural similarity The separation of **enantiomers** illustrates the resolution of structurally similar components. In the case of carbon, enantiomers differ only by the spatial orientation of the four groups attached. A stationary phase called Chirasil-Val®, for example, is capable of separating the enantiomers of amino acids. **Cholestane** has a total of eight **chiral centres** and although separation of any one of the 128 enantiomeric pairs requires a chiral stationary phase, the separation of many of the **diastereoisomers** may be accomplished on common non-polar stationary phases and is a slightly less stringent example.

Polarity The separation of components of different polarity is exemplified by the analysis of, for example, aliphatic hydrocarbons and carboxylic acids. The former are non-polar and are straightforward to analyse using a number of different conditions and stationary phases (i.e. non-polar, which usually gives better results, as well as polar), whilst underivatised carboxylic acids may only be analysed successfully on highly polar phases. The polarity of carboxylic acids may be reduced by derivatisation (e.g. formation of methyl esters) and they may then be analysed on non-polar to medium-polarity stationary phases.

Analytical pyrolysis

The application of pyrolysis to the degradation of organic matter has already been mentioned with respect to the Rock Eval® instrument. However, pyrolysis may also be coupled either directly to a mass spectrometer (py-MS) or to a gas chromatograph followed by an appropriate detector (py–GC–FID or py–GC–MS). The principle underlying analytical pyrolysis is that organic matter not directly amenable to chromatographic analysis is degraded by the input of thermal energy. Thus, relatively weak bonds (e.g. C—O or C—S bonds) of kerogen or asphaltene are preferentially broken to release smaller molecules that may then be analysed. Two methods for supplying the thermal energy are commonly employed: resistive heating and Curie point heating.

Analytical pyrolysers employing resistive heating typically consist of a coil or ribbon of platinum and are used for solid and liquid samples, respectively. Solid samples (about 0.2–0.5 mg) are placed in a small quartz tube between two loose glass wool plugs and the tube inserted into the platinum coil of the pyrolysis probe. In order to avoid contamination of the tube or glass wool with finger lipids, cotton gloves should be worn and tweezers used for handling. The probe is then inserted into the pyrolysis interface or injector and the pyrolysis programme initiated. It is common to heat the sample almost instantaneously to the final pyrolysis temperature (e.g. 600°C; about 8 or 600 ms for ribbon or coil, respectively) at which it is held for a short period (e.g. 20 s). Alternatively, the temperature may be increased more slowly at a specified, controlled rate.

The Curie point of a ferromagnetic material is the temperature at which it loses its ferromagnetism and displays only paramagnetic properties. This temperature depends only on the material and for nickel, iron and cobalt it is 358, 770 and 1131°C, respectively. In addition, since alloys of different ferromagnetic metals also have a characteristic Curie point, there is a choice of pyrolysis temperature available (e.g. 480, 510, 610 and 925°C). By subjecting a wire of the selected alloy to an RF field in which it is heated by magnetic induction

(Box 4.2), the temperature of the wire may be rapidly raised (about 300–400 ms) to its characteristic Curie point and, since no further energy will be absorbed above this temperature, it will remain at the Curie point for the required pyrolysis time. Although the rise–time and especially the temperature are well defined, it is not possible to obtain a specific temperature programme for the pyrolysis. The sample may be applied to the pyrolysis wire as a solution followed by evaporation of the solvent or, if it is a solid, a small amount may be pressed onto the wire using a hydraulic press.

A major drawback with py–GC or py–GC–MS has been the difficulty in calibrating the detector response to obtain quantitative data. However, this has partly been solved by the use of a polymer, poly-*t*-butylstyrene, that pyrolyses to form predominantly *t*-butylstyrene which may be used for quantification. The polymer itself is soluble in dichloromethane and a known volume (e.g. 5 μl) of solution, and therefore amount of poly-*t*-butylstyrene, is added to the sample prior to pyrolysis. It has been found effective to add the standard directly to the sample on the ribbon or Curie point wire or in the quartz tube, immediately prior to pyrolysis. Alternatively, the solid sample and solution of standard may be thoroughly mixed and then placed in the quartz tube or pressed onto the Curie point wire. The amount of a pyrolysis product may then be calculated by comparing its chromatographic peak area with that of *t*-butylstyrene; however, since the relative response factor is often unknown, the results are typically semiquantitative.

Analytical high-performance liquid chromatography (HPLC)

In addition to group-type separation of crude oils and sediment extracts as already discussed, HPLC may be used for the analysis of components that are either thermally labile or non-volatile and are therefore not readily amenable to analysis by GC. As a result of the wide range of stationary and mobile phases available, as well as the ability of modern instrumentation to perform reproducible, accurate and complex flow and mobile phase composition programming, HPLC is an extremely versatile analytical technique.

The major use of HPLC in the organic geochemistry laboratory is for the analysis of pigments (i.e. carotenoids and tetrapyrroles and their degradation products). Tetrapyrrole pigments include the many types of chlorophylls, their immediate degradation products and the non-functionalised porphyrins. The latter are believed to be derived from the chlorophylls during diagenesis by reduction of all functional groups, aromatisation and exchange of the chelating metal ion. Analysis of these components by GC has not generally been successful. The carotenoid and chlorophyll distribution may yield taxonomic information (i.e. it may be characteristic of a particular organism) and, therefore, it is common to determine these components in a single chromatographic analysis. Both normal and reverse phase HPLC have been used but the latter is more common. It is common to use a fluorescence and an absorbance detector in series to measure the carotenoids, chlorophylls and their degradation products since they all exhibit absorbance at 440 nm and the latter fluoresce when excited by light of approximately 430 nm.

Porphyrins occur in the geosphere predominantly as the nickel and vanadyl (V=O) chelates. Although metallated porphyrins have been successfully analysed using reverse phase HPLC it is more common to analyse the demetallated, or free-base, porphyrins using normal phase HPLC. However, if the nickel and vanadyl porphyrin distributions are required, they should be separated (e.g. by TLC) prior to demetallation and analysis.

Whether normal or reverse phase HPLC is employed, it is usual to measure the absorbance at approximately 400 nm.

Mass spectrometry (MS)

Application of mass spectrometric techniques to elemental analysis and the determination of isotope ratios in the field of inorganic geochemistry is discussed in Chapters 8–11 and the theory of mass spectrometry is covered in Boxes 8.1, 8.2, 10.1 and 11.1. The mass spectrometer is also used in organic geochemistry, usually coupled to a separatory method such as gas chromatography. The coupling of gas chromatography to mass spectrometry (i.e. GC–MS) has enabled organic geochemists to identify and quantify many hundreds of compounds. Although sophisticated and expensive instruments that may be applied to a number of different tasks are commercially available, the advent of small and relatively inexpensive instruments has meant that there are few laboratories that do not have immediate access to GC–MS. These 'benchtop' instruments do not have the mass range or capability of different ionisation modes of most large instruments but they produce data adequate for the majority of applications and are reliable and simple to operate. The practicalities of organic MS are discussed in detail by Chapman (1993).

The mass spectrometer is commonly used as a means of detecting and quantifying known analytes but the organic geochemist is often interested in identifying novel, unknown organic molecules. In many cases, therefore, it is important to obtain a mass spectrum of the analyte, from which the structure of the compound may be determined (or at least proposed) by interpretation of the fragment ions. The interpretation of organic mass spectra will not be discussed here and the reader is directed to McLafferty and Tureček (1993).

Direct insertion probe MS   The direct insertion probe, or simply 'probe', inlet was one of the first methods devised to introduce an analyte into the mass spectrometer and both non–volatile as well as labile organic molecules may be introduced. Typically the sample is loaded into a small glass crucible that is seated in the probe tip and the probe is pushed through vacuum seals and a ball valve into the source volume. The probe may be electrically heated to volatilise non–volatile, stable analytes using a known temperature programme. Alternatively, labile analytes may be cooled by a flow of gas through the probe. The probe inlet is useful for the analysis of pure compounds and if the mass spectrometer is capable of medium or high resolution (Box 11.1) then an accurate molecular mass may be measured and the molecular formula of the compound calculated. Under low-resolution conditions, a mass spectrum may be obtained and structure elucidation attempted. Analytes that are mixtures of similar compounds may also be analysed and it is usual to arrange the conditions so that there is the minimum of fragmentation.

The analyte may be ionised by either electron impact (EI, Fig. 9.2) or chemical ionisation (CI) methods. The latter technique relies on the presence of a reagent gas in the ion source (typically methane, iso–butane, hydrogen or ammonia) which is ionised by the electron beam and in turn ionises the analyte. It is usually observed that CI causes less fragmentation than EI (i.e. it is a 'softer' ionisation technique). The carbon number range of porphyrins isolated from sediments or crude oils is often determined by EI probe MS.

Isotope ratio mass spectrometry (IRMS)

Isotope ratios of hydrogen, carbon, nitrogen, oxygen and sulphur have all been determined in crude oil and sedimentary organic matter. The technique is discussed fully in Chapter 9 and only those aspects pertinent to the measurement of the carbon isotope ratio of organic matter will be covered in this chapter. The interested reader is also directed to Faure (1986) or Coleman and Fry (1991) for further information.

The carbon isotope ratio standard (i.e. PDB; Table 9.1) has been exhausted for a number of years and additional standards whose isotope ratio is known accurately with respect to PDB are used instead. The most useful of these to the organic geochemist is a lubricating oil, NBS-22, which is reported to have $\delta^{13}C$ of $-29.81 \pm 0.06‰$ (Schoell *et al.*, 1983). There is a finite quantity of even these standards and most laboratories use them to calibrate a secondary working standard, which is then used for day-to-day analysis.

To determine the $\delta^{13}C$ ratio of a sample the carbon must be converted into CO_2 (Table 9.1 lists the analyte gases for other elements). Two methods have been used to achieve this: dynamic combustion and sealed-tube combustion. Although both methods yield comparable data, the latter is quicker and less expensive and is most commonly used (see Coleman and Fry, 1991). Briefly, an appropriate amount of sample and CuO wire are placed in a borosilicate glass tube closed at one end, which is then evacuated and sealed. Both the tubes and CuO should be preignited to reduce organic contamination. The tubes are then combusted ($550°C$; about $1–5$ h) in a preheated muffle furnace. The $\delta^{13}C$ value of the CO_2 and, therefore, of the sample may be determined using a dual-inlet mass spectrometer (see Chapter 9 and Fig. 9.4).

Gas chromatography– mass spectrometry (GC–MS)

Gas chromatography–mass spectrometry (Fig. 16.5) combines the resolving power of capillary GC with the sensitivity of MS and is one of the most useful analytical techniques available today. Since adoption by organic geochemists in the mid 1960s it is used in almost every study involving biological markers. The technique is so useful because it enables the separation, quantification and most importantly identification of the many components present in geological organic matter. The practice and theory of GC–MS are discussed by Message (1984) and Chapman (1993).

In order to maintain chromatographic resolution as well as sensitivity, analytes must be transferred from the gas chromatograph to the mass spectrometer efficiently. Modern vacuum technology enables the capillary column to be run directly into the source volume and the transfer of the sample from the injector to the mass spectrometer will be highly efficient if it is ensured that the region of the column between the chromatograph and the mass spectrometer (i.e. the transfer line) is maintained at or slightly above the maximum temperature that the column experiences. This ensures there are no 'cold spots' which will tend to reduce the transfer of less volatile analytes into the mass spectrometer (especially those eluting at the maximum chromatographic temperature).

Many variables affect the distribution of ions in a mass spectrum and it is important to note that there is no *single* 'true' mass spectrum. In order that the spectra of known analytes appear as expected (i.e. similar to those in the many mass spectral databases), the operating parameters of the mass spectrometer must be adjusted. This may be conveniently achieved by introducing a small amount of 'calibration' compound (e.g. perfluorokerosene) into the source of the mass spectrometer. The mass spectrometer is then 'tuned' so that the spectrum measured closely matches that of the reference spectrum for the calibration compound, which entails ensuring that all the correct m/z values are present and that they occur in the correct relative intensity. One of the variables that has a significant effect on the fragmentation pattern of an analyte molecule is the energy of the ionising electrons (i.e. the electron

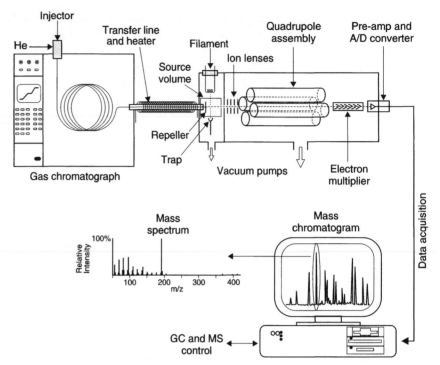

Figure 16.5   Schematic of a gas chromatograph–mass spectrometer. Analytes in the sample are separated in the capillary column and are introduced directly into the source volume of the mass spectrometer via a heated transfer line. Electrons from a heated filament ionise the analyte molecules which are expelled from the source by a repeller plate and focused by a series of slits into the quadrupole mass analyser. For clarity, the column effluent, electron beam and ionised molecules are shown in the same plane; however, they are often orthogonal to each other. The quadrupole consists of four parallel rods having a hyperbolic or circular internal surface. Opposite rods are electrically connected with one pair having a positive and the other a negative voltage applied, both of which comprise DC and RF components. The RF component of the two pairs of rods is 180°C out of phase, which causes ions of successive m/z value to be transmitted. Ions passing through the mass analyser impinge on an electron multiplier which produces a cascade of electrons and amplifies the initially small current ($\leq 10^{-9}$ A) by a factor of up to 10^6. This current is converted into a voltage by a preamplifier, digitised and sent to a computer for manipulation and storage. Typically, the computer will acquire and process the data as well as controlling the mass spectrometer and sometimes the gas chromatograph as well. By summing the data for all ions detected, during one scan of the mass range, a total ion chromatogram may be displayed. In addition, the full mass spectrum of any peak may be displayed.

energy). The ionisation probability of most analytes reaches a maximum with electrons in the energy range 50–100 eV and therefore an electron energy of 70 eV is widely used; small fluctuations around this energy also have little effect on the fragmentation pattern.

GC–MS enables the mass spectrum of each GC peak to be measured. Doing so may generate a very large amount of information. Because of the high resolution of modern capillary gas chromatography, the mass spectrometer must be capable of quickly, accurately and reproducibly scanning a mass range of, for example, 50 to 550 amu. In order to characterise a chromatographic peak, it is necessary to acquire a number of data points (i.e. mass spectra) over the peak profile and it is convenient to scan the mass range every second. Consequently, the mass spectrometer must be capable of scanning a **decade** or more in less than 1 s, since it is necessary to allow time for stabilisation between individual scans. On the whole, quadrupole mass analysers (Box 10.1) have been found to be best suited for this application although modern laminated magnet construction and electronic control have enabled magnetic sector mass spectrometers to be used successfully as well.

Although the entire mass range of interest may be scanned for each GC peak (Fig. 16.6a), a process often called full data collection, it is also possible to collect data for a small number of selected m/z values only (Fig. 16.6b and c). This method, known as selected ion monitoring (SIM), is useful for routine analysis, since it is accompanied by an increase in sensitivity and a decrease in the amount of data storage required. It is widely used for the analysis of biological markers and some of the more common ions monitored by SIM and their origin are listed in Table 16.1. SIM is of limited utility for identifying unknown components. A more complete list of biological markers and their characteristic fragmentation is given by Peters and Moldowan (1993).

Whether full data collection or SIM is performed, a chromatogram of a single m/z (i.e. an ion chromatogram) may be used to display the distribution of a class of analytes. The utility of ion chromatography is illustrated by the following example. A fingerprint of an

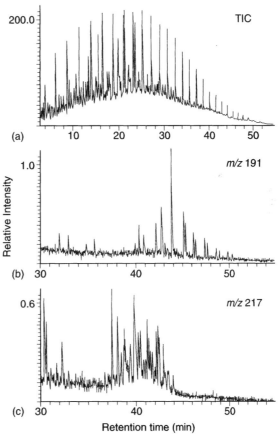

Figure 16.6   Chromatograms of (a) the total ion current (TIC), (b) $m/z = 191$ (pentacyclic triterpanes), and (c) $m/z = 217$ (steranes) of an oil, illustrating that a small amount of polycyclic biological markers may be readily detected in the presence of a substantially larger amount of other components (i.e. *n*-alkanes) as indicated by the relative intensity of the TIC, $m/z = 191$ and $m/z = 217$ chromatograms (200:1:0.6).

Table 16.1   Common biological markers and their characteristic mass spectral fragment ions. R represents an alkyl group: in these examples, methyl ($-CH_3$), ethyl ($-CH_2CH_3$) or *iso*-propyl ($-CH(CH_3)_2$)

Ion (m/z)	Compound Type	Structure and Fragmentation
177	triterpanes (*e.g.* C_{27} hopanes)	177
191	triterpanes (*e.g.* hopanes)	191 191 R $R=C_3H_7$
205	triterpanes (*e.g.* C_{32} hopanes)	205
217/218	steranes	R 217/218
231	triaromatic steroids	R 231
253	monoaromatic steroids	R 253

aliphatic fraction of a crude oil or sediment extract may be obtained by GC alone but the components revealed (e.g. *n*-alkanes, methyl-branched alkanes and **isoprenoids**) are structurally simple. This simplicity means that many of the processes (e.g. thermal maturation, biodegradation and **migration**) that affect these components are not recorded in a unique manner and, consequently, they are not readily apparent from the distributions observed. The components which often have the greatest capacity to record these processes (i.e. the structurally complex polycyclic biological markers) are usually present in low concentration and are effectively hidden in a background of interfering chemical signals. However, GC–MS may be used for their analysis and quantification by using the mass spectrometer to detect their characteristic fragmentation ions (Table 16.1). For example, to investigate the distribution of **pentacyclic triterpanes** and **steranes**, ions at $m/z = 191$ and $m/z = 217$ (or $m/z = 218$), respectively, may be plotted. Figure 16.6 shows a plot of the total ion current (TIC; i.e. the sum of the intensity of all ions monitored during full data collection) and of the ions at $m/z = 191$ and 217. Note that the intensity of the TIC signal is substantially greater than that for either of the individual ions, demonstrating that even minor components may be detected using full data collection.

GC–MS analysis of geological organic matter is often performed qualitatively. Thus, the distribution of the biological markers present may be determined and the ratios of selected components calculated. These ratios are used to assess the degree to which a sediment or crude oil has been altered by a particular process. A considerable number of ratios have been devised to determine, for example, the source or thermal maturity of sedimentary organic matter or the degree of biodegradation of a crude oil, and the interested reader should consult Peters and Moldowan (1993) for further information.

It is often more useful to measure the absolute concentration of analytes, rather than the ratio of one analyte to another. The use of an internal standard is the best method for performing quantitative GC and the same is true for GC–MS. However, the problem of determining the relative response factor is exacerbated since the species detected and quantified by GC–MS is typically a fragment ion of the analyte. The internal standard should ideally have a similar structure to the analyte and undergo fragmentation to produce ions of the same m/z ratio. Although the fragment ion chosen is usually the largest single ion in the mass spectrum of the analyte and internal standard (this ion is called the base peak), it is rarely the case (at least for aliphatic biological markers) that it represents more than 50% of the total ion current. Small changes in the operating conditions of the mass spectrometer may alter the abundance of the base peak, relative to all the other ions, without dramatically altering the mass spectrum and this will in turn affect the amount of analyte measured. In order to minimise this problem, the mass spectrometric conditions should be kept as constant as possible and the quantification standards should be routinely analysed with each batch of samples. Although the procedure is rather time consuming, and it is not always possible to satisfy completely all the necessary criteria, the use of internal standard quantification is to be encouraged since it enables the effect of processes such as thermal maturation, biodegradation and migration on the concentration of biological markers to be studied.

Advanced mass spectrometric techniques
In addition to GC–MS, a number of other techniques are used for research. These include gas chromatography–high-resolution mass spectrometry, gas chromatography–mass spectrometry–mass spectrometry (GC–MS–MS), high-performance liquid chromatography–mass spectrometry (usually abbreviated LC–MS), high-performance liquid chromatography–inductively coupled plasma–mass spectrometry (HPLC–ICP–MS) and gas chromatography–isotope ratio mass spectrometry (GC–IRMS). Further details are provided by Chapman (1993).

Any mass spectrometer that has two mass analysers, whether electrostatic, magnetic or quadrupole, is technically an 'MS–MS' instrument. However, medium- and high-resolution instruments, which employ both an electrostatic and a magnetic analyser, have traditionally been called double-focusing mass spectrometers (Box 11.1) and the term MS–MS is generally reserved for instruments having tandem quadrupole mass analysers. Whereas the ions exiting from the mass analyser of a single quadrupole instrument impinge on the detector, in an MS–MS the ions exiting from the first mass analyser pass through a 'collision cell', where they undergo dissociation into product ions. These product ions then pass into the second mass analyser and thence to the detector. The selectivity of GC–MS–MS is thus enhanced over that of GC–MS since not only is a chromatographic separation achieved but, by choosing a specific dissociation of a parent ion into a product ion, a mass spectrometric separation is achieved as well. An assortment of experiments may be performed. In one example, the first quadrupole is scanned over the desired m/z range and a specific product ion, produced by dissociation in the collision cell, is transmitted by the second quadrupole to the detector (called 'parent scanning'). In this way, it is possible to analyse a product ion that is common to the various parent ions of a particular compound class. In the case of the sterane biological markers, for example, the base peak product ion at $m/z = 217$ (see Table 16.1), transmitted by the second quadrupole mass analyser, is related to the parent ions at $m/z = 372$, 386 or 400 (corresponding to the molecular ions of the C_{27}, C_{28} or C_{29} steranes, respectively), mass analysed by the first quadrupole. Thus, a selective analysis for C_{27}, C_{28} or C_{29} steranes *only* is achieved.

The coupling of HPLC and MS (usually called LC–MS) is not as straightforward as GC–MS. One of the major problems that must be overcome is the efficient removal of the mobile phase whilst transferring the analyte into the source of the mass spectrometer. Typically this involves an input of power to volatilise the mobile phase and an efficient pumping system to maintain the required vacuum in the source. LC–MS has already proved to be useful for the analysis of biological and environmental samples (e.g. for proteins and dyes, respectively) and has been applied to the analysis of chlorophylls and their degradation products in geological samples. Since HPLC is ideally suited to the analysis of polar, non-volatile or labile analytes, the use of LC–MS will increase in the future.

The application of ICP–MS to the determination of elements in inorganic matrices is discussed in Chapter 10. Typically the sample, as an aqueous solution or a slurry, is nebulised and introduced directly into the argon plasma without any chromatographic separation. By coupling HPLC to ICP–MS it is possible to separate, detect and quantify organometallic compounds. The coupling is straightforward since the effluent from the column may be nebulised and introduced into the argon plasma; however, a number of problems may still arise. For example, large amounts of organic solvent in the mobile phase may extinguish the plasma or polyatomic interferences may occur. These may be overcome by desolvation techniques or by the introduction of an additional gas (e.g. N_2) into the nebuliser, respectively. The presence of metals, especially nickel and vanadium, in crude oil is well known. Although a proportion of these metals is chelated to porphyrins, the speciation of the remainder (associated with the more polar fractions) is poorly understood. In the last 15 years, produced crude oil has become increasingly heavy (Altgelt and Boduszynski, 1994) and since polar material is more abundant in heavy oils a higher concentration of metals is also present. Metals poison the catalysts used in the refining of crude oil and therefore it is important to determine their chemical form and then devise methods to remove them. HPLC–ICP–MS has been used to analyse the metalloporphyrins in geological materials and, once appropriate chromatographic conditions are determined, the methodology may be used to analyse the organometallic components in the polar fraction of crude oils.

Within the last 5 years, instruments which are capable of measuring the carbon isotope ratio of individual components of a mixture have become commercially available. Both compound-specific isotope analysis (CSIA) and GC–IRMS have been used to describe this technique. Whereas IRMS yields an average $\delta^{13}C$ value for a sample, GC–IRMS enables the $\delta^{13}C$ value of individual analytes to be determined. In order to achieve this, the sample is first separated by gas chromatography and, after elution from the column, each analyte is combusted to CO_2, the $\delta^{13}C$ of which is then measured by an isotope ratio mass spectrometer. The $\delta^{13}C$ values measured by GC–IRMS are typically slightly less accurate and precise than those measured by IRMS but this disadvantage is outweighed by being able to determine the isotope ratio of structurally similar analytes. The $\delta^{13}C$ value of biological organic matter may depend on the trophic level of an organism (e.g. autotrophic or heterotrophic) since the source of carbon and the biosynthetic pathways may be different. Thus, it may be possible to determine the trophic level of the organisms from which individual lipids are derived. For example, the **pristane** and **phytane** present in most sediments are believed to be mainly derived from the side chain of chlorophyll *a*. However, by measuring the δ^{13}C value of these isoprenoids it has been demonstrated that in some environments the former is derived from photosynthetic phytoplankton (i.e. chlorophyll *a*), whereas the latter is derived from methanogenic bacteria.

Applications

Biodegradation

Biogradation is a process by which the chemical composition of crude oil is altered by bacteria. This may occur in a reservoir and it is important to determine the extent to which the oil has been degraded before production since the hydrocarbon components most useful to industry, and which lend the oil suitable physical characteristics for production and refining, may have been removed, with the result that the oil will be heavier and polar components and metals more concentrated. Although there is no single scale of biodegradation, structurally simple components are initially removed (e.g. *n*-alkanes and then isoprenoids) and as the extent of biodegradation increases so more complex components (e.g. polycyclic biological markers) are degraded. The initial stages of biodegradation may be determined by GC (i.e. loss of *n*-alkanes and isoprenoids); however, changes in the distribution of the polycyclic biological markers, as a result of more severe biodegradation, require the use of GC–MS. Generally it is observed that the distribution of steranes is significantly altered at intermediate levels of biodegradation.

Figure 16.7 shows the $m/z = 217$ chromatograms of a tar–sand bitumen and a crude oil from a reservoir at a depth of approximately 1300 m from southern Oklahoma, USA. On the basis of the distribution of aromatic steroids and porphyrins and other geochemical data, the tar–sand bitumen and the crude oil are believed to be related (Michael *et al.*, 1989). However, the $m/z = 217$ chromatograms indicate that the tar–sand bitumen (Fig. 16.7a) has been biodegraded to the extent that all the C_{27}–C_{29} steranes have been removed, whilst the distribution of the steranes of the crude oil (Fig. 16.7b) does not appear to have been affected. This is clearly shown by the absence of the C_{29} steranes (shaded in Fig. 16.7b) in the $m/z = 217$ chromatogram of the tar–sand bitumen. The components that remain are C_{30} steranes and those having a rearranged steroid nucleus (horizontally shaded and stippled in Fig. 16.7a, respectively).

Depositional environment

In theory, the biochemical composition of all organisms is unique and it is likely that, at the time of deposition, each different **depositional environment** will have a different assemblage of organisms contributing to the sedimentary organic matter. Consequently, the initial composition of sedimentary organic matter may be characteristic of the environment in

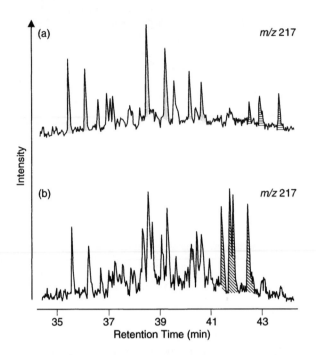

Figure 16.7   (a) Chromatogram of $m/z = 217$ of a bitumen isolated from a tar sand from southern Oklahoma, USA. Note the presence of rearranged steranes and C_{30} steranes (stippled and horizontally shaded, respectively) and the absence of C_{29} steranes (cf. Fig. 16.7b), the latter indicating that this sample has experienced biodegradation. (b) Chromatogram of $m/z = 217$ of a crude oil from southern Oklahoma, USA. The presence of steranes (e.g. C_{29}, shaded) indicates that the degree of biodegradation experienced by crude oil has not affected the sterane distribution.

which the sediment was deposited. If the sediment ultimately generates crude oil then some of these characteristics may be passed on to the oil. All geochemical processes (e.g. thermal maturation, migration and biodegradation) alter the composition of sedimentary organic matter and crude oil and therefore the initial characteristics become increasingly obscure; however, by using a combination of compound classes, as well as sedimentological and petrological data, it is usually possible to make a broad assessment of the depositional environment.

Figure 16.8 shows the $m/z = 191$ and 217 chromatograms of three types of crude oil from Sumatra (Sosrowidjojo *et al.*, 1994). These ions are characteristic of pentacyclic triterpanes (of which there are many types including **hopanes**) and steranes, respectively, and for each type of oil the distribution of the components is quite different. Type 1 oils are characterised by a number of demethylated hopanes (30-nor-17α-hopanes; solid shading, Fig. 16.8a) and C_{30} steranes (24-n-propylcholestanes; Fig. 16.8b), which indicate formation from marine carbonate source rocks. Apart from the complexity of the $m/z = 217$ chromatogram of the Type 2 oils, which in part derives from the presence of 4-methylsteranes, neither of the chromatograms reveals much about the depositional environment of the source rock. However, the presence of **botryococcane** (not illustrated herein) indicates that Type 2 oils are derived from a source rock deposited under fresh to brackish water lucustrine conditions. The $m/z = 191$ and 217 chromatograms of the Type 3 oils are dominated by **oleanane** (O; Fig. 16.8a) and the **bicadinanes** (W, T and R; horizontal shading,

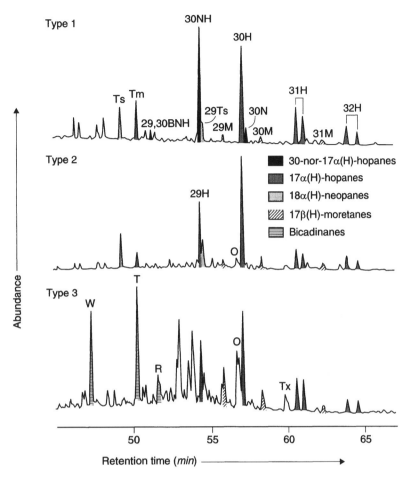

Figure 16.8(a) Chromatograms of $m/z = 191$ of three classes of crude oil illustrating how the pentacyclic triterpane biological markers may be used to differentiate the depositional environment (modified from Sosrowidjojo *et al.*, 1994. The biomarker composition of some crude oils from Sumatra. *Org. Geochem.* **21**, 303–312. With kind permission of Elsevier Science Ltd. The Boulevard, Longford Lane, Kidlington OX5 1GB, UK).

Fig. 16.8) which indicate that these oils are formed from sediments deposited in a deltaic environment. These compounds or their functionalised precursors occur in the higher plants and higher plant resins found in contemporary deltas (e.g. the Niger delta, Nigeria) and are believed to be characteristic of this environment. Thus, the depositional environment of a sediment or that of the source rock which generated a crude oil may be determined by considering the distribution of the biological markers, of which the pentacyclic triterpanes and steranes are but two.

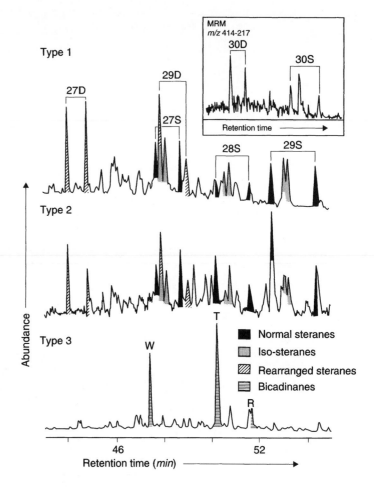

Figure 16.8(b)   Chromatograms of $m/z = 217$ of three classes of crude oil illustrating how the sterane biological markers may be used to differentiate the depositional environment (modified from Sosrowidjojo *et al.*, 1994. The biomarker composition of some crude oils from Sumatra. *Org. Geochem.* **21**, 303–312. With kind permission of Elsevier Science Ltd. The Boulevard, Longford Lane, Kidlington OX5 1GB, UK).

Acknowledgements

I wish to thank Professor Steve Rowland for the help and advice he has given me, not only whilst writing this chapter but since I began working with him. Members of the Petroleum and Environmental Geochemistry Group, both past and present, are also thanked for their assistance.

Vacuum technology

G.A. Ingram

Introduction

Any analytical instrument that employs a beam of charged particles (e.g. a mass spectrometer or an electron microscope) or that measures parts of the electromagnetic spectrum that suffer strong absorption in air (e.g. soft X-rays) requires a high vacuum if it is to function effectively. Though some modern instruments have automated vacuum systems, it is desirable for every analyst to understand the basics of vacuum technology.

Units

In the SI system (Appendix D) the quality of a vacuum is measured as the pressure (in pascals, $Pa = N\ m^{-2}$) of residual gases in the system. Historically, however, vacuum has been recorded in various other units, which may still be found on older equipment. The conversion factors between them are given in Table A.1.

Rotary pump

The simplest and most widely used type of vacuum pump is the rotary-vane pump, which pumps air by mechanical displacement. This type of pump is used in most vacuum systems, either as the sole means of evacuation for low-vacuum applications (reducing pressure down to about 0.1 Pa (10^{-3} torr)) or as a 'backing pump' to maintain a working vacuum for

Table A.1   Units for expressing vacuum pressure

	Pascal	Torr	Atmospheres	Millibar
1 pascal (Pa)	1	7.53×10^{-3}	9.87×10^{-6}	10^{-2}
1 torr*	133	1	1.31×10^{-3}	1.33
1 atmosphere	101,325	760	1	1013
1 millibar (mbar)	100	0.753	9.87×10^{-4}	1

*Pressure expressed as millimetres of mercury in a notional manometer.

a high-vacuum pump (see next section) that can only operate at reduced pressure. It is also used as a 'roughing pump' for initial evacuation of any part of an instrument that is vented to air, for example when changing specimens.

The pumping mechanism is immersed in a reservoir of vacuum oil (Fig. A.1a) which acts both as lubricant and coolant. A cylindrical rotor with spring-loaded vanes rotates within a cylindrical chamber, about an eccentric axis. It is driven by an electric motor and belt operating outside the vacuum. The vanes press against the inner surface of the chamber, providing a gas-tight seal. As the rotor turns, vane A sweeps out a volume into which the air in the volume being pumped can initially expand. As vane B sweeps across the inlet port, this pocket of air is isolated, compressed by the eccentric mechanism as rotation continues and eventually vented through the exhaust port (a non-return valve) into the surrounding oil bath, thence escaping to the atmosphere. Rotation of the vanes repeats this expansion–compression cycle and progressively displaces the air in the pumped chamber to the atmosphere. Such pumps are designed to be run continuously and are capable of realising a vacuum of 0.1 Pa, though it may take some hours to evacuate a large volume. A rotary pump must always be isolated from the vacuum chamber and air admitted to the inlet port before switching off, as oil from the pump will otherwise seep into and contaminate the chamber owing to the difference in pressure.

High-vacuum pumps

An instrument like a mass spectrometer requires a higher vacuum than a rotary pump can provide. Such instruments utilise various high-vacuum pumps that operate in conjunction

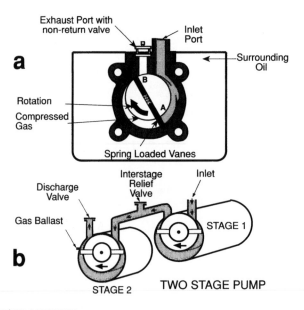

Figure A.1   Schematic of a rotary-vane pump.

with a rotary 'backing' pump. As such pumps can only be exposed to atmosphere when switched off, they must be protected by appropriate isolation valves.

Oil diffusion pump

A diffusion pump operates on the principle of transferring momentum from high-velocity oil-vapour molecules to the gas molecules that are to be pumped from the vacuum chamber. Diffusion pump oil must have high molecular weight, low vapour pressure at high temperature, and chemical inertness. Oils based on phenyl ethers are commonly used, for example Santovac 5.

The operation of the pump depends on oil being heated electrically until a vapour emerges at high speed from downward-pointing jets (Fig. A.2). The oil-vapour jets entrain residual air molecules and sweep them down to the lower part of the pump. The oil condenses on the cold walls (water-cooling coils are fitted to the ouside of the pump body – Fig. A.2) and drains back to the heater. The entrained air is pumped away by a rotary pump. This pumping mechanism requires backing pressures below 10 Pa (partly to avoid oxidation of hot oil by residual air). 'Back-streaming' of oil vapour from the pump into the vacuum chamber may be minimised by interposing between them carefully designed baffles (Fig. A.3), and by deploying a cold trap cooled externally by liquid nitrogen (77 K). A diffusion pump is capable of attaining a high vacuum of 10^{-5} Pa (10^{-7} torr), or 10^{-7} to 10^{-8} Pa (10^{-9} to 10^{-10} torr) where used in tandem.

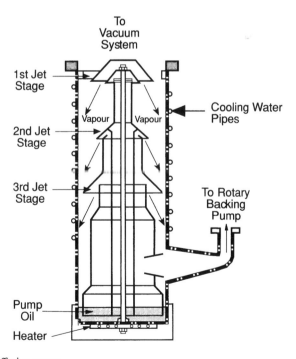

Figure A.2   Three-stage diffusion pump.

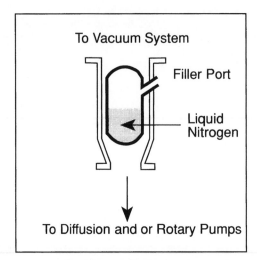

LIQUID NITROGEN TRAP

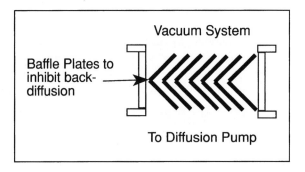

Figure A.3   Schematic of a cold trap and a baffle.

Turbomolecular pump

The turbomolecular pump consists of a stack of rotating turbine blades that compress residual gas by momentum transfer. The blades are rotated at a very high speed ($< 75{,}000$ rpm); molecules impacting on them are impelled into the backing line. Like a diffusion pump, a 'turbo pump' needs to be backed by a rotary pump, and operates in the same vacuum range as a diffusion pump but has the advantage of much faster pumping speeds. It may also require water or air for cooling.

Sputter ion pump

The sputter ion pump, or simply 'ion pump', consists of a cellular anode between two titanium plates as cathodes (Fig. A.4) in a strong magnetic field applied normal to the cathode. A 5–7 kV potential difference is applied between the cathodes and anode. Electrons initially produced by an electron gun follow spiral trajectories under the influence of the magnetic

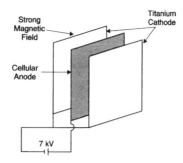

Figure A.4   Sputter ion pump.

field, thereby increasing path length and increasing the probability of collision with residual gas molecules. Such collisions generate negatively charged molecular ions, which accelerate towards the cathode plates where on impact they **sputter** titanium from the cathode. Sputtering causes turnover of the electrode surface, eroding some parts and depositing clean metal on other parts, permanently trapping the gas ions. Ion pumps utilise four different removal mechanisms, the relative importance of which varies with the gas species. Reactive gases such as N_2, O_2 and H_2 are pumped by **gettering**, as in the sublimation pump (below), inert gases like Ar and Kr are pumped by ion burial, and light gases H_2 and He are by ion burial and diffusion into the titanium. This pump is only used at vacuums of 10^{-5} mbar or less.

Sublimation pump

This type of pump makes use of **gettering**, the chemical combination of the gas molecules with reactive ions or atoms which then condense from the vacuum. The commonest getter material used for vacuum pumps is titanium, as it is easily sublimed and readily obtainable.

A titanium–molybdenum alloy filament is heated by an electric current of about 40 A. It attains a temperature of 1400°C at which the titanium sublimes to vapour. The Ti atoms combine with reactive gas molecules, and by condensing on surrounding surfaces form a thin film on the inside of the pump in which residual gas molecules are trapped. Titanium sublimes only at pressures below 0.1 Pa (10^{-3} torr), and the sublimation pump is used at pressures below 10^{-3} Pa (10^{-5} torr). It cannot be used to pump hydrocarbons; for this reason sublimation pumps should not be used in combination with an oil diffusion pump.

Vacuum Measurement

Table A.2

Type of gauge	Vacuum range
Thermocouple	$10^3 - 10^{-2}$ Pa ($10 - 10^{-4}$ torr)
Pirani	$10^3 - 10^{-2}$ Pa ($10 - 10^{-4}$ torr)
Penning	$1 - 10^{-5}$ Pa ($10^{-2} - 10^{-7}$ torr)
Bayard–Alpert	$10^{-1} - 10^{-9}$ Pa ($10^{-3} - 10^{-11}$ torr)

Pirani gauge

This consists of a fine wire of Pt or W passing down the centre of a tube of 1 cm diameter open to the vacuum. A known potential difference is applied across the wire. The current generates heat in the wire, which is conducted away by the surrounding gas. At pressures close to atmospheric, thermal conductivity is independent of pressure, but below 10^3 Pa (10 torr), gas conductivity varies with the number of molecules present, and therefore with pressure. The temperature, and therefore the resistance, of the filament will increase, and the current will decrease, as the thermal conductivity of the gas falls. A meter measures the current as the resistance changes; the scale is usually graduated directly in pressure units. Below 1 Pa (10^{-3} mbar) the conductivity becomes very small and heat loss from the wire is mainly by radiation, and thus the Pirani gauge can only be used between 1 and 1000 Pa.

Thermocouple gauge

The thermocouple gauge differs from the Pirani gauge in that the temperature of the filament is measured directly by a thermocouple. It is more compact than the Pirani, which requires a long filament. It operates over the same pressure range.

Calibration is achieved by using dry nitrogen, but the sensitivity of these gauges depends on the identity of the gas being measured. The gauges should not be deployed close to the inlet of a rotary pump, as back-streaming oil vapour will cause the gauge to give erroneously high readings.

Penning gauge

The cold-cathode Penning gauge (Fig. A.5) consists of two parallel cathode plates that are arranged on either side of an anode ring. A 2 kV voltage is applied between the cathodes and anode and a magnetic field is applied perpendicular to the cathode plates. Electrons ionise the gas molecules and hence a pressure-dependent ion current is produced which can be measured. The gauge operates only in a pressure range where the electric discharge can be maintained, from 1 Pa (10^{-2} torr) down to pressures of 10^{-5} Pa (10^{-7} torr).

Bayert–Alpert ion gauge

In this gauge electrons are supplied by a hot filament, in much greater numbers than with the Penning gauge. The electrons are accelerated by a potential difference of 150 V towards a spiral wire 'grid' (Fig. A.6). Gas molecules ionised near the grid are collected by a central anode wire which is at a lower potential than the grid, and the ion current which is produced is measured. This gauge will measure down to 10^{-9} Pa (10^{-11} torr).

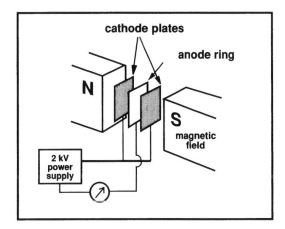

Figure A.5   Schematic diagram of a Penning gauge.

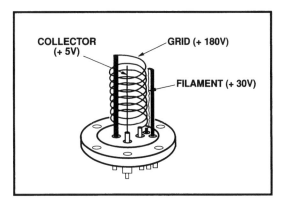

Figure A.6   Bayert–Alpert head.

Materials, connections and baking

Nearly all vacuum systems are made out of stainless steel or glass, with ceramic feed-throughs for electrical contacts. These materials are chosen for their strength and rigidity (chamber walls will have to bear loads of $10,335 \text{ kg m}^{-2}$), for their very low vapour pressures and low outgassing rates, and for their capacity to be baked to discharge adsorbed gases.

Various types of seal are used to seal couplings and entry ports. For a chamber that is regularly opened, such as a mass spectrometer source chamber, the lid or door will have a **viton** seal ('O–ring'); the seal is created by pressing the O-ring between two flat surfaces. For couplings that only require occasional opening a copper gasket between two knife edges (known as conflats) provides the seal. For seals that are broken infrequently, 24 carat gold is used in the form of a wire ring and is squashed against the flat surface of a flange.

All vacuum systems become contaminated from the repeated opening of the system to introduce samples, from degassing of samples, and from the decomposition of vacuum oil.

Pumping speeds decline and the ultimate vacuum deteriorates owing to gases sorbed on the interior surfaces of the system. When the deterioration has reached an unacceptable level, it is necessary to bake the system that is under vacuum at approx. 150°C, using either tailor-made ovens or heating tape. It may be necessary to switch off the high vacuum pumps, but this is system dependent.

Glossary

Words printed in **bold** letters refer the reader to other Glossary entries. Parentheses at the end of some definitions indicate the units in which a quantity is normally reported.

Statistical definitions with initials in brackets after them are derived from the definitions recognised by the International Organization for Standardization (see ISO, 1992), the International Union of Pure and Applied Chemistry (see IUPAC, 1978) and the Analytical Methods Committee of the Royal Chemical Society (see AMC, 1987).

AAS	abbreviation for atomic absorption spectrometry (Chapter 5).
absorbance	a measure of the attentuation of light intensity in passing through a sample. Absorbance is given by: $$\text{Absorbance} = \log_{10}(I_0/I)) = -\log(1-D)$$ where I_0 is the intensity of the incident light and I is the intensity of light emerging from the specimen. (D is optical density.) (dimensionless)
absorption	attenuation of spectral intensity by the analyte itself, by other elements present in the sample, by the air or by components of the analytical instrument.
absorption edge	discontinuity in the X-ray absorption spectrum of an element: photon energies on the short-wavelength side are just sufficient to ionise the relevant (K, L, M, N) shell of the atom whereas those on the long-wavelength side are not, resulting in an abrupt increase in absorption coefficient in passing to the short-wavelength side.
accuracy	the closeness of agreement between the **true value** and an individual measured value [ISO], embracing both random and systematic error (**bias**).
ADC	'analogue to digital converter': a circuit that translates an analogue signal (usually a voltage or current) into digital form for computer processing.
aliphatic	describing an organic compound that is devoid of aromatic (e.g. benzene) rings. Aliphatic hydrocarbons may be acyclic (e.g. n-alkanes) or mono- or polycyclic (e.g. steranes, Table 16.3).
alpha particle (α)	a positively charged nuclear particle consisting of two protons and two neutrons. Identical to a ^{4}He nucleus.
α-decay	radioactive decay resulting in the emission of **alpha particles**.

amu	atomic mass units.
analyte	the element or compound whose concentration an analysis is intended to determine.
anonymous, anonymity	describes replicate readings of the same **value** (i.e. analyses of the same sample) carried out according to a protocol that prevents them being identified as replicates at the time of analysis.
aqua regia	1:3 mixture by volume of concentrated nitric and hydrochloric acids.
aromatic	describing an **unsaturated** organic compound that is based on one or more benzene rings, or that has properties similar to those of benzene.
asphaltene	the fraction of a crude oil or sediment extract that is soluble in some organic solvents (e.g. dichloromethane) but not others (e.g. n-pentane) (cf. **maltene**).
astigmatism	an optical defect in which a lens exhibits different focal lengths for rays passing through different sectors of the lens (Box 14.1).
atomic mass unit (amu)	unit in which atomic masses are conventionally expressed = mass of ^{12}C atom$/12.0000 = 1.66056 \times 10^{-27}$ kg.
atomic number Z	the cardinal property of a chemical element that identifies the number of protons in its nucleus. Equal to the number of electrons in the *neutral* atom.
background	the notional intensity of a spectrum at the position of an analyte peak if that peak were absent. It is measured under exactly the same conditions as the analyte peak. It is usually understood to *exclude* spectral overlap and non-sample analyte contributions.
Beer–Lambert law	(or Beer's law). $I = I_0 \exp(-\mu\rho x)$ where I_0 is the intensity of a beam of electromagnetic radiation (e.g. X-rays) entering a block of absorbing material of density ρ, I is the intensity after traversing a path length x, and μ is the **mass absorption coefficient** of the material.
β (beta) particle	a high-energy electron ejected from an atomic nucleus.
β-decay	radioactive decay resulting in the emission of **beta particles**.
bias	of an analysis refers to the closeness of agreement between the mean of a series of analyte concentration measurements and the **true value** of the concentration (in practice, the best available estimate of the true value); cf. **accuracy**.
bicadinanes	a group of C_{30} pentacyclic hydrocarbons whose biosynthetic origin is presently unknown. They appear to be characteristic of organic matter derived from higher plants.

binding energy	the minimum energy (in keV) that must be given to an electron in a specific energy level in an atom for it to escape from the atom altogether.
biological marker	'Any organic compound detected in the geosphere whose basic skeleton suggests an unambiguous link with a known contemporary natural product' (Mackenzie, 1984). Not to be confused with ecotoxicological usage which refers to biological change induced by altered physiology and/or biochemistry.
biomarker	synonym for **biological marker**.
bitumen fraction	see **lipid**.
blank	a term whose precise meaning depends on the user. An *instrument blank* is an artificial sample containing zero concentration of the **analyte**, used to determine an instrument's baseline response. A *reagent* or *procedure blank* is a preparation for analysis containing no sample but prepared using the same reagents and/or the same method as unknown samples. A *blank reading* is an apparent analyte signal determined on such a sample.
bremsstrahlung	the German name (literally 'braking rays') for the X-ray continuum emitted when a high-energy charged-particle beam is used to generate **characteristic** X-rays in a material. The particles are decelerated by interaction with the electrostatic fields of atomic nuclei that they pass close to, causing them to emit X-ray photons whose energies are unrelated to atomic energy levels and which give rise to a background continuum (Box 6.1).
brilliance	(of an X-ray beam) number of photons per second, per mm^2 of source area, per milliradian of solid angle per unit energy bandwidth.
calibrate, calibration	the act of relating an instrumental signal to the magnitude of the physical parameter it represents; for example, the **intensity** of a spectral line needs to be calibrated (usually by measuring a series of samples of known composition) before it can be converted into an estimate of concentration.
carotenoid	large class of organic compounds (typically C_{40}) that may be classed as **isoprenoids**. They are highly coloured (often yellow/orange, e.g. β-carotene) and play a role in photosynthesis. With chlorophylls, they are sometimes classed as pigments.
cathodoluminescence	emission of visible light resulting from electron bombardment.
CCD	see **charge-coupled device**.
to **centre**, to **peak-centre**	(mass spectrometry) to determine and set the exact magnetic flux density for an ion beam to pass symmetrically into a collector.

certified reference material (CRM)
a **reference material** one or more of whose compositional properties are certified by a recognised procedure, that is accompanied by a certificate issued by a certifying body [after ISO].

chain of custody
system of sample management and documentation (used in many commercial testing laboratories) that records every person who has in turn taken responsibility for a specified sample or batch, from the moment it arrives in the laboratory to its final destination (e.g. disposal, return to client, delivery to specified third party). Each transfer of responsibility is documented by the signatures of those relinquishing and receiving the material, with the date and time.

channeltron
a form of electron multiplier, consisting of a small horn-shaped ceramic body whose interior surface is coated with semiconducting PbO, biased across its length ~ -3 kV. The device amplifies individual ion arrivals in a similar fashion to an electron multiplier (Fig. 6.3.2); the coating acts as a 'continuous dynode' in place of the individual dynodes of a conventional electron multiplier.

characteristic X-ray
X-ray emission at a specific wavelength (or series of wavelengths) associated with a specific element in the sample.

charge-coupled device
(CCD) a self-scanning array of photodiodes used in spectrometry, image detection and astronomy. See Smith and Thompson (1988). Such devices can operate at low light intensities and can record peak profiles and shifts as well as peak intensities.

chemical shift
a slight shift in the characteristic wavelength emitted by an element owing to a change in the atom's atomic environment and/or bonding.

chiral centre
a carbon atom in a complex organic molecule that is connected to four different groups (i.e. possesses no element of symmetry).

cholestane
a C_{27} tetracyclic steroid alkane (i.e. a sterane; see Table 16.1) having a carbon skeleton similar to cholesterol.

chromatography
class of chemical technique in which the components of a mixed solution or a mixed gas are separated by selective adsorption on or distribution between two chemical phases, one stationary and the other mobile.

coefficient of variation
an estimate of the precision of a measurement equal to $2 \times 100s/\bar{x}\%$ among repeated measurements, where $\bar{x}$ is the mean value of the repeated measurements and s is standard deviation.

collimator
a device for trimming a beam of electromagnetic radiation or ions into a narrower beam of limited divergence.

Compton scattering
inelastic scattering of photons by atoms or electrons, in which the scattered photon has a longer wavelength (lower photon energy) than the incident photon; cf. **Rayleigh scattering**.

concentration	in chemical usage, a measure of how much of a particular chemical component (in mass or molar units) is present in unit *volume* of the substance in which it resides. Also used loosely in geochemistry to signify the amount of analyte per unit *mass* of sample (e.g. %, μg g^{-1}, ppb).
conduction band	an empty or partly filled band of electron energy levels in a crystal that allows electrons promoted to it to migrate in response to an applied electric field.
confidence interval	See Box 4.5.
counter	synonym for radiation or particle detector.
dead time	the interval of time (μs) after the arrival of a photon during which a detector is disabled and incapable of detecting a subsequent photon. Sometimes refers to the accumulated detector dead time over a counting interval as a percentage of elapsed time.
decay constant	the rate constant of radioactive decay. The amount m of a radioactive parent nuclide remaining after time t is given by $$m = m_0 \mathrm{e}^{-\lambda t}$$ where m_0 is the initial amount of the nuclide (at $t = 0$) and λ is the decay constant. (s^{-1}, year^{-1})
degrees of freedom	where a sample mean is calculated from n readings, the number of degrees of freedom is n–1. See Miller and Miller (1993) for explanation.
delay time	pause to allow instrument to stabilise before performing a measurement. In automated instruments, this may be a programmable parameter.
detection limit	the lowest concentration of an **analyte** giving a signal that can be distinguished with reasonable confidence from **blank** or **background** measurements. The working definition used in this book is *the concentration or amount of analyte corresponding to a measurement level* $3\sigma_\mathrm{B}$ *units above the value for zero analyte,* where σ_B is the standard deviation of readings on the blank [after AMC]. The statistical principles underlying the detection limit are explained in Box 4.4.
diffraction	See Box 4.3.
drift	gradual change of analyte sensitivity or instrument parameters with time.
dynamic range	the ratio of the highest concentration measurable by a particular technique to the **detection limit** for the same element by the same technique. Sometimes restricted to that part of the dynamic range over which the **calibration** is linear. Also used for the range of signal magnitude over which an amplifier or ADC responds in a linear manner.

dynode	one of a series of electrodes emitting secondary electrons in an **electron multiplier**.
earth	UK term for electrical **ground** (North America). A terminal having a low-impedance connection to the ground.
electromagnetic radiation, electromagnetic wave	encompasses all transverse waves propagating energy through space by coupled oscillation of electric and magnetic fields, of which light is the most familiar example.
electron multiplier	a device for amplifying a low dc signal (see Box 6.3).
electron-volt (eV)	unit of energy widely used in spectroscopy and atomic physics. *The energy acquired by an electron accelerated by a potential difference of 1 volt.* Equal to 1.602×10^{-19} joules.
enantiomers	two alternative forms of a complex molecule that are non-superimposable mirror images of each other.
energy level	one of a restricted number of **quantised** energies that an electron can have in an atom according to wave mechanical principles.
epithermal neutrons	neutrons having energies in the range 10^{-2}–10^2eV, somewhat more energetic than **thermal neutrons**.
escape peak	if an incoming photon causes the detector medium (Si, Ge or Ar) to **fluoresce** and a fluorescent photon escapes from the detector, the detector will register the *difference* between the original photon energy and the escaping photon energy, resulting in a spurious low-energy peak in the energy spectrum, called an escape peak.
excitation	the process of promoting electrons in an atom to higher **energy levels**, creating vacancies in lower levels (an excited state).
excitation energy	the energy required to excite an electron from the ground to a given excited state (cf. **ionisation energy**). (eV *or* kJ mol^{-1})
Faraday constant F	Quantity of electric charge residing on 1 mole of any singly charged ionic species. Equal to Avogadro's number $\times$ charge on an electron. (C mol^{-1})
FET	field–effect transistor, a high-impedance, low-noise charge amplifier.
fitness for purpose	an aspect of **quality assurance** referring to the capacity of an analysis to serve well the purpose for which it was carried out (with regard to **precision, bias, detection limit**, turnaround, **traceability**, etc.).
fluorescence	emission of electromagnetic radiation ('secondary radiation') arising from **excitation** of atoms by **electromagnetic radiation** of higher **photon** energy ('primary radiation').

fragmentation	the breakdown of a complex organic molecule into simpler fragment ions in the course of mass spectrometric analysis. See Table 16.1.
frequency	*physics:* the number of oscillations or events taking place per second (Hz); *statistics:* the number of observations whose values fall within a specified value interval.
fulvic acid	organic constituents that are soluble in aqueous base and aqueous acid.
FWHM	'full width at half maximum', a convention for measuring the spectral width of emission lines in wavelength space (nm).
γ(gamma)-ray	a high-energy electromagnetic **photon** having a frequency greater than 10^{20} Hz or a wavelength less than 3 pm (quantum energy exceeding 10^5 eV). γ-rays emitted from atoms are produced by *nuclear*, not electronic, transitions (cf. **X-ray**).
Gaussian distribution	see normal distribution.
GC–MS	abbreviation for gas chromatography–mass spectrometry.
geostandard	in a geochemical context, a **reference material** comprising a rock or mineral powder (finely ground and carefully homogenised) or other material that has been analysed independently for numerous elements by numerous methods in numerous laboratories, for which there exists an authoritative published compilation of chemical analyses.
gettering, getter	the chemical combination of the residual gas molecules in a vacuum with sublimed atoms of the getter, which, by condensing on cooler parts of the pump, remove them from the vacuum, the basis of the sublimation vacuum pump (Appendix A). 'Getter' refers to the reactive, sublimed material used for the purpose, usually titanium.
ground	North American term for electrical **earth**.
half-life $t_{1/2}$	the time required for the radioactivity of a specific radionuclide to decay to half its initial value. $t_{1/2} = (\ln 2)/\lambda$. (s, day, year)
humic acid	comprises organic constituents soluble in aqueous bases but not aqueous acids.
humin	the fraction of organic matter in unconsolidated sediments and soils that is not soluble in aqueous base or aqueous acid (cf. **kerogen**).
ICP–AES	abbreviation for inductively coupled **plasma**–atomic emission spectrometry.
ICP–MS	abbreviation for inductively coupled **plasma**-mass spectrometry.

in phase
refers to two wave profiles coinciding in space or time (see Fig. 4.3.1b).

induction heating
a method of heating a conducting material without physical contact. The sample is surrounded by a coil carrying a radio-frequency alternating current. The sample behaves as the secondary coil of a transformer, and the secondary current flowing in it causes resistive heating.

instability
significant variation in ion beam intensity, electromagnetic intensity or instrument parameters during a time interval less than that used for measurement.

integration time
time taken (or preset) for an instrument to complete a single spectral measurement without changing settings.

intensity
of a particle or photon beam: energy flux per unit time per unit area normal to the direction of propagation; *of an individual wave:* the square of the wave amplitude.

interference (analytical)
a systematic error introduced into the analysis of one element, **nuclide** or compound in a sample by the presence of another. This may arise from circumstantial overlap of spectral lines (**spectral interference**) or by the effect of other sample constituents on analyte **sensitivity** (**matrix effects**).

interference (optical)
an interaction between two waves of identical wavelength in the same region of space. If they are **in phase**, their amplitudes reinforce each other to produce a stronger signal (constructive interference); if they are perfectly **out of phase** (Fig. 4.3.1c), the amplitudes counteract one another and the signal is diminished or eliminated (destructive interference).

interference correction
a numerical correction applied to the measured spectral intensity of an analyte to compensate for overlapping spectra of other elements present in the sample.

internal standard
an element or isotope in a sample that can be used to **calibrate** other elements being analysed. Such an element has been added to the sample in a known amount (much larger than that already present), or its concentration in the sample has been determined by a separate analysis.

international standard
see **geostandard**.

ion exchange
technique of element separation in solution using the selective adsorption properties of certain resins or crystals in the presence of solvents of varying composition (e.g. pH). See Harland (1994).

ion exchange resin
an organic (acrylic or styrene) co-polymer resin to which functional groups have been added to provide ion exchange capacity. Addition of acidic (sulphonic ($-SO_3H$) or carboxylic ($-COOH$)) functional groups generates a *cation exchange resin*, whereas the

	introduction of basic functionality (amine ($-NH_2$) groups) produces an *anion exchange resin*.
ion optics	techniques for accelerating, mass selecting and focusing an ion beam by means of magnetic and electric fields.
ionisation efficiency	the ratio between the number of ions of an element produced in an ion source and the number of atoms introduced.
ionisation energy	the energy required to remove the most easily detached electron from an atom to a state of rest at infinity. (eV *or* kJ mol^{-1})
ionisation potential	obsolete synonym for **ionisation energy**. (eV *or* kJ mol^{-1})
isobar, isobaric	a different **nuclide** having the same **mass number** as the nuclide being analysed.
isobaric interference	an **interference** effect arising in mass spectrometry from the presence of an **isobaric nuclide** having the same mass/charge ratio.
isotope	atoms of an element whose nuclei contain the same number of neutrons N. Most elements consist of several isotopes, each characterised by a different value of N.
isotope dilution (ID)	technique of mass spectrometric analysis in which the sample is **spiked** with a known amount of the analyte element artificially enriched in one particular isotope. The initial concentration of the analyte is calculated from the difference in isotope ratios between the mixed and natural samples. (See Box 8.3)
isotope/isotopic ratio	the ratio of abundances (in numbers of atoms) of two isotopes of the same element in a sample. (dimensionless)
iterative, iteration	cyclic calculation process involving successively closer approximation to the required value. Applied to correction of matrix effects in **XRFS** analysis, where correction of major element concentrations for mass absorption requires prior knowledge of the major element composition that is being determined.
K, L, M 'shells'	Electron 'shells' comprising orbitals with principal quantum number values: $n = 1$ (K shell = 1s orbital only), $n = 2$ (L shell = 2s + 2p orbitals) and $n = 3$ (M shell, 3s + 3p + 3d). See Box 6.1.
kerogen	the fraction of organic matter in consolidated rocks that is wholly insoluble in organic solvents (cf. **humin**).
leaching	chemical pretreatment of a sample in order to remove selectively spurious components that may bias an analysis.
light	electromagnetic radiation in the wavelength range visible to the human eye (400–700 nm, equivalent to 7.5–4.2 $\times 10^2$ THz).
line broadening	an atomic emission line produced by an electron transition between atomic **energy levels** is very narrow (**FWHM** $\sim 10^{-5}$ nm). In practice, linewidths are substantially wider because random atomic motions introduce small positive and

negative Doppler shifts. This 'Doppler broadening' increases with source temperature. Line broadening may also be brought about by high pressure or a strong magnetic field.

linear	defining a straight line.
lipid	soluble fraction of organic matter in sedimentary rocks and petroleum.
magnetic susceptibility	the ratio of the intensity of magnetisation of a material to the magnetic field giving rise to it.
mass absorption coefficient	the constant in the **Beer–Lambert law** that relates loss of intensity to the path length in, and the density of, the absorbing medium. The mass absorption coefficient (MAC) of a compound is the weighted mean of the MACs of the individual elements present. ($cm^2\,g^{-1}$)
mass bias	the ratio between the measured and true **isotopic ratios** of two isotopes 1 amu apart. (dimensionless)
mass defect Δ	the difference between the observed **relative atomic mass** of an atom and the sum of the masses of the constituent protons and neutrons. (**atomic mass units**)
mass fractionation	(mass spectrometry) a kinetic phenomenon peculiar to thermal ionisation sources, in which light atoms are preferentially released, usually leading to systematic change in measured **isotope ratio** with time.
mass number A	the total number of protons and neutrons in a nucleus $= Z + N$.
mass resolution	the ratio $M/\Delta M$, where ΔM is the mass difference between the closest pair of peaks that a mass spectrometer can resolve (or between which a quadrupole can discriminate), and M is the mean mass of the two peaks.
matrix effect	a form of **interference** in which the **sensitivity** of the analyte depends on **concentrations** of other elements present in the sample (the 'matrix').
metamorphic	describes a class of rocks that have recrystallised in the solid state due to heating and/or stress.
milliequivalents (meq)	units of molar concentration expressed in terms of the notional charge on all ions of a given species (e.g. Fe^{2+} present in solution at a molality of $0.064\,mol\,kg^{-1}$ amounts to $2 \times 0.064\,meq\,kg^{-1}$).
mole	(abbreviated 'mol') an amount of a compound (or element) whose mass expressed in grams is numerically equal to its **RMM** (or **RAM**).
molecular flow	flow regime in a gas at low pressure ($<10\,Pa$) in which the mean free path of the gas molecules exceeds the diameter of the container, so that molecule–wall collisions outnumber molecule–

	molecule collisions. Such flow can result in significant isotope fractionation; cf. **viscous flow**.
molecular sieve	compound (often a variety of the mineral zeolite) whose crystal structure incorporates large cavities or channels that can selectively accommodate small molecules; widely used in the chemical industry to separate ions and molecules according to size.
monochromatic	*of light or other radiation*: consisting of a single wavelength.
Mylar®	trade name for thin polyester film (usually 1 μm or 6 μm) used as X-ray-transparent window separating parts of an X-ray spectrometer at differing pressure.
neutron flux $_n$	the number of neutrons passing through unit area (perpendicular to the neutron beam) per unit time. Units $m^{-2}\,s^{-1}$.
neutron number N	the number of neutrons in a nucleus.
noise	thermal ('Johnson') noise: background electrical fluctuations limiting the ultimate sensitivity of electronic apparatus; arises from random (Brownian) motion of electrons. Amplitude proportional to $(kT)^{0.5}$.
normal distribution	see Box 1.1.
normalised	(mass spectrometry) refers to a measurement of an **isotope ratio** that has been corrected for mass fractionation (based on the observed mass fractionation in another isotope ratio of known value).
nucleon	a general term for a nuclear particle, embracing both protons and neutrons.
nuclide	a substance consisting of atoms with particular values of Z and N. A specific isotope of a specific element.
NSO compound	organic molecules containing N, S or O atoms.
ohmic heating	heating arising from passage of an electric current through a resistive medium.
out of phase	two waves having the same wavelength or period are 'out of phase' when the maxima of one wave coincide with the minima of the other (see Fig. 4.3.1c).
PAH	polynuclear aromatic hydrocarbon.
parts per billion (ppb)	unit of trace element concentration = μg of element per kg of sample.
parts per million (ppm)	unit of trace element concentration = mg of element per kg of sample.
period	(of a wave) the time taken for equivalent points on successive cycles of a travelling wave to pass the same point. (s)

phase	refers to different points on a wave profile. See **in phase** and **out of phase**.
photoelectric effect	the emission of electrons from a material resulting from excitation by energetic photons.
photoelectron	a free electron produced by the **photoelectric effect**.
photomultiplier	see Box 6.3.
photon	a **quantum** of electromagnetic radiation (e.g. light).
plasma	high-temperature gas in which a sufficient proportion of the atoms are ionised to make the gas electrically conducting.
Poisson distribution	see Box 1.1.
polar, polarity	refers to a covalent bond linking atoms of very different electro-negativity. A compound containing such a bond possesses a partial electric dipole.
population mean	the mean of all possible measurements of a particular **value** (cf. **sample mean**).
power density	*in laser mass spectrometry*, the energy deposited per unit area of target per unit duration of pulse; *in SSMS* (Chapter 11), the energy flux passing through the plasma per unit area of cross-section per unit time. In either case the working unit is $W\ cm^{-2}$.
ppb	see **parts per billion**.
ppm	see **parts per million**.
precise, precision	refers to the closeness of agreement between results obtained by applying the analytical procedure several times under prescribed conditions; the smaller the random uncertainties affecting the results, the *more* precise the measurement[1] [after ISO]. Precision is a qualitative statement solely about *random* error, without reference to the **true value** of the concentration.
probe current	the electron current (measured in nanoamperes or microamperes) arriving at the specimen surface in an electron microprobe or scanning electron microscope.
protocol	a precisely specified sample preparation or analytical procedure designed to minimise inter-laboratory **bias**.
pyrolytic graphite tube	(or pyrolytically coated graphite) preformed graphite tube that has been heated to $2500°C$ in an atmosphere of methane, forming a dense surface layer of graphite that is unreactive and impervious to sample solution.

---

[1] Though it is acceptable to describe a technique as having 'high precision' (meaning a low random error), it is preferable to cite the actual standard deviation of the measurements.

quality assurance	'planned activities designed to ensure that a **quality control** programme is being properly implemented' (after Garfield, 1991).
quality control	in the context of an analytical laboratory, a planned, systematic programme of measures to verify and document the precision and accuracy of analyses produced in the laboratory, introducing **anonymously** into the work stream appropriate replicates and **reference materials** and maintaining a continuous record of the results obtained.
quantised	a physical attribute is said to be quantised when it can adopt only certain specific values (like the frequencies of a guitar string) instead of ranging across a continuum. Electrons in atoms are thus confined to a limited number of prescribed **energy levels**.
quantum	the smallest, indivisible unit of any form of physical energy.
radiofrequency	a frequency in the range equivalent to radio waves (10 kHz to 1 GHz).
radiogenic	refers to a **nuclide** whose abundance in a sample is the result of, or has been enhanced by, radioactive decay of another nuclide. Thus ^{87}Sr is the radiogenic isotope of strontium, being the daughter nuclide of the decay of ^{87}Rb.
radioisotope	a radioactive **isotope**.
radionuclide	a radioactive **nuclide**.
random error	an error which varies in an unpredictable manner, in absolute value and in sign, when a large number of measurements of the same **value** of a quantity are made under effectively identical conditions [ISO].
rare earth elements	the lanthanide elements $_{57}$La (lanthanum) to $_{71}$Lu (lutetium). The related element $_{39}$Y (yttrium) is sometimes included. See **REE**.
Rayleigh scattering	scattering of **photons** by atoms or electrons, in which the scattered photon has the same wavelength (no loss of photon energy) as the incident photon (cf. **Compton scattering**).
REE	abbreviation for **rare earth element**. Sub-divided by geochemists into LREEs (light REEs, La–Eu) and HREEs (heavy REEs, Gd–Lu).
reference material	a material or substance one or more properties of which are sufficiently well established to be used for the calibration of an analytical determination or the assessment of a measurement method [ISO].
relative atomic mass (RAM) M	the mass of an atomic species expressed in **atomic mass units**. Formerly 'atomic weight'.
relative molecular mass (RMM) $\underline{M}$	the mass of a molecular species expressed in **atomic mass units**. Formerly 'molecular weight'.

relative sensitivity factor	the sensitivity of an analyte as a ratio of the sensitivity of another element, or of the same element under different conditions (e.g. in the absence of matrix effects).
relaxation	the process in which an atom or molecule returns from an excited state to its ground state: an electron loses energy in moving from a high **energy level** to fill a vacancy in a lower level by emitting a **photon**. See Box 4.1.
repeatability	the closeness of agreement between the results of successive measurements of the same **value** carried out over a *short period of time* by the same method on the same instrument by the same analyst in the same laboratory [after ISO].
representative	a sample is representative if its composition can be said to reflect without bias the mean composition of the entire body of material from which it has been taken.
reproducibility	the closeness of agreement between the results of measurements of the same **value** carried out by the same method, instrument, analyst and laboratory over a period of time that is *long* in comparison to the duration of a single measurement. [Note: this conflicts in detail with ISO usage.] Compare with **repeatability**.
resolution	the capability of a spectrometer to separate adjacent peaks in a spectrum. The ratio $\lambda/\Delta\lambda$, where $\Delta\lambda$ is the wavelength difference between the closest pair of peaks that a spectrometer can resolve, and λ is the mean wavelength of the two peaks.
RSD	relative standard deviation = $100\sigma/\text{mean}$. (%)
sample	a portion of material selected to be **representative** of a larger quantity of material. Often used loosely for a material to be analysed.
sample mean $\bar{x}$	the mean of a finite sub-set of measurements, providing an *estimate* of the mean of the entire population (see **population mean**): $$\bar{x} = \frac{\sum\limits_{i=1}^{n} x_i}{n}$$
sample standard deviation s	see under **standard deviation**.
scintillation	minute flash of light produced as an ionising particle or photon interacts with certain kinds of crystal or organic liquid (a scintillator or phosphor). Scintillation detector – see Box 6.3.
self-reversal	(self-absorption) an atomic emission line from a high-temperature source may undergo absorption by ground-state atoms in the cooler periphery of the device, suppressing the central portions of the line profile. In combination with **line broadening**, this may cause splitting of emission lines.

sensitivity	signal intensity observed per unit abundance of the analyte in the sample.
SIMS	abbreviation for secondary ion mass spectrometry, a synonym for the ion microprobe (Chapter 15).
specific gravity	the density of a substance expressed as a multiple of the density of pure water at $4°C$. (dimensionless)
spectral intensity	that part of the total **intensity** lying within a specific wavelength (or mass) interval.
spectral interference	interference arising from incomplete isolation of radiation emitted or absorbed by the analyte from other radiation detected by the instrument [IUPAC].
spectroscopy	a form of spectrometry in which shifts in wavelength ('chemical shifts') are used to study differences in chemical bonding or atomic environment.
spherical aberration	a defect of a lens causing it to have a different focal length for the periphery of the beam compared with that for the axial part of the beam. This form of distortion is particularly acute with magnetic lenses used to focus electron beams.
spike, spiked	a substantial amount of an element added to a sample to determine spectral interference and/or to calibrate a wavelength scale. *In isotope dilution:* the accurately known amount of the analyte element, artificially enriched in one isotope, that is introduced into the sample being analysed (see Box 8.3).
spontaneous emission	emission of radiation due to electronic transition that is not **stimulated** (cf. Box 9.3).
sputter	extract ions from a specimen surface by means of ion bombardment.
Sr isotope stratigraphy	method of comparing relative ages of marine sedimentary rocks using known variations in $(^{87}Sr/^{86}Sr)_{seawater}$ through geological time.
SSMS	abbreviation of spark source mass spectrometry (Chapter 11).
stable isotope	in principle, any non-radioactive isotope. Used in geochemistry to refer to isotopes of certain low-Z elements (H, C, N, O, S, etc.) whose isotopes undergo mass fractionation to a measurable extent in various geological processes, and are therefore of interest to geochemists ('stable' as distinct from the radioactive/radiogenic isotope systems of elements like Sr, Nd and Pb whose isotopic compositions change with time).
standard	material of accurately known composition used to **calibrate** an analysis. Usually several standards covering a range of compositions are used to prepare a **calibration** curve.

standard addition	a technique for overcoming **matrix effects**. Known amounts of analyte are added to aliquots of the unknown sample (as powder, solution or during the course of fusion) and homogenised. When the signal intensities for the spiked aliquots are plotted against the amounts of analyte added, a straight line is obtained whose (negative) intercept on the x axis indicates the amount of analyte originally present in the unknown.
standard deviation σ, s	the most widely used measure of the dispersion of readings about a mean value. The standard deviation for an effectively infinite population of readings is

$$\sigma = \sqrt{\frac{\sum_{i=1}^{n}(x_i - \mu)^2}{n - 1}}$$

where x_i represents each individual reading in a large population of n readings of the same variable, and μ is the mean of the entire population. In practical terms, the standard deviation of a population is estimated from a small sub-set of replicate readings, yielding the *sample standard deviation, s*

$$s = \sqrt{\frac{\sum_{i=1}^{n}(x_i - \bar{x})^2}{n - 1}}$$

where $\bar{x}$ represents the **sample mean** of the sub-set.

standard error on the mean *se*	the 'standard error on the mean' is the **standard deviation** of a series of n measurements divided by $\sqrt{n}$, and reflects the increasing confidence in the mean value as n increases. It represents the uncertainty on the mean value of a series of measurements, but is *not* a measure of reproducibility or precision of individual readings.
stimulated emission	emission from an excited state prompted by irradiation (Box 9.3).
systematic error	an error that is not diminished by repeated measurement of the **value** concerned.
Teflon®	trade name of polytetrafluoroethylene.
thermal neutrons	'slow' neutrons with energies in the lowest range ($<10^{-2}$ eV), so called because their energies correspond to those of thermal vibration.
thermionic emission	emission of electrons from the surface of an electrically heated conductor.
TIMS	abbreviation of thermal ionisation-mass spectrometry.

traceability	property of an analysis or analytical **value** whereby it can be related, with a specified uncertainty, to a specified **reference material**, usually a **certified reference material**, through an unbroken chain of comparison.
transmission	the ratio between the number of ions received by a collector per unit time and the number of ions of that mass produced in the mass spectrometer source during the same unit of time.
true value	the hypothetical **value** of a quantity obtained by measurements that are not subject to error, either systematic or random. The true value is an ideal concept and cannot in general be known.
ultraviolet light	**electromagnetic radiation** in the wavelength range between visible **light** and **X-rays**: 3–400 nm.
urea	Organic compound $$O{=}C{\overset{\displaystyle \diagup NH_2}{\diagdown NH_2}}$$ that forms clathrates with n-alkanes.
unsaturated	describing an organic compound in which some carbon atoms are linked by double or triple bonds.
value	magnitude of a quantity expressed as the product of a number and the unit of measurement.
vapour	a gas at a temperature below its critical point, which can therefore be condensed to a liquid by an increase in applied pressure.
variable	a quantity whose variation is being examined. In a table of geochemical analyses, each element or compound whose concentration is reported is a separate variable.
variance σ^2	the square of the standard deviation. Variance is an *additive* function (the total variance is the sum of variances of contributory sources of error) whereas standard deviation is not.
variate	a member of a statistical population. In a table of geochemical analyses, each **sample** analysed is a separate variate.
viscous flow	flow regime in a gas (pressure > 10 Pa) in which the mean free path of the gas molecules is less than the diameter of the container; flow is accomplished mainly through inter-molecular collisions. Viscous flow does not cause isotope fractionation; cf. **molecular flow**.
viton	synthetic rubber (elastomer) used for vacuum seals because it retains its flexibility over a wide temperature range.
wavelength λ	the linear distance between equivalent points on successive cycles of a wave. (m, mm, µm, nm)

work function W the energy required to remove the most easily detached electron from the bound state in a metal surface to a state of rest at infinity. (eV *or* kJ mol^{-1})

working value   concentrations assigned to elements in geostandards. They are classified in terms of perceived reliability as *recommended* (probably close to true), *proposed* and *information* (order of magnitude) values (Govindaraju, 1994).

X-ray   a high-energy electromagnetic **photon** having a **frequency** between 10^{17} and 10^{20} Hz or a **wavelength** between 3 pm and 3 nm (**quantum energy** between 10^2 and 10^5 eV). X-rays are the highest-energy form of electromagnetic radiation produced by *electronic* transitions in atoms.

X-ray diffraction   see Box 6.2.

XRFS   abbreviation for X-ray fluorescence spectrometry (Chapter 6).

Symbols and constants used in this book

Symbol	Meaning	Value or comment	Units
a	Repeat distance in diffraction grating	See Fig. 4.3.1	m
A	Mass number	$Z + N$	–
A_t	Activity of irradiated nucleus	–	Bq
$\alpha_{A,B}$	Isotope fractionation factor	see Box 9.1	–
B	Magnetic induction	Force exerted per ampere meter	$Wb\,m^{-2}$
C_γ	Count rate in γ-ray counting	–	s^{-1}
C_i	Concentration of constituent i		%, ppm, $kg\,m^{-3}$
c	Speed of light	2.998×10^8	$m\,s^{-1}$
d	Inter-planar spacing in diffracting crystal	–	nm
	Diameter of particle		μm
ΔE	Energy difference between electron energy levels	–	eV, $kJ\,mol^{-1}$
Δ	Mass defect	$[Z(m_p + m_e) + Nm_n] - M$	kg
D	Optical density	$(I_0 - I/I_0) = 1 - \tau$	–
$\delta^{18}O$ etc	δ notation for isotope fractionation	see Box 9.1	parts per 10^3 (‰)
e	Base of Naperian (natural) logarithms	2.17823	–
e^+	Charge of proton	1.602×10^{-19}	C
e^-	Charge of electron	-1.602×10^{-19}	C
E	First ionisation energy	–	eV or $kJ\,mol^{-1}$
E	Electric field strength	Force exerted per unit charge	$N\,C^{-1}$
E_q	Quantum energy or photon energy	–	eV or $J\,mol^{-1}$
ϵ	Detector efficiency	–	–
f_i	Relative sensitivity factor for element i	–	–
F	Faraday constant	9.649×10^4	$C\,mol^{-1}$
ϕ	Angle of incidence of light ray	–	rad or °
ϕ_n	Neutron flux		$m^{-2}\,s^{-1}$
h	Planck constant	6.626×10^{-34}	$J\,s$
H	Magnetic field strength	B/μ_0	$A\,m^{-1}$
i	General integer	–	–
I	Intensity of light beam or spectral line	–	(arbitrary units)
j	General integer	–	–
$\mathcal{J}$	Neutron flux parameter	see Box 9.2	–
k	Boltzmann constant	1.381×10^{-23}	$J\,K^{-1}$
	General constant proportionality	–	–
ln	Natural logarithm (to base e)	$= \log_e$	–
log	Logarithm to base 10	$= \log_{10}$	–
L	Avogadro number (or constant)	6.0221×10^{23}	mol^{-1}
λ	Decay constant	–	s^{-1}, $year^{-1}$
	Wavelength	–	nm, μm
m	Mass	–	kg
m_i	Mass of element of species i	–	kg
m_e	Mass of electron	$9.110 \times 10^{-31} kg = 0.5486 \times 10^{-3}$	amu
m_p	Mass of proton	$1.673 \times 10^{-27} kg = 1.0073$	amu

Symbol	Meaning	Value or comment	Units
m_n	Mass of neutron	1.675×10^{-27} kg = 1.0087	amu
M	Relative atomic mass	formerly 'atomic weight'	amu
$\underline{M}$	Relative molecular mass	Formerly 'molecular weight'	amu
μ	Hypothetical population mean of all possible measurements of a given value	–	*
μ	Total mass absorption coefficient (at a specified wavelength) of a compound material (e.g. rock sample)	–	$cm^2\,g^{-1}$
μ_i	Mass absorption coefficient of component i at specified wavelength	–	$cm^2\,g^{-1}$
n	General integer		
	Number of measurements, samples	–	–
	Principal quantum number	$1, 2, 3, 4, 5, \ldots$	–
N	Neutron number of an isotope	–	–
N	Number of target nuclei present in sample	–	–
ν	Frequency	–	$Hz\,(=s^{-1})$
pH	Acidity or alkalinity of solution	$-\log_{10}[H^+]$	dimensionless
P	Pressure	–	Pa
P_b	Branching ratio of radioactive decay		
π	Circumference: diameter ratio of a circle	3.141592	–
q	Charge		
Q	Intensity of ion beam	–	(arbitrary units)
r	Radius	–	m
r_e	Radius of curvature of ion moving in an electric field		m
r_m	Radius of curvature of ion moving in a magnetic field		m
R	Gas constant	$8.314\,J\,K^{-1}\,mol^{-1}$	
R_{ij}	Abundance ratio of isotopes i and j	–	Dimensionless
ρ	Density	mass/volume	$g\,cm^{-3}$, $kg\,dm^{-3}$
s^2	variance		
σ	Standard deviation of the hypothetical population distribution of all possible readings of a value	–	*
σ^2	Variance of population	–	*
σ_n	Neutron-capture cross-section		m^2
σ_s, $\sigma_s{}^2$	Sampling precision, sampling variance		
se	Standard error on the mean	–	*
Σ	Summation of a number of terms		
$\sum_{i=1}^{n}$	Summation of a series of n terms, each identified by successive values of the integer i varying from 1 to n		
ΣFeO	The total iron content of a rock sample expressed in the form of % ferrous oxide		
t	Time	–	s
	t-statistic for significance testing	see Box 2.1	
t_d	Time between irradiation and counting	–	s
t_i	Irradiation time	–	s
t_D	Dead time	–	ns
$t_{1/2}$	Half-life	$= (\ln 2)/\lambda$	s, year
T	Temperature	–	K or °C
τ	Optical transparency	$1 - D$	–
θ	Angle of diffraction		rad or °
θ_i	Fraction of specific isotope in element	–	–
v	Velocity	–	$m\,s^{-1}$
V	Volume	–	m^3
U	Electric potential (or potential difference)	–	V

Symbol	Meaning	Value or comment	Units
W	work function	–	$eV, kJ\,mol^{-1}$
$\bar{x}$	Sample mean of a finite series of measurements of the same value	–	–
X	Molar or atomic fraction	–	–
$X_{j,s}$	Abundance of isotope j in element s	–	–
z	Net charge on an ion (+ve or −ve)	(in electron–charge units)	–
$ze^{+/-}$	Net charge on an ion (+ve or −ve)	–	C
Z	Atomic number	–	–
$\bar{Z}$	Mean atomic number of substance	–	amu

*Units are those of the variable being measured.

Bold type denotes a vector quantity. Symbols denoting variables and constants are printed in *italic* (e.g. V for volume) to distinguish them from units, which are not italicised (e.g. V for volt).

SI units

Symbol	Unit	Physical quantity	Definition/equivalent units
Basic SI units			
m	metre	length	
kg	kilogram	mass	
s[†]	second	time	
A	ampere	electric current	
K	kelvin	temperature	
mol	mole	amount of substance	
rad	radian	angle	θ rad $= (\pi/180)$ $\theta° = 0.01745\theta°$
Derived SI Units			
J	joule	energy, work	$\mathrm{kg\,m^2\,s^{-2}}$ *
N	newton	force	$\mathrm{kg\,m\,s^{-2}} = \mathrm{J\,m^{-1}}$
Pa	pascal	pressure	$\mathrm{kg\,m^{-1}\,s^{-2}} = \mathrm{N\,m^{-2}}$
W	watt	power	$\mathrm{kg\,m^2\,s^{-3}} = \mathrm{J\,s^{-1}}$
C	coulomb	electric charge	$\mathrm{A\,s}$
V	volt	electrical potential (and difference)	$\mathrm{kg\,m^2\,s^{-3}\,A^{-1}} = \mathrm{W\,A^{-1}}$
Ω	ohm	electric resistance	$\mathrm{kg\,m^2\,s^{-3}\,A^{-2}} = \mathrm{V\,A^{-1}}$
Wb	weber	magnetic flux	$\mathrm{kg\,m^2\,s^{-2}\,A^{-1}} = \mathrm{V\,s}$
S	siemens	electrical conductance	$\Omega^{-1} = \mathrm{A^2\,s^3\,kg^{-1}\,m^{-2}}$
T	tesla	magnetic flux density	$\mathrm{kg\,s^{-2}\,A^{-1}} = \mathrm{Wb\,m^{-2}}$
H	henry	inductance	$\mathrm{kg\,m^2\,s^{-2}\,A^{-2}} = \mathrm{V\,s\,A^{-1}}$
Hz	hertz	frequency	(cycles, pulses, photons) $\mathrm{s^{-1}}$
Bq	becquerel	radioactivity	(disintegrations) $\mathrm{s^{-1}}$
Gy	gray	radiation dose	
Sv	sievert	biologically equivalent radiation dose	

Non-SI units and abbreviations retained in this book

Abbreviation	*Unit*	*Variable measured*	*SI equivalent*
%	percent by mass	mass fraction ('concentration')	$\mathrm{kg\,(100\,kg)^{-1}}$
Å	angstrom unit	length	$0.1\,\mathrm{nm} = 10^{-10}\,\mathrm{m}$
eV	electron-volt	energy state of individual atom	1.602×10^{-19} J or 96.5 kJ mol^{-1}
ml	millitre[‡]	volume (of solution)	$\approx 10^{-6}\,\mathrm{m^3} = 10^{-3}\,\mathrm{dm^3} = 1\,\mathrm{cm^3}$
ppm	parts per million	concentration (mass units)	$\mathrm{\mu g\,g^{-1}}$
ppb	parts per billion	concentration (mass units)	$\mathrm{ng\,g^{-1}}$
a	year (annum)[†]	time	3.154×10^6 s

*The SI system prefers the notation $\mathrm{m\,s^{-2}}$ to $\mathrm{m/s^2}$.

[†]Though the second is the fundamental SI unit of time, earth scientists retain the year as the unit of geological time. The preferred abbreviation is 'a' (for annum); hence 'Ga' (for gigayear $= 10^9$ a), 'Ma', etc.

[‡]In some literature the abbreviation 'mL' is used.

Prefixes for SI units

	Smaller				Larger		
Symbol	Prefix	Factor	Example	Symbol	Prefix	Factor	Example
m	milli–	10^{-3}	mV (millivolt)	k	kilo–	10^{3}	kJ (kilojoule)
μ	micro–	10^{-6}	μA (microampere)	M	mega–	10^{6}	MΩ (megaohm)
n	nano–	10^{-9}	nC (nanocoulomb)	G	giga–	10^{9}	GPa (gigapascal)
p	pico–	10^{-12}	pm (picometre)	T	tera–	10^{12}	THz (terahertz)

Bibliography

Chapter 1

Appelo, C.A.J. and Postma, D. 1994. *Geochemistry, groundwater and pollution*. Rotterdam: Balkema, xvi + 536pp.

Frontasyeva, M.V., Nazarov, V.M. and Steinnes, E. 1994. Moss as monitor of heavy metal deposition: comparison of different multi-element analytical techniques. *J. Radioanal. Nucl. Chem.* **181**, 363–371.

Gill, R. 1995. *Chemical Fundamentals of Geology*, 2nd edn. London: Chapman & Hall.

Gill, R.C.O., Nielsen, T.F.D., Brooks, C.K. and Ingram, G. 1988. Tertiary volcanism in the Kangerdlugssuaq region, E. Greenland: trace element geochemistry of the Lower Basalts and tholeiitic dyke swarms. In: Morton, A.C. and Parson, L.M. (eds): *Early Tertiary Volcanism and the Opening of the NE Atlantic*, Geol. Soc. London Spec. Publ. **39**, 161–179.

Govindaraju, K. 1994. 1994 compilation of working values and sample description for 383 geostandards. *Geostandards Newsl.* **18**, 1–58.

Miller, J.C. and Miller, J.N. 1993. *Statistics for Analytical Chemistry*, 3rd edn. New York: Ellis Horwood.

Potts, P.J., Tindle, A.G. and Webb, P.C. 1992. *Geochemical Reference Material Compositions. Rocks, minerals, sediments, soils, carbonates, refractories and ores used in research and industry*. Boca Raton, FL: CRC Press.

Thompson, M. and Howarth, R.J. 1976. Duplicate analysis in geochemical practice. Part 1: theoretical approach and estimation of analytical reproducibility. *Analyst* **101**, 690–698.

Chapter 2

Allman, M. and Lawrence, D.F. 1972. *Geological laboratory techniques*. London: Blandford.

Argyraki, A., Ramsey, M.H. and Thompson, M. 1995. Proficiency testing in sampling: pilot study on contaminated land. *Analyst* **120**, 2799–2805.

Barcelona, M.J. 1988. Overview of the sampling process. In: Keith, L.H. (ed.): *Principles of Environmental Sampling*, Washington, DC: American Chemical Society, pp 3–23.

Brooks, R.R. 1983. *Biological methods of prospecting for minerals*. New York: Wiley.

Chester, R. 1990. *Marine Geochemistry*. London: Unwin Hyman.

Clifton, E., Hunter, R.E., Swanson, F.J. and Phillips, R.L. 1969. Sampling size and meaningful gold analysis. *US Geol. Sur. Prof. Pap.* **625-C**.

Davis, O.L.1954. *Statistical Methods in Research and Production*, 2nd edn. Edinburgh: Oliver and Boyd.

DoE. 1994. *Sampling strategies for contaminated land*, CLR Report No 4. London: UK Department of the Environment.

Ferguson, C.C. 1992. The statistical basis for spatial sampling of contaminated land *Ground Eng.* **25**, 34.

Fitton, J.G. and Gill, R.C.O. 1970. The oxidation of ferrous iron in rocks and minerals during mechanical grinding. *Geochim. Cosmochim. Acta* **34**, 518–524.

Gregory, M.R. and Johnston, K.A. 1987. A nontoxic substitute for hazardous heavy liquids – aqueous sodium polytungstate ($3Na_2WO_4.9WO_3.H_2O$) solution. *New Zealand J. Geol. Geophys.* **30**, 317–320.

Gy, P.M. 1979. *Sampling of particulate materials. Theory and practice.* New York: Elsevier.

Hutchison, C.S. 1974. *Laboratory handbook of petrographic techniques.* New York: Wiley.

Ingamells, C.O. and Switzer P. 1973. A proposed sampling constant for use in geochemical analysis. *Talanta* **20**, 547–568.

Keith, L.H. (ed.) 1988. *Principles of Environmental Sampling.* Washington, DC: American Chemical Society.

Keith, L.H. 1991. *Environmental Sampling and Analysis: A practical guide*, Chelsea MI: Lewis Publishers.

Kratochvil, B., Wallace D. and Taylor, J.K. 1984. Sampling for chemical analysis. *Anal. Chem.* **56**, 113R–129R.

Markert, B. (ed.) 1994. *Sampling of environmental materials for trace analysis.* Weinheim: VCH.

Miller, J.C. and Miller, J.N. 1988. *Statistics for analytical chemistry*, 2nd edn. Chichester: Ellis Horwood.

Miller, J.C. and Miller, J.N. 1993. *Statistics for analytical chemistry*, 3rd edn. Chichester: Ellis Horwood.

Ministry of Agriculture, Fisheries and Food 1986. *The analysis of agricultural materials*, 3rd edn. London: HMSO.

Munsell Colour Company Inc. 1971. *Munsell Soil Colour Chart.* Baltimore, Maryland 21218, USA.

Paetz, A. and Crobmann, G. 1994. Problem and results in the development of international standards for sampling and pretreatment of soils. In: Markert, B. (ed.): *Sampling of environmental materials for trace analysis.* Weinheim: VCH.

Potts, P.J. 1987. *A Handbook of Silicate Rock Analysis.* Glasgow: Blackie, pp. 19–24.

Ramsey, M.H. 1994. Error estimation in environmental sampling and analysis. In: Markert, B. (ed.): *Sampling of environmental materials for trace analysis.* Weinheim: VCH.

Ramsey, M.H., Dong, D., Thornton, I., Watt, J. and Giddens, R. 1991. Discrimination between aluminium held within vegetation and that contributed by soil contamination - using a combination of electron probe micro analysis (EPMA) and inductively coupled plasma-atomic emission spectrometry (ICP-AES). *J. Environ. Geochem. Health* **13**, 114–118.

Smith, R. and James, G.V. 1981. *The sampling of bulk materials.* London: The Royal Society of Chemistry.

Wager, L.R. and Brown, G.M. 1960. Collection and preparation of material for analysis. In: Smales, A.A. and Wager, L.R. (eds): *Methods in Geochemistry.* New York: Interscience, 4–33.

Webster, R. 1977. *Quantitative and numerical methods in soil classification and survey.* Oxford: Oxford University Press.

Chapter 3

Arden, J.W. and Gale, N.H. 1974. New electrochemical technique for the separation of lead at trace levels from natural silicates. *Anal. Chem.* **46**, 2–9.

Bock, R. 1979. *Handbook of decomposition methods in analytical chemistry.* Glasgow: Blackie.

Cremer, M. and Schloker J. 1976. Lithium borate decomposition of rocks, minerals and ores. *Am. Mineral.* **61**, 318–321.

Govindaraju, K and Mevelle, G. 1987. Fully automated dissolution and separation methods for inductively coupled plasma atomic emission spectrometry rock analysis. Application to the determination of rare earth elements. *J. Anal. At. Spectrom.* **2**, 615–621.

Harland, C.E. 1994. *Ion exchange: theory and practice.* Cambridge: Royal Society of Chemistry.

Hooker, P.J., O'Nions, R.K. and Pankhurst, R.J. 1975. Determination of rare earth elements in USGS standard rocks by mixed-solvent ion exchange and mass spectrometric isotope dilution. *Chem. Geol.* **16**, 189–196.

Ingamells, C.O. 1970. Lithium metaborate flux in silicate analysis. *Anal. Chim. Acta* **52**, 323–334.

Jarvis, I. 1992. Sample preparation for ICP-Ms. In: Jarvis, K.E. *et al.* (eds): *Handbook of Inductively Coupled Plasma–Mass Spectrometry*. Glasgow: Blackie, pp. 172–224.

Jeffery, 1970. *Chemical methods of rock analysis*. Oxford: Pergamon, 509pp.

Johnson, W.M and Maxwell, J.A. 1981. *Rock and Mineral Analysis*. New York: Wiley, 489pp.

Jolly, S.C. (ed.) 1963. Metallic impurities in organic matter. In: *Official, Standardised and Recommended Methods of Analysis*. Cambridge: Heffer, 3–19.

Krogh, T.E. 1973. A low-contamination method for hydrothermal decomposition of zircon and extraction of U and Pb for isotopic age determinations. *Geochim. Cosmochim. Acta* 37, 485–494.

Meier, A.L. 1980. *J. Geochem. Explor.* 13, 77.

Potts, P. J. 1987. *A Handbook of Silicate Rock Analysis*. Glasgow: Blackie, 622pp.

Sulcek, Z. and Povondra, P. 1989. *Methods of Decomposition in Inorganic Analysis*. Boca Raton, FL: CRC Press, 325pp.

Thompson, M. and Walsh, J. N. 1989. *A Handbook of Inductively Coupled Plasma Spectrometry*. Glasgow: Blackie, 316pp.

Chapter 4

Atkins, P.W. 1994. *Physical Chemistry*, 5th edn. Oxford: Oxford University Press.

Boumans, P.W.J.M. 1980. *Line Coincidence Tables for Inductively Coupled Plasma Atomic Emission Spectroscopy*. Oxford: Pergamon.

Ebdon, L. 1982. *An Introduction to Atomic Absorption Spectrometry*. Philadelphia: Heyden, 138pp.

Gill, R. 1995. *Chemical Fundamentals of Geology*, 2nd edn. London: Chapman and Hall.

Greenfield, S., Jones, I.Ll. and Berry, C.T. 1964. High-pressure plasmas as spectroscopic emission sources. *Analyst* 89, 713–720.

Nakamura, N. 1974. Determination of REE, Ba, Fe, Mg, Na and K in carbonaceous and ordinary chondrites. *Geochem. Cosmochim. Acta* 38, 575–775.

Parsons, M.L., Forster, A.R. and Anderson, D. 1980. *An Atlas of Spectral Interferences in ICP Spectroscopy*. New York: Plenum.

Potts, P.J. 1987. *A Handbook of Silicate Rock Analysis*. Glasgow: Blackie.

Reed, T.B. 1961. Induction-coupled plasma torch. *J. Appl. Phys.* 32, 821–824.

Rosler, H.J. and Lange, H. 1972. *Geochemical Tables*. Amsterdam: Elsevier, 468pp.

Shriver, D.F., Atkins, P.W. and Langford, C.H. 1994. *Inorganic Chemistry*, 2nd edn. Oxford: Oxford University Press.

Thompson, M. and Walsh, J.N. 1989. *Handbook of Inductively Coupled Plasma Spectrometry*. London: Chapman & Hall, 316pp.

Walsh, J.N. 1992. Use of multiple internal standards for high-precision, routine analysis of geological samples by inductively coupled plasma–atomic emission spectrometry. *Chem. Geol.* 95, 113–121.

Walsh, J.N., Buckley, F. and Barker, J. 1981. The simultaneous determination of the rare earth elements in rocks using inductively coupled plasma source spectrometry. *Chem. Geol.* 33, 141–153.

Wendt, R.H. and Fassel,V.A. 1965. Induction coupled plasma spectrometric excitation source. *Anal. Chem.* 37, 920–922.

Winge, R.K., Fassel, V.A., Peterson, V.J. and Floyd, M.A. 1984. *Inductively Coupled Plasma–Atomic Emission Spectroscopy. An Atlas of Spectral Information*. Amsterdam: Elsevier.

Chapter 5

Allen, S.E. (ed.) 1995. *Chemical Analysis of Ecological Materials*. Oxford: Blackwell.

Cape, J.N. 1987. Non-suppressed ion chromatography in acid rain analysis. In: Rowland, A.P. (ed.): *Chemical Analysis in Environmental Research*. Huntingdon: Institute of Terrestrial Ecology.

Cresser, M.S. 1994. *Flame Spectrometry in Environmental Chemical Analysis: A Practical Guide*. Cambridge: The Royal Society of Chemistry.

Cresser, M.S., Armstrong, J., Dean, J.R., Watkins, P. and Cave, M. 1992. Atomic spectrometry update – environmental analysis. *J. Anal. At. Spectrom.* 8, 1R–66R.

Cresser, M.S., Armstrong, J., Cook, J.M., Dean, J.R., Watkins, P. and Cave, M. 1995. Atomic spectrometry update – environmental analysis. *J. Anal. At. Spectrom.* 10, 9R–60R.

Davidson, W. 1987. Measuring the pH of fresh water. In: Rowland, A.P. (ed.): *Chemical Analysis in Environmental Research*. Huntingdon: Institute of Terrestrial Ecology.

Drever, J.I. 1988. *The Geochemistry of Natural Waters*, 2nd edn. Englewood Cliffs, NJ: Prentice Hall.

Driscol, C.T. 1984. A procedure for the fractionation of aqueous aluminium in acidic waters. *Int. J. Environ. Anal. Chem.* 16, 267–283.

Fang, Z. 1995 *Flow-injection atomic absorption spectrometry*. Chichester: Wiley.

Farago, M.E. and Mehra, A. 1994. Analytical techniques for plant analysis. In: Farago, M.E. (ed.): *Plants and the Chemical Elements: Biochemistry, uptake, tolerance and toxicity*. Weinheim: VCH.

Farah, K.S. and Sneddon, J. 1993. Simultaneous determination of copper, iron, manganese and zinc in bovine liver and estuarine liver using flame atomic absorption spectrometry with background correction. *Anal. Lett.* 26, 709–719.

Fuller, C.W. 1979. *Electrothermal Atomisation for Atomic Absorption Spectrometry*. London: The Chemical Society.

Gran, G. 1950. Determination of the equivalence point in potentiometric titrations. *Acta Chem. Scand.* 4, 559–567.

Gran, G. 1952. Determination of the equivalence point in potentiometric titrations Part II. *Analyst* 77, 661–671.

Greenberg, A.E., Clesceri, L.S. and Eaton, A.D. (eds) 1992. *Standard Methods for the Examination of Water and Wastewater*, 18th edn. Washington, DC: APHA, AWWA and WEF.

Grimshaw, H.M. 1989. Analysis of soils. In: Allen, S.E. (ed.): *Chemical Analysis of Ecological Materials*. Oxford: Blackwell.

Haygarth, P.M., Rowland, A.P., Sturup, S. and Jones, K.C. 1993. Comparison of instrumental methods for the determination of total selenium in environmental samples. *Analyst* 118, 1303–1308.

Hem, J.D. 1982. Conductance: a collective measure of dissolved ions. In: Minear, R.A. and Keith, L.H. (eds): *Water Analysis*, Vol. 1, Inorganic Species, Part 1. New York: Academic Press.

Hill, S.B., Dawson, J.B., Price, W.J., Riby, P., Shuttler, I.L. and Tyson, J.F. 1994. Atomic spectrometry update – advances in atomic absorption and fluorescence spectrometry and related techniques. *J. Anal. At. Spectrom.* 9, 213R–265R.

Howard, A.G. and Arbab-Zavar, M.H. 1981. Determination of 'inorganic' arsenic(III) and arsenic(V), 'methylarsenic' and 'dimethylarsenic' species by selective hydride evolution atomic-absorption spectroscopy. *Analyst* 106, 213–220.

Howard, A.G. and Comber, S.D.W. 1992. Hydride-trapping techniques for the speciation of arsenic. *Mikrochim. Acta* 109, 27–33.

Hunt, L.E. and Howard, A.G. 1994. Arsenic speciation and distribution in the Carnon estuary following the acute discharge of contaminated water from a disused mine. *Mar. Pollut. Bull.* 28, 33–38.

Jackson, L.L., Baedecker, P.A., Fries, T.L. and Lamothe, P.J. 1995. Geological and inorganic materials. *Anal. Chem.* 67, 71R–85R.

Kane, J.S. 1988. Optimisation of flame parameters for simultaneous multi-element atomic absorption spectrometric determination of trace elements in rocks. *J. Anal. At. Spectrom.* 3, 1039–1045.

Kramer, J.R. 1982. Alkalinity and acidity. In: Minear, R.A. and Keith, L.H. (eds): *Water Analysis*, Vol. 1, Inorganic Species, Part 1. New York: Academic Press.

Lin, S.-I. and Hwang, H.-P. 1993. The design of an online flow-injection system with a gravitational phase separator for flame atomic absorption spectrometer and its analytical performance. *Talanta* 40, 1077–1083.

Lopez Garcia, I., Arroyo Cortez, J. and Hernandez Cordoba, M. 1993. Flow injection flame atomic absorption spectrometry for slurry atomization. Determination of iron, calcium and magnesium in samples with high silica content. *Talanta* 40, 1677–1685.

Luecke, W. 1992. Anionic matrix interference on alkali elements in atomic absorption spectrometry – its significance to silicate analyses. *Chem. Geol.* 98, 323.

Lust, A. 1994. Atomic spectroscopy bibliography, January–June 1994. *At. Spectrosc.* 15, 169–201.

Methods for the Examination of Waters and Associated Materials (MEWAM) 1979. *The Measurement of Electrical Conductivity and the Laboratory Determination of the pH value of Natural, Treated and Waste Waters 1978*. London: HMSO.

Methods for the Examination of Waters and Associated Materials (MEWAM) 1980. *Atomic Absorption Spectrophotometry 1979 version; an essay review*. London: HMSO.

Methods for the Examination of Waters and Associated Materials (MEWAM) 1981. *The Determination of Alkalinity and Acidity in Water 1981*. London: HMSO.

Methods for the Examination of Waters and Associated Materials (MEWAM) 1987. *Lead and Cadmium in Fresh Waters by Atomic Absorption Spectrophotometry (Second Edition). A General Introduction to Electrothermal Atomisation Atomic Absorption Spectrophotometry*. London: HMSO.

Methods for the Examination of Waters and Associated Materials (MEWAM) 1988. *The Determination of pH in Low Ionic Strength Waters 1988*. London: HMSO.

Methods for the Examination of Waters and Associated Materials (MEWAM) 1990. *The Determination of Anions and Cations, Transition Metals, Other Complex Ions and Organic Acids and Bases in Water by Chromatography 1990*. London: HMSO.

Potts, P.J. 1987. *A Handbook of Silicate Rock Analysis*. Glasgow: Blackie.

Sen Gupta, J.G. 1993. Twenty-five years of research and application of atomic absorption spectroscopy to geoanalysis in the Geological survey of Canada. *Can. J. Appl. Spectrosc.* **38**, 145–149.

Smith, S.B. and Hieftje, G.M. 1983. A new background correction method for atomic absorption spectrometry. *At. Spectrosc.* **37**, 419–424.

Stockwell, P.B. and Corns, W.T. 1994. Environmental sensors based on atomic fluorescence. *Analyst* **119**, 1641–1645.

Vander Voet, A.H.M. and Riddle, C. 1993. *The Analysis of Geological Materials Volume 1: A Practical Guide*. Ontario Geological Survey Miscellaneous Paper 149.

Whiteside, P.J. 1979. *An introduction to atomic absorption spectrometry*. Cambridge: Pye Unicam.

Wilson, A.D. 1955. Determination of ferrous iron in rocks and minerals. *Bull. Geol. Surv. Great Britain* **9**, 56–58.

Yao, J. and Huang, W. 1986. Use of different graphite tubes in the determination of Be in rocks by graphite furnace atomic absorption spectrometry. *Fenxi Huaxue* **14**, 273.

Chapter 6

Anderman, G. and Kemp, J.W. 1958. Scattered X-rays as internal standards in X-ray emission spectroscopy. *Anal. Chem.* **30**, 1306–1309.

Gill, R. 1995. *Chemical Fundamentals of Geology*, 2nd edn. London: Chapman & Hall.

Govindaraju, K. 1994. 1994 compilation of working values and sample description for 383 geostandards. *Geostandards Newsl.* **18**, 1–158.

Klockenkaemper, R. and von Bohlen, A. 1992. Total reflection X-ray fluorescence – an efficient method for micro-, trace and surface layer analysis. *J. Anal. At. Spetrom.* **7**, 273–279.

Klockenkaemper, R., Knoth, J., Prange, A. and Schwenke, H. 1992. Total reflection X-ray fluorescence spectrometry. *Anal. Chem.* **64**, 1115A-1122A.

Norrish, K. and Hutton, J.T. 1969. An accurate X-ray spectrographic method for the analysis of a wide range of geological samples. *Geochim. Cosmochim. Acta* **33**, 431–453.

Reynolds, R.C. 1963. Matrix corrections in trace element analysis by X-ray fluorescence: estimation of the mass absorption coefficient by Compton scattering. *Am. Mineral.* **48**, 1133–1143.

Smith, J.V. and Rivers, M.L. 1995. Synchrotron X-ray microanalysis. In: Potts, P.J. *et al.* (eds): *Microprobe Techniques in the Earth Sciences*. London: Chapman & Hall, pp.163–233.

Tertian, R. and Claisse, F. 1994. *Principles of Quantitative X-ray Fluorescence Analysis*. Chichester: Wiley, 385 pp.

Chapter 7

Aarnio, P.A., Hakulinen, T.T. and Routti, J.T. 1992. Expert system for nuclide identification and interpretation of gamma spectrum analysis. *J Radioanal. Nucl. Chem.* **160**, 245–252.
Alfassi, Z.B. 1985. Epithermal neutron activation analysis. *J. Radioanal. Nucl. Chem.* **96**, 151–165.
Alfassi, Z.B. (ed.) 1989. *Activation Analysis.* Boca Raton, Fl: CRC Press.
Amiel, S. (ed.) 1981a. *Non-destructive Activation Analysis.* Amsterdam: Elsevier.
Amiel, S. 1981b. Neutron counting in activation analysis. In: Amiel, S. (ed.): *Non-destructive Activation Analysis.* Amsterdam: Elsevier, pp. 43–52.
Browne, E. and Firestone, R.B. 1986. *Table of Radioactive Isotopes.* New York: Wiley.
Busche, F.D. 1989. Using plants as an exploration tool for gold. *J. Geochem. Explor.* **32**, 199–209.
Dams, R., Robbins, J.A., Rahn, K.A. and Winchester, J.W. 1970. Non-destructive neutron activation analysis of air pollution particulates. *Anal. Chem.* **42**, 861–868.
Das, H.A. Faanhof, A., and van der Sloot, H.A. 1989a. *Environmental Radioanalysis.* Amsterdam: Elsevier.
Das, H.A., Faanhof, A. and van der Sloot, H.A. 1989b. *Radioanalysis in Geochemistry.* Amsterdam: Elsevier.
Ehmann, W.D. and Vance, D.E. 1991. *Radiochemistry and Nuclear Methods of Analysis.* New York: Wiley.
Guinn, V. P. 1980. Cyclic nuclear activation analysis. *Radiochem. Radioanal. Lett.* **44**, 133–138.
Hoffman, E.L. 1992. Instrumental neutron activation in geoanalysis. *J. Geochem. Explor.* **44**, 297–319.
McConnell, J. and Davenport, P. 1989. Gold and associated trace elements in Newfoundland lake sediment: their application to gold exploration. *J Geochem Explor.* **32**, 33–50.
Parry, S.J. 1991. *Activation Spectrometry in Chemical Analysis.* New York: Wiley.
Peisach, M. 1981. Prompt techniques. In: Amiel, S. (ed.): *Non-destructive Activation Analysis.* Amsterdam: Elsevier, pp. 93–111.
Schaug, J., Rambaek, J.P., Steinnes, E. and Henry, R.C. 1990. Multivariate analysis of trace element data from moss samples used to monitor atmospheric deposition. *Atmos. Environ.* **24A**, 2625–2631.
Spyrou, N.M. 1981. Cyclic activation analysis – a review. *J. Radioanal. Chem.* **61**, 211–242.
Steinnes, E. 1980. Atmospheric deposition of heavy metals in Norway studied by the analysis of moss samples using neutron activation analysis and atomic absorption spectrometry. *J. Radioanal. Chem.* **58**, 387–391.
Tolgyessy, J. and Klehr, E.H. 1987. *Nuclear Environmental Chemical Analysis.* Chichester: Ellis Horwood.
Wyttenbach, A., Bajo, S. and Tobler, L. 1987. Aerosols deposited on spruce needles. *J. Radioanal. Nucl. Chem.* **114**, 137–145
Zikovsky, L. and Badillo, M. 1987. An indirect study of air pollution by neutron activation analysis of snow. *J. Radioanal. Nucl. Chem.* **114**, 147–153.

Chapter 8

Belshaw, N.S. and O'Nions, R.K. 1990. The Cambridge ISOLAB 120: An assessment of abundance sensitivity. *Adv. Inorg. Mass Spectrom.* **1**, 1–3.
Belshaw, N.S., O'Nions, R.K. and Vonblanckenburg, F. 1995. A SIMS method for Be-10/Be-9 ratio measurement in environmental materials. *Int. J. Mass Spectrom. Ion Process.* **142**, 55–67.
Cameron, A.E., Smith, D.H. and Waler, R.L. 1969. Mass spectrometry of nanogram-size samples of lead. *Anal. Chem.* **41**, 525.
Christensen, J.N., Halliday, A.N., Lee, D.C. and Hall, C.M. 1995. In-situ Sr isotopic analysis by laser-ablation. *Earth Planet. Sci. Lett.* **136**, 79–85.

Compston, W. and Oversby, V.M. 1969. Lead isotope analysis using a double spike. *J. Geophys. Res.* **74**, 4338–4348.

Dickin, A.P. 1995. *Radiogenic isotope geology*. Cambridge: Cambridge University Press.

Faure, G. 1987. *Principles of Isotope Geology*, 2nd edn, New York: Wiley.

Kober, B. 1987. Single-zircon evaporation combined with Pb+ emitter bedding for $^{207}Pb/^{206}Pb$-age investigations using thermal ion mass spectrometry, and implications to zirconology. *Contrib. Miner. Petrol.* **96**, 63–71.

Lee, D.C. and Halliday, A.N. 1995. Precise determinations of the isotopic compositions and atomic weights of molybdenum, tellurium, tin and tungsten using ICP magnetic-sector multiple collector mass-spectrometry. *Int. J. Mass Spectrom. Ion Process.* **146**, 35–46.

Makishima, A. and Nakamura E. 1991. Calibration of Faraday cup efficiency in a multicollector mass spectrometer. *Chem. Geol. (Isot. Geosci. Sect.)* **14**, 105–110.

Martin, C.E. 1991. Osmium isotopic characteristics of mantle-derived rocks. *Geochim. Cosmochim. Acta* **55**, 1421–1434.

Nier, A.O. 1940. A mass spectrometer for routine isotope abundance measurements. *Rev. Sc. Instrum.* **11**, 212–216.

Palacz, Z., Haines, C. and Turner, P. 1995. Effective lifetime of new Faraday cups. *VG Isotech Techn. Note* 307.

Saxton, J.M., Lyon, I.C. and Turner, G. 1995. High-precision, in-situ oxygen-isotope ratio measurements obtained from geological and extraterrestrial materials using an Isolab-54 ion microprobe. *Analyst* **120**, 1321–1326.

Steiger, R.H. and Jaeger, E. 1977. Subcommission on geochronology: convention on the use of decay constants in geo- and cosmochronology. *Earth Planet. Sci. Lett.* **36**, 359–362.

Thirlwall, M.F. 1982. A triple filament method for rapid and precise analysis of rare earth elements by isotope dilution. *Chem. Geol.* **35**, 155–166.

Thirlwall, M.F. 1991a. High precision multicollector isotope analysis of low levels of Nd as oxide. *Chem. Geol. (Isot. Geosci. Sect.)* **94**, 13–22.

Thirlwall, M.F. 1991b. Long-term reproducibility of multicollector Sr and Nd isotope ratio analysis. *Chem. Geol. (Isot. Geosci. Sect.)* **94**, 85–104.

Thirlwall, M.F. and Walder, A.J., 1995. In situ hafnium isotope ratio analysis of zircon by inductively coupled plasma multiple collector mass spectrometry. *Chem. Geol. (Isot. Geosci. Sect.)* **122**, 241–247.

Walder, A.J., Platzner, I. and Freedman, P.A. 1993. Isotope ratio measurement of lead, neodymium and neodymium-samarium mixtures, hafnium and hafnium-lutetium mixtures with a double focusing multiple collector inductively coupled plasma mass spectrometer. *J. Anal. At. Spectrom.* **8**, 19–23.

Wasserburg, G.J., Jacobsen, S.B., DePaolo, D.J., McCulloch, M.T. and Wen, T. 1981. Precise determination of Sm/Nd ratios, Sm and Nd isotopic abundances in standard solutions. *Geochim. Cosmochim. Acta* **45**, 2311–2323.

Woodhead, J.D., Volker, F. and McCulloch, M.T. 1995. Routine lead-isotope determinations using a Pb-207 Pb-204 double spike – a long-term assessment of analytical precision and accuracy. *Analyst* **120**, 35–39.

Chapter 9

Baker, J., Snee, L. and Menzies, M. 1996. A brief Oligocene period of flood volcanism in Yemen; implications for the duration and rate of continental flood volcanism at the Afro-Arabian triple junction. *Earth Planet. Sci. Lett.*, **138**, 39–55.

Craig, H. 1957. Isotopic standards for carbon and oxygen and correction factors for mass spectrometric analysis of carbon dioxide. *Geochim. Cosmochim. Acta* **12**, 133–149.

Faure, G. 1986. *Principles of isotope geology*, 2nd edn. New York: Wiley, pp. 405–428.

Friedman, I. and O'Neil., J.R. 1977. Compilation of stable isotope fractionation factors of geological interest. In: Fleischer, M. (ed): *Data of Geochemistry*, 6th edn, Chapter KK, US Geol. Survey. Prof. Pap. 440-KK, 12pp. and 49 figs.

Hoefs, J. 1987. *Stable Isotope Geochemistry* 3rd edn. Berlin: Springer-Verlag.

Pillinger, C.T. 1984. Light element stable isotopes in meteorites – from grams to picograms. *Geochim. Cosmochim. Acta*, **48**, 2739–2766.

Sharp, Z.D. 1992. In situ laser microprobe techniques for stable isotope analysis. *Chem. Geol. (Isot. Geosci. Sect.)*, **101**, 3–9.

Valley, J.W., Taylor H.P. Jr and O'Neil, J.R. (eds) 1990. *Reviews in Mineralogy Volume 16: Stable isotopes in high-temperature geological processes*. Washington, DC: Min. Soc. Am.

Chapter 10

Bakowska, E., Falkner, K., Barnes, R.M. and Edmond, J.M. 1989. Sample handling of gold at low concentration in limited volume solution preconcentrated from seawater for inductively coupled plasma mass spectrometry. *Appl. Spectrosc.*, **43**, 1283–1286.

Colodner, D.C., Boyle, E.A., Edmond, J.M. and Thompson, J. 1992. Post depositional mobility of platinum, iridium and rhenium in marine sediments. *Nature* **358**, 402–404.

Gregoire, D.C. 1987. Influence of instrument parameters on non-spectroscopic interferences in inductively coupled plasma mass spectrometry. *Appl. Spectrosc.* **41**, 897–903.

Hall, G.E.M., Pelchat, J.C. and Loop, J. 1990. Determination of zirconium, niobium, hafnium and tantalum at low levels in geological materials by inductively coupled plasma mass spectrometry. *J. Anal. At. Spectrom.* **5**, 339–349.

Houk, R.S. and Thompson, J.J. 1988. Inductively coupled plasma mass spectrometry. In: Gross, H.L. (ed.): *Mass Spectrometry Reviews* 7, New York: Wiley, pp. 425–461.

Jackson, S.E., Fryer, B.J., Gosse, W., Healey, D.C., Longerich, H.P. and Strong, D.F. 1990. Determination of the precious metals in geological materials by inductively coupled plasma mass spectrometry (ICP-MS) with nickel sulphide fire-assay collection and tellurium co-precipitation. *Chem. Geol.* **83**, 119–132.

Jarvis, I. 1992. Sample preparation for ICP-MS. In: Jarvis, K.E. *et al.* (eds): *Handbook of Inductively Coupled Plasma Mass Spectrometry*, Glasgow: Blackie, pp. 173–224.

Jarvis, K.E. 1989. Determination of the rare earth elements in geological samples by inductively coupled plasma mass spectrometry. *J. Anal. At. Spectrom.* **4**, 563–570.

Jarvis, K.E. and Williams, J.G. 1993. Laser ablation inductively coupled plasma mass spectrometry (LA-ICP-MS): a rapid technique for the direct, quantitative determination of major, trace and rare earth elements in geological samples. *Chem. Geol.* **106**, 251–262.

Jarvis, K.E., Gray, A.L. and Houk, R.S. (eds) 1992. *A Handbook of Inductively Coupled Plasma Mass Spectrometry*, Glasgow: Blackie, 380pp.

Jarvis, K.E., Gray, A.L. and McCurdy, E. 1989. Avoidance of spectral interference on europium in inductively coupled plasma mass spectrometry by sensitive measurement of the doubly charged ion. *J. Anal. At. Spectrom.* **4**, 743–747.

Juvonen, R., Kallio, E. and Lakonaa, T. 1994. Determination of precious metals in rocks by inductively coupled plasma mass spectrometry using nickel sulphide concentration. Comparison with other pre-treatment methods. *Analyst* **119**, 617–621.

Lichte, F.E., Wilson, S.M., Brooks, R.R., Holzbecher, J. and Ryan, D.E. 1986. New method for the measurement of osmium isotopes applied to a New Zealand Cretaceous/Tertiary boundary shale. *Nature* **332**, 816–817.

McLaren, J.W., Lam, J.W.H., Berman, S.S., Akatsuka, K. and Azerado, M.A. 1993. On-line method for the analysis of seawater for trace elements by inductively coupled plasma mass spectrometry. *J. Anal. At. Spectrom.* **8**, 279–286.

Sun, M., Jain, J.S., Zhou, M.F. and Kerrich, R. 1993. A procedural modification for enhanced recovery of precious metals (Au, PGE) following nickel sulphide fire assay and Te-coprecipitation. Application for analysis of geological samples by ICP-MS. *Can. J. Appl. Spectrosc.* **38**, 103–108.

Williams, J.G. 1992. Instrument options. In: Jarvis, K.E., Gray, A.L. and Houk, R.S. (eds): *Handbook of Inductively Coupled Plasma Mass Spectrometry*, Glasgow: Blackie, pp. 58–80.

Chapter 11

Ahearn, A.J. (ed.) 1972. *Trace analysis by mass spectrometry*. New York: Academic Press, 460pp.

Anders, E, and Grevesse, N. 1989. Abundances of the elements: meteoritic and solar. *Geochim. Cosmochim. Acta* **53**, 197–214.

Govindaraju, K. 1994. 1994 compilation of working values and sample description for 383 geostandards. *Geostand. Newsl.* **18**, 1–58.

Hofmann, A.W. 1988. Chemical differentiation of the Earth: the relationship between mantle, continental crust, and oceanic crust. *Earth Planet. Sci. Lett.* **90**, 297–314.

Hofmann, A.W., Jochum, K.P., Seufert. M. and White, W.M. 1986. Nb and Pb in oceanic basalts: new constraints on mantle evolution. *Earth Planet. Sci. Lett.* **79**, 33–45.

Jochum, K.P., Seufert, H.M., Midinet-Best, S., Rettmann, E., Schönberger, K. and Zimmer, M. 1988. Multi-element analysis by isotope dilution-spark source mass spectrometry (ID-SSMS). *Fresenius Z. Anal. Chem.* **331**, 104–110.

Jochum, K.P., Seufert, H.M. and Thirlwall, M.F. 1990. Multi-element analysis of 15 international standard rocks by isotope-dilution spark source mass spectrometry. *Geostand. Newsl.* **14**, 469–473.

Jochum, K.P., Arndt, N.T. and Hofmann, A.W. 1991. Nb-Th-La in komatiites and basalts: constraints on komatiite petrogenesis and mantle evolution. *Earth Planet. Sci. Lett.* **107**, 272–289.

Jochum, K.P., Laue, H.-J., Seufert, H.M. and Hofmann, A.W. 1994. First analytical results using a multi-ion counting system for a spark source mass spectrometer. *Fresenius Z. Anal. Chem.* **350**, 642–644.

Matus, L., Seufert, H.M. and Jochum, K.P., 1994. Microanalysis of geological samples by laser plasma ionisation mass spectrometry (LIMS). *Fresenius J. Anal Chem.* **350**, 330–337.

Ramendik, G., Verlinden, J. and Gijbels, R. 1988. Spark source mass spectrometry. In: Adams, F., Gijbels, R. and van Grieken, R. (eds): *Inorganic Mass Spectrometry*, Chichester: Wiley, pp. 17–84.

Taylor, S.R. and Gorton, M.P. 1977. Geochemical application of spark source mass spectrography– III. Element sensitivity, precision and accuracy. *Geochim. Cosmochim. Acta* **41**, 1375–1380.

Taylor, S.R. and McLennan, S.M. 1985. *The continental crust: its composition and evolution*. Oxford: Blackwell, 312pp.

Chapter 12

Brown, L. 1984. Applications of accelerator mass spectrometry. *Annu. Rev. Earth Planet. Sci.* **12**, 39–59.

Brown, L. 1987. [10]Be: recent applications in Earth sciences. *Philos. Trans. R. Soc.* **A323**, 75–86.

Currie, L.A. 1992. Mankind's perturbations of particulate carbon. In: Taylor, R.E. (ed.): *Four Decades of Radiocarbon Studies: An Interdisciplinary Perspective*, Berlin: Springer-Verlag, Ch. 31, pp. 535–568.

Elmore. D, and Phillips, F.M. 1987. Accelerator mass spectrometry for measurement of long-lived radioisotopes. *Science* **236**, 543–550.

Faure, G. 1986 *Principles of isotope geology*, 2nd edn. New York: Wiley, pp. 405–428
Lal, D. and Peters, B. 1967. Cosmic-ray-produced radioactivity on the Earth. *Handb. Phys.* 4/2, 551–612, Berlin: Springer Verlag.
Long, A. (ed.) 1989. 13th International Radiocarbon Conference. *Radiocarbon* 31, 229–1082.
Morris, J. D. 1991. Applications of cosmogenic [10]Be to problems in the Earth sciences. *Annu. Rev. Earth Planet. Sci.* 19, 313–350.

Chapter 13

Broderick, B.E. 1991. Laboratory accreditation: the operation of an established scheme. *Mikrochim. Acta* III, 17–21.
Eby, G.N. 1972. Determination of the rare earths, yttrium and scandium abundances in rocks and minerals by an ion exchange X-ray fluorescence procedure. *Anal. Chem,* 44, 2137–2143.
Garfield, F.M. 1991. *Quality Assurance Principles for Analytical Laboratories.* Arlington, VA: ACOC.
Gill, R. 1995. *Chemical Fundamentals of Geology*, 2nd edn. London: Chapman & Hall.
Interim Canadian Environmental Quality Criteria for Contaminated Sites, drinking water standard, 1991.
Potts, P.J. 1987. *A handbook of silicate rock analysis.* Glasgow: Blackie, Ch. 10, pp. 326–382.
Prichard, E. 1995. *Quality in the Analytical Chemistry Laboratory.* Chichester: Wiley.
Taylor, S.R. 1982. *Planetary science: a lunar perspective.* Houston: Lunar and Planetary Institute.
Walsh, J.N., Buckley, F. and Barker, J. 1981. The simultaneous determination of the rare earth elements in rocks using inductively coupled plasma source spectrometry. *Chem. Geol.* 33, 141–153.

Chapter 14

Champness, P.E. 1995. Analytical electron microscopy. In: Potts, P.J. *et al.* (eds): *Microprobe Techniques in the Earth Sciences.* London: Chapman & Hall, pp. 91–139.
Droop, G.T.R., 1985. Alpine metamorphism in the south-east Tauern Window, Austria: 1. P-T variations in space and time. *J. Metamorph. Petrol.* 3, 371–402.
Ferry, J.M. and Spear, F.S. 1978. Experimental calibration of the partition of Fe and Mg between garnet and biotite. *Contrib. Miner. Petrol.* 66, 113–7.
Goldstein, J.I., Newbury, D.E., Echlin, P., Joy, D.C., Romig, A.D. Jr, Lyman, C.E., Fiori, C. and Lifshin, E. 1992. *Scanning Electron Microscopy and X-ray Microanalysis*, 2nd edn, New York: Plenum Press.
Nott, J.A. 1993. X-ray microanalysis in pollution studies. In: Sigee, D.C., *et al.* (eds): *X-ray Microanalysis in Biology.* Cambridge: Cambridge University Press, pp. 257–281.
Potts, P.J. 1987. *A Handbook of Silicate Rock Analysis.* Glasgow: Blackie.
Ray, S. and McLeese, D.W. 1987. Biological cycling of cadmium in the marine environment. In: Nriagu, J.O. and Sprague, J.B. (eds): *Cadmium in the Aquatic Environment.* New York: Wiley, pp. 199–220.
Reed, S.J.B. 1993. *Electron Microprobe Analysis*, 2nd edn. Cambridge: Cambridge University Press.
Read, S.J.B. 1995. Electron microprobe analysis. In: Potts, P.J. *et al.* (eds): *Microprobe Techniques in the Earth Sciences.* London: Chapman & Hall. pp. 49–89.
Sweatman, T.R. and Long, J.V.P. 1969. Quantitative electron-probe microanalysis of rock-forming minerals. *J. Petrol.* 10, 332–379.

Walker, G., Rainbow, P.S., Foster, P. and Crisp, D.J. 1975. Barnacles: possible indicators of zinc pollution? *Mar. Biol.* **30**, 57–65.

Yardley. B.W.D. 1989. *An Introduction to Metamorphic Petrology*. Harlow: Longman.

Chapter 15

Bowring, S.A., Williams, I.S. and Compston, W. 1989. 3.96 Ga gneisses from the Slave Province, N.W.T., Canada. *Geology* **17**, 971–975.

Clement, S.W., Compston, W. and Newstead, G. 1977. Design of a large, high resolution ion microprobe. *Proc. First Int. Conf. SIMS, Muenster.*

Eldridge, C.S., Compston, W., Williams, I. S., Walshe, J.L. and Both, R.A. 1987. In situ microanalysis for $^{34}S/^{32}S$ ratios using the ion microprobe SHRIMP. *Int. J. Mass Spectrom. Ion Process.*, **76**, 65–83.

Fahey, A.J., Goswami, J.N., McKeegan, K.D. and Zinner, E. 1985. Evidence for extreme ^{50}Ti enrichments in primitive meteorites. *Astrophys. J. Lett.*, **296**, L17–L20.

Farver, J.R. and Giletti, B.J. 1985. Oxygen diffusion in amphiboles. *Geochim. Cosmochim. Acta* **49**, 1403–1411.

Froude, D.O., Ireland, T.R., Kinny, P.D., Williams, I.S., Compston, W., Williams, I.R. and Myers, J.S. 1983. Ion microprobe identification of 4,100–4,200 Myr-old terrestrial zircons. *Nature* **304**, 616–618.

Giletti, B.J. and Yund, R.A. 1984. Oxygen diffusion in quartz. *J. Geophys. Res.* **89**, 4039–4046.

Giletti, B.J., Semet, M.P. and Yund, R.A. 1978. Studies in diffusion, III. Oxygen in feldspars: an ion microprobe determination. *Geochim. Cosmochim. Acta* **42**, 45–57.

Hart, S.R. and Dunn T. 1993. Experimental cpx/melt partitioning of 24 trace elements. *Contrib. Mineral. Petrol.* **113**, 1–8.

Hervig, R.L. and Dunbar, N.W. 1992. Cause of chemical zoning in the Bishop (California) and Bandelier (New Mexico) magma chambers. *Earth Planet. Sci. Lett.* **111**, 97–108.

Ireland, T.R., Compston, W. and Heydegger, H.R. 1985. Titanium isotopic anomalies in hibonites from the Murchison carbonaceous chondrite. *Geochim. Cosmochim. Acta* **49**, 1989–1993.

Irving, A.J. and Frey, F.A. 1984. Trace element abundances in megacrysts and their host basalts: constraints on partition coefficients and megacryst genesis. *Geochim. Cosmochim. Acta* **48**, 1201–1221.

Johnson, K.T.M. and Dick, H.J.B. 1992. Open system melting and temporal and spatial variations of peridotite and basalt at the Atlantis II fracture zone. *J. Geophys. Res.* **97**, 9219–9241.

Johnson, K.T.M., Dick, H.J.B. and Shimizu, N. 1990. Melting in the oceanic upper mantle: an ion microprobe study of diopsides in abyssal peridotites. *J. Geophys. Res.* **95**, 2661–2678.

Kelemen, P.B., Dick H.J.B. and Quick, J.E. 1992. Formation of harzburgite by pervasive melt/rock reaction in the upper mantle. *Nature* **358**, 635–641.

Layne, G.D. and Stix, J. 1990. Volatile and light lithophile element evolution of the Cerro Toledo rhyolite, Jemez Mountains, New Mexico: an ion microprobe study, *EOS* **71**, 651.

Lundberg, L.L., Crozaz, G. and McSween, H.Y. Jr. 1990. Rare earth elements in minerals of the ALHA77005 shergottite and implications for its parent magma and crystallization history. *Geochim. Cosmochim. Acta* **54**, 2535–2547.

Macfarlane, A.W. and Shimizu, N. 1991. SIMS measurements of $\delta^{34}S$ in sulfide minerals from adjacent vein and stratabound ores. *Geochim. Cosmochim. Acta* **55**, 525–541.

Mason, B. and Allen, R.O. 1973. Minor and trace elements in augite, hornblende, and pyrope megacrysts from Kakanui, New Zealand. *NZ J. Geol. Geophys.* **16**, 935–947.

Moore, R.O., Gurney, J.J., Griffin, W.L. and Shimizu, N. 1991. Ultra-high pressure garnet inclusions in Monastery diamonds: trace element abundance patterns and conditions of origin. *Eur. J. Mineral.* **3**, 213–230.

Shimizu, N. 1975. Rare earth elements in garnets and clinopyroxenes from garnet lherzolite nodules in kimberlites. *Earth Planet. Sci. Lett.* **25**, 26–32.

Shimizu, N. 1986. Silicon-induced enhancement in secondary ion emission from silicates, *Int. J. Mass Spectrom. Ion Process.* **69**, 325–338.

Shimizu, N. and Hart, S. R. 1982a. Isotope fractionation in secondary ion mass spectrometry *J. Appl. Phys.* **52**, 1303–1311.

Shimizu, N. and Hart, S. R. 1982b. Applications of the ion microprobe to geochemistry and cosmochemistry. *Annu. Rev. Earth Planet. Sci.* **10**, 483–526.

Shimizu, N. and Richardson, S.H. 1987. Trace element abundance patterns of garnet inclusions in peridotite-suite diamonds. *Geochim. Cosmochim. Acta* **51**, 755–758.

Shimizu, N., Semet, M. P. and Allegre, C. J. 1978. Geochemical applications of quantitative ion-microprobe analysis. *Geochim. Cosmochim. Acta* **42**, 1321–1334.

Shimizu, N., Gurney, J.J. and Moore, R. 1989. Trace element geochemistry of garnet inclusions in diamonds from the Finsch and Koffiefontein kimberlite pipes. *Abstract. Workshop on Diamonds, IGC, Washington, DC*, pp. 100–101.

Sigmund, P. 1969. Theory of sputtering I. Sputtering yield of amorphous and polycrystalline targets. *Phys. Rev.* **184**, 383–416.

Slodzian, G., Lorin, J. C. and Havette, A. 1980. Isotopic effect on the ionization probabilities in secondary ion emission *J. Phys.* **23**, 555–558.

Slodzian, G., Chaintreau, M. and Dennebouy, R. 1986. The emission objective lens working as an electron mirror: self regulated potential at the surface of an insulating sample. *Proc. SIMS V, Washington, DC*, pp. 158–160.

Sobolev, A.V. and Shimizu, N. 1993. Ultra-depleted primary melt included in an olivine from the Mid-Atlantic Ridge. *Nature*, **363**, 151–154.

Takazawa, E., Frey, F.A. Shimizu, N., Obata, M. and Bodinier, J.L. 1992. Geochemical evidence for melt migration and reaction in the upper mantle. *Nature*, **359**, 55–58.

Zinner, E., Fahey, A.J., Goswami, J.N., Ireland, T.R. and McKeegan, K.D. 1986. Large [48]Ca anomalies are associated with [50]Ti anomalies in Murchison and Murray hibonites. *Astrophys. J. Lett.* **311**, L103–L107.

Chapter 16

Altgelt, K.H. and Boduszynski, M.M. 1994. *Composition and Analysis of Heavy Petroleum Fractions.* New York: Marcel Dekker.

Bordenave, M.L. 1993. (ed.) *Applied Petroleum Geochemistry*. Paris: Éditions Technip.

Chapman, J.R. 1993. *Practical Organic Mass Spectrometry: A Guide for Chemical and Biochemical Analysis*, 2nd edn. Chichester: Wiley.

Coleman, D.C. and Fry, B. 1991. (eds) *Carbon Isotope Techniques*. San Diego: Academic Press.

Durand, B. and Nicaise, G. 1980. Procedures for kerogen isolation. In: Durand, B. (ed): *Kerogen: Insoluble Organic Matter from Sedimentary Rocks*. Paris: Éditions Technip, pp. 35–53.

Eglinton, G. and Murphy, M.T.J. 1969. (eds) *Organic Geochemistry: Methods and Results*. Berlin: Springer-Verlag.

Engel, M.H. and Macko, S.A. 1993. (eds) *Organic Geochemistry: Principles and Applications*. New York: Plenum.

Faure, G. 1986. *Principles of Isotope Geology*, 2nd edn. New York: Wiley.

Fried, B. and Sherma, J. 1986. *Thin-Layer Chromatography: Techniques and Applications*, 2nd edn. New York: Marcel Dekker.

Harwood, L.M. and Moody, C.J. 1989. *Experimental Organic Chemistry: Principles and Practice*. Oxford: Blackwell Scientific.

Hunt, J.M. 1979. *Petroleum Geochemistry and Geology*. San Francisco: W.H. Freeman.

Jennings, W. 1987. *Analytical Gas Chromatography*. San Diego: Academic Press.

Killops, S.D. and Killops, V.J. 1993. *An Introduction to Organic Geochemistry*. Harlow: Longman.

McLafferty, F.W. and Tureček, F. 1993. *Interpretation of Mass Spectra*, 4th edn. Mill Valley, CA: University Science Books.

McMaster, M.C. 1994. *HPLC: A Practical User's Guide*. New York: VCH.

Message, G.M. 1984. *Practical Aspects of Gas Chromatography/Mass Spectrometry*. New York: Wiley.

Michael, G.E., Lin, L.H., Philp, R.P., Lewis, C.A. and Jones, P.J. 1989. Biodegradation of tar-sand bitumens from the Ardmore/Anadarko Basins, Oklahoma-II. Correlation of oils, tar sands and source rocks. *Org. Geochem.* **14**, 619–633.

Middleditch, B.S. 1989. *Analytical Artifacts: GC, MS, HPLC, TLC and PC*. Amsterdam: Elsevier.

Peters, K.E., and Moldowan, J.M. 1993. *The Biomarker Guide: Interpreting Molecular Fossils in Petroleum and Ancient Sediments*. Englewood Cliffs, NJ: Prentice Hall.

Poole, C.F. and Schuette, S.A. 1984. *Contemporary Practice of Chromatography*. Amsterdam: Elsevier.

Ranný, M. 1987. *Thin-Layer Chromatography with Flame Ionization Detection*. Dordrecht: Reidel.

Rossini, F.D., Mair, B.J. and Streiff, A.J. 1953. *Hydrocarbons from Petroleum: The Fractionation, Analysis, Isolation, Purification, and Properties of Petroleum Hydrocarbons*. New York: Reinhold.

Schoell, M., Faber, E. and Coleman, M.L. 1983. Carbon and hydrogen isotope compositions of the NBS 22 and NBS 21 stable isotope reference materials: an inter-laboratory comparison. *Org. Geochem.* **5**, 3–6.

Sosrowidjojo, I.B., Alexander, R. and Kagi, R.I. 1994. The biomarker composition of some crude oils from Sumatra. *Org. Geochem.* **21**, 303–312.

Stahl, E. 1969. *Thin Layer Chromatography*, 2nd edn. Berlin: Springer-Verlag.

Tissot, B.P. and Welte, D.H. 1984. *Petroleum Formation and Occurrence*, 2nd edn. Berlin: Springer-Verlag.

Tyson, R.V. 1995. Sedimentary Organic Matter: Organic facies and Palynofacies. London: Chapman and Hall.

Appendix A

O'Hanlon, John F. 1989. *A User's Guide To Vacuum Technology*. New York: John Wiley.

Appendix B

AMC (Analytical Methods Committee of the Royal Society of Chemistry) 1987. Recommendations for the definition, estimation and use of the detection limit. *Analyst* **112**, 199–204.

Garfield, F.M. 1991. *Quality Assurance Principles for Analytical Laboratories*. Arlington, VA: ACOC.

Govindaraju, K. 1994. 1994 compilation of working values and sample description for 383 geostandards. *Geostandards Newsl.* **18**, 1–58.

Harland, C.E. 1994. *Ion exchange: theory and practice*, 2nd edn. Cambridge: Royal Society of Chemistry.

ISO 1992. *Guide 30: Terms and definitions used in connection with reference materials*. Geneva: International Organization for Standardization.

IUPAC (International Union of Pure and Applied Chemistry) 1978. Nomeclature, symbols, units and their usage in spectrochemical analysis – III. *Spectrochim. Acta*, **34B**, 261.

MacKenzie, A.S. 1984. Applications of biological markers in petroleum geochemistry. In: J. Brooks and Welte D. (eds): *Advances in Petroleum Geochemistry*, Vol. 1, pp.115–214. New York: Academic Press.

Miller, J.C. and Miller, J.N. 1993 *Statistics for Analytical Chemistry*, 3rd edn. New York: Ellis Horwood.

Smith, F.G. and Thomson, J.H. 1988. *Optics*, 2nd edn. Chichester: Wiley.

Index

Terms printed in **bold** are defined in the Glossary (Appendix B, pp 281-298). Decimal numbers (e.g. '6.10') in normal type refer to tables; those printed in *italic* refer to figures; those in **bold** to boxes. Greek lettered references appear at the start of each entry for the corresponding Roman letter. Analyte elements are listed by symbol only.